NATURALISTIC EPISTEMOLOGY:
A SYMPOSIUM OF TWO DECADES

BOSTON STUDIES IN THE PHILOSOPHY OF SCIENCE



VOLUME 100

# NATURALISTIC EPISTEMOLOGY

## *A Symposium of Two Decades*

*Edited by*

ABNER SHIMONY
*Boston University*

*and*

DEBRA NAILS
*University of the Witwatersrand*

D. REIDEL PUBLISHING COMPANY

A MEMBER OF THE KLUWER ACADEMIC PUBLISHERS GROUP

DORDRECHT / BOSTON / LANCASTER / TOKYO

**Library of Congress Cataloging-in-Publication Data**

CIP

Naturalistic epistemology.

(Boston studies in the philosophy of science ; v. 100)
Bibliography: p.
Includes index.
1. Knowledge, Theory of. 2. Naturalism. I. Shimony, Abner. II. Nails, Debra, 1950– . III. Series.
Q174.B67 Vol. 100 [BD161] 001′.01 s [121] 86–31600
ISBN 90–277–2337–0

Published by D. Reidel Publishing Company,
P.O. Box 17, 3300 AA Dordrecht, Holland

Sold and distributed in the U.S.A. and Canada
by Kluwer Academic Publishers
101 Philip Drive, Norwell, MA 02061, U.S.A.

In all other countries, sold and distributed
by Kluwer Academic Publishers Group
P.O. Box 322, 3300 AH Dordrecht, Holland



Printed in The Netherlands

# TABLE OF CONTENTS

## PART III / CRITICISMS OF NATURALISTIC EPISTEMOLOGY

ABNER SHIMONY

# INTRODUCTION

## I. AIMS OF THE INTRODUCTION

The systematic assessment of claims to knowledge is the central task of epistemology. According to naturalistic epistemologists, this task cannot be well performed unless proper attention is paid to the place of the knowing subject in nature. All philosophers who can appropriately be called 'naturalistic epistemologists' subscribe to two theses: (a) human beings, including their cognitive faculties, are entities in nature, interacting with other entities studied by the natural sciences; and (b) the results of natural scientific investigations of human beings, particularly of biology and empirical psychology, are relevant and probably crucial to the epistemological enterprise.

Naturalistic epistemologists differ in their explications of theses (a) and (b) and also in their conceptions of the proper admixture of other components needed for an adequate treatment of human knowledge — e.g., linguistic analysis, logic, decision theory, and theory of value. Those contributors to this volume who consider themselves to be naturalistic epistemologists (the majority) differ greatly in these respects. It is not my intention in this introduction to give a taxonomy of naturalistic epistemologies. I intend only to provide an overview which will stimulate a critical reading of the articles in the body of this volume, by facilitating a recognition of the authors' assumptions, emphases, and omissions.

My procedure will be to summarize four major objections which have been raised against theses (a) and (b) and then to examine briefly the ways in which the objections have been or could be answered. I shall not abstain from indicating my own point of view but shall not present it in detail (and the need to do so is mitigated by the inclusion of my article later in the volume).

## II. CRITICISMS OF NATURALISTIC EPISTEMOLOGY

1. Thesis (a) has been criticized for crudely conceiving nature as an



*Abner Shimony and Debra Nails (eds.), Naturalistic Epistemology*, 1—13.

aggregate of entities, without recognizing that an entity may belong to nature in some respects and not in others. Hegel, for example, regarded nature as the domain of externality and spirit as the domain of inwardness, and accordingly a human being belongs to nature in bodily aspects but not spiritually. Other philosophers have interpreted all natural phenomena physicalistically, and therefore thoughts and feelings are outside nature, unless, of course, they too are construed physicalistically. Kant, by contrast, does not exclude all mental phenomena from the domain of nature, and yet his conception of nature is incompatible with thesis (a). "By nature . . . taken substantivally is meant the sum of appearances in so far as they stand, by virtue of an inner principle of causality, in thoroughgoing interconnection" (*Critique of Pure Reason* B446). Since causality, according to Kant, is a category imposed upon experience by the Understanding, that faculty is the source of lawfulness in nature. A naturalistic account of the operation of the Understanding would, accordingly, be a nonsensical inversion. Kant acknowledges that there is an empirical self which falls within the domain of experience, since it is represented to the inner sense, and it can be investigated scientifically by empirical psychology (B158, B152). But neither empirical psychology nor any other natural science yields information about the transcendental self and its faculties, the Sensibility and the Understanding, which are the source of formal principles of order in experience.

2. Thesis (b) has been criticized on the grounds that the findings of the natural sciences concerning mental phenomena are relevant only to descriptive epistemology, which studies the way in which thought processes actually work. The essential part of the enterprise of epistemology, however, is not descriptive but normative, prescribing the thought processes which ought to be followed in order to make correct inferences and in general to achieve knowledge. The failure to discriminate normative from descriptive epistemology would be a conflation of *ought* and *is*, which has been shown by meta-ethics to be a fundamental error.

3. Thesis (b) has been criticized on the grounds that the knowledge which the natural sciences provide concerning human beings is indirect and inferential and is therefore not trustworthy unless scientific methodology has been justified. But scientific methodology is a part of epistemology. Consequently, any program making essential use of the

results of the natural sciences in epistemology, for the purpose of assessing claims to knowledge, is blatantly circular. Fear of circularity is the reason for the persistence of a long philosophical tradition, from Descartes to Lorenzen, which accepts thesis (a) that human knowers are natural entities but abstains from using the results of the natural sciences in epistemological investigations. According to this tradition, circularity can be avoided only by a stepwise construction, in which the lowest level is in some sense directly founded (e.g., by the *cogito* argument, or by the immediacy of sense data), and each successive layer rests firmly upon its predecessors. The working out of the details of this stepwise construction is the correct way to carry out the enterprise of epistemology. (Of course, skepticism results if the same architectural model of human knowledge is retained, but doubt is cast upon the validity of some indispensable principle of stepwise construction, for example, by Hume's criticism of induction.)

4. Thesis (b) has been criticized on the grounds that a naturalistic study of human beings seriously underestimates the historical and cultural character of human cognition. When human beings are studied by the natural sciences, the emphasis is upon biologically heritable traits. To be sure, among these traits is the capacity for culture, which makes it possible for human beings to exhibit non-genetic transmittable diversity. Acknowledging the biological basis of the capacity for culture, however, does not justify subsuming the study of culture under the natural sciences. If the genetic component in human knowledge is relatively unimportant compared to the cultural component, then little scope remains to naturalistic epistemology. The anatomical and physiological features of human beings, and even some of the primitive elements of human mentality, may have remained almost unchanged during the past one hundred thousand years, but the advanced aspects of human mentality have been so radically transformed during the historical epoch that the very concept of a fixed 'human nature' must be rejected. Alternatively, if one wishes to retain the expression 'human nature', one must treat it as a cultural variable. Different versions of this argument against naturalistic epistemology and in favor of a historical or cultural epistemology have been presented by many philosophers, including Vico, Herder, Marx, Dilthey, and Mannheim. Habermas and other members of the critical school have given related arguments for the separation of epistemology from the natural sciences, though an

important element in their position is the analysis of the normative aspects of cognition, and to this extent their objections to naturalistic epistemology overlap with the second objection above.

## III. ANSWERS TO CRITICISMS

### *Objection 1*

This objection is actually a cluster of different arguments against treating the knowing subject as an entity in nature. I shall restrict my attention to two members of the cluster. In particular, I shall not attempt to survey answers to Hegel, since that would require an extensive metaphilosophical investigation of logic and language. Furthermore, because of Hegel's insistence upon a historical analysis of spirit, his contraposition of spirit to nature anticipates later historicist treatments of cognition, and these will be discussed below.

To the objection that the physical character of nature precludes the classification of the knowing subject as a natural entity, one straightforward answer is to accept a physicalist reduction of psychology. This is the position of Quine, who is such a central figure that many students are unaware of the possibility (and, in my opinion, desirability) of disentangling naturalistic epistemology from a physicalist conception of mind. It is, however, beyond the scope of this Introduction to review the arguments for and against a physicalist philosophy of mind.

Another straightforward answer is to deny the physicalist characterization of nature and to postulate, as Whitehead does, that proto-mentality is exhibited throughout nature, even at the level of elementary particles. Whitehead's proposal remains conjectural and programmatic, however, not only because proto-mental traits have not been detected in primitive physical systems, but because it is far from clear how they could be detected.

A third answer is to accept the wealth of known correlations between mental and physical phenomena, which are observed in ordinary life and confirmed by psychophysical and medical investigations, as overwhelming evidence that the human subject is enmeshed causally with the subject's own body and with other physical objects. These objects act upon the subject so as to give rise to experiences, often in predictable ways, and conversely the subject is somehow able to effect changes in the configurations of physical objects, again often in predictable ways. Much of the literature of naturalistic epistemology is concerned with such psychophysical correlations. Lower organisms learn by conditioning, and higher organisms by processes involving consciousness as well as by conditioning, that certain stimuli of the sense organs are reliable indications of the existence and properties of remote objects. When naturalistic epistemologists explain the reliability of cognitive processes, they typically make reference to correlations, without any pretense to the kind of fundamental understanding which Whitehead aims at. It might be objected that explanations in terms of correlations alone constitute promissory notes for fundamental understanding, and if this is not forthcoming in the long run then naturalistic epistemology would be bankrupt. An advocate of instrumentalist interpretations might reply that nothing more can be reasonably demanded than appropriate correlations. In my opinion, however, it would be most unsatisfactory to rely upon instrumentalism in order to defend naturalistic epistemology. A central motivation of naturalistic epistemology is to avoid falling into philosophical anthropocentrism, and resorting to an instrumentalist interpretation of scientific theories in preference to a realist interpretation is *prima facie* a move towards anthropocentrism.

The Kantian objection to thesis (a) is compelling only if Kant is correct in maintaining that our minds can form synthetic *a priori* judgments, which are necessary and universal. It would be hard to give an account of such judgments as cogent as Kant's — that they result from the imposition by the Sensibility and the Understanding of forms upon experience. However, the development of a physically applicable non-Euclidean geometry and of a non-deterministic physics has had the effect of casting great doubt upon the possibility of making synthetic *a priori* judgments, and thereby upon the Kantian thesis that the transcendental self legislates to nature. Furthermore, the biological theory of evolution has provided naturalistic epistemologists with a powerful tool for explaining those elements of the child's cognitive equipment which seem to be *a priori* in the sense of not being learned during the child's brief span of experience (e.g., the organizing principles of the Gestalt psychologists, the schemata of Piaget, and the universal rules of language of Chomsky). Evolutionary epistemologists from Spencer to Popper and Campbell have interpreted such elements, whatever they may be, as phylogenetically *a posteriori* even though they are ontogenetically *a priori*. Explaining the *a priori* elements of human cognitive equipment in this way does not presuppose a Lamarckian theory of the inheritance of acquired characteristics, for their incorporation into the human genome can be plausibly attributed to natural selection.

*Objection 2*

A straightforward answer to the objection that naturalistic epistemology can be descriptive but not normative is provided by the familiar distinction between categorical and hypothetical norms. Factual knowledge of human nature may well be inadequate to supply an 'ought' in the sense of a categorical norm and yet fully adequate to justify a hypothetical norm. The latter has the form: follow a specified prescription as a means for achieving a certain end (without any commitment to the ultimate desirability of that end). It is clear upon reflection that normative epistemology aspires only to hypothetical norms: follow specified prescriptions for the purpose of obtaining the truth or a good approximation to the truth, at least with high probability, concerning matters of interest. This answer clearly avoids any conflation of 'ought' and 'is'.

This answer, however, gives rise to other problems. How is the concept of *truth* to be construed — as coherence of beliefs, as correspondence with facts having an ontological status independent of belief, as pragmatic reliability, or in some other way? Here is one of the fundamental problems of analytic epistemology, which is a different enterprise from normative epistemology, but also apparently different from descriptive epistemology. It is not evident, without actually engaging in analysis, that a naturalistic treatment of the knowing subject will answer this question or even throw some light upon it. Furthermore, if a prescription is recommended on the grounds that it is a means for obtaining the truth or approximations to the truth, then the question arises of justifying that it is a *good* means. If a justification is offered which draws upon the results of the natural sciences, then there is a circularity unless grounds independent of the results of the natural sciences can be provided for relying upon scientific methodology. Consequently, the straightforward answer given to the objection concerning normative epistemology has the effect of aggravating the objection that naturalistic epistemology is circular.

*Objection 3*

The charge of circularity against naturalistic epistemology may be answered by denying that a full justification needs to be supplied for each instrument used in the epistemological enterprise. The demand for a full justification of the results of the natural sciences and of the scientific method itself is engendered by a foundationalist or architectural model of human knowledge, which was shared by early rationalists like Descartes and early empiricists like Hume. After three centuries of futile attempts to realize a foundationalist program, we should have learned to accept its impossibility as an aspect of the human condition. This is the answer given by Quine (1969, p. 76):

> such scruples against circularity have little point once we have stopped dreaming of deducing science from observations. If we are out simply to understand the link between observation and science, we are well advised to use any available information, including that provided by the very science whose link with observation we are seeking to understand.

It is a genuine question, according to Quine, to ask why science works so well, but he insists that the answer should be supplied by science itself. Physics, geology, biology, etc. describe the natural environment and state the laws which govern it; and psychology, neurology, etc. study the ways in which human beings are stimulated by the environment, undergo consequent internal transformations, and finally arrive at remarkably reliable representations and predictions. This application of the natural sciences to the process of acquiring knowledge about nature is the surrogate offered by Quine for the discredited foundationalist program.

There is obviously a close connection between Quine's proposal to validate science by scientific investigation and the thesis that coherence of beliefs is the primary criterion of their truth. The appeal to coherence is congenial to Quine, in view of his well known assertions that scientific theories are largely underdetermined by data and that all statements, including those of mathematics and logic, are subject to revision in the interest of optimizing the harmony of the entire 'web of belief'.

The great weakness of the criterion of coherence is that indefinitely many competing systems of beliefs — including those which appeal to allegedly authoritative books, individuals, and institutions — may satisfy it equally well. Some philosophers sympathetic both with naturalistic epistemology and with the appeal to coherence have therefore looked for other criteria for discriminating among competing coherent systems of beliefs. An interesting proposal of additional criteria is Sosa's 'perspectival coherentism', which requires that the subject be able to place himself within the world depicted by his set of beliefs and to exhibit his means of access of that world.

Another way to reply to the objection of circularity is to distinguish between vicious and non-vicious circles and to argue that the latter are permissible and unavoidable in any philosophy that aims to be systematic and comprehensive. Black, for example, has offered inductive arguments for the validity of the principle of induction, and he bases his claim for non-viciousness of his kind of circularity on the fact that the conclusion of the inductive inference is an assertion of the validity of the rule of inference used in the induction — an assertion that is not one of the premisses of the inference. His claim of non-viciousness has been widely criticized, however, for example by Achinstein. Other philosophers (e.g., Shimony) argue that a circular procedure is non-vicious if it does not insulate the propositions under investigation from the possibility of refutation and thus respects Peirce's injunction not to "block the road to inquiry."

Related to considerations of non-vicious circularity, but less concerned with the formal structure of reasoning, is the proposal of a dialectical formulation of naturalistic epistemology. It may be possible to construe Quine as advocating such a formulation, even though the word 'dialectical' is probably not to his taste. In the dialectic as practiced by Plato, the method of investigation is not precisely articulated initially. The starting point consists of the beliefs of some of the interlocutors and an attitude of questioning on the part of another. Progress in articulating the rules of method is achieved in tandem with the clarification of concepts and the increase of the body of knowledge. (The *Sophist* and Book VI of *The Republic* state some methodological principles explicitly, but these have been preceded by exercises of the informal Socratic dialectic.) Furthermore, it is characteristic of the Platonic dialectic to be reflective about itself and self-critical. We can transpose these Platonic conceptions to the enterprise of naturalistic epistemology, and can even figuratively use the expression 'dialogue with nature'.[1] In a historical reconstruction of human intellectual achievements we can imagine an initial stage in which the intellectual equipment consisted of a crude world view, a set of crude tacit epistemological assumptions, a primitive methodology developed during earlier epochs by natural selection, and innate mechanisms for processing sensory data. A remarkable feature of this intellectual equipment was its susceptibility to 'bootstrapping' (to use and broaden Glymour's term). The primitive methodology sufficed to recognize exemplary cases of successful predictions and exemplary cases of failure to predict, and reflections upon these cases were historically the occasions of refinements of methodology and of world view. Further argument is needed, of course, to establish that exemplary cases of scientific advance were not merely the stimuli for methodological improvements which could have been made *a priori* by an ideally clear mind, but actually supplied indispensable inductive evidence for them. Once the stage of documented intellectual history is reached, imaginative reconstructions can and should be replaced by a study of the actual interplay between scientific discoveries and philosophical reflections on them. But even the rational reconstruction of human intellectual achievements prior to documentation is best carried out by an imaginative dialectic, if it is accepted that the program of foundationalism is unrealizable.

The dialectic formulation of naturalistic epistemology offers an avenue of escape from the problem of circularity. It regards primitive scientific discoveries to be justified by a primitive and largely innate methodology, and refinements of methodology to be justified by the scientific discoveries. In this way there is a mutual correction of a scientific world picture and of procedures for arriving at a scientific world picture, and the starting point of the process is not hedged against revision. What has just been said about methodology also applies to other parts of epistemology, such as the treatment of the relation between primary and secondary qualities.

One of the virtues of the dialectic formulation of naturalistic epistemology is its efficacy in dealing with the deep problem posed by the fact that human beings are, in a sense, the prisoners of their own faculties. Investigating the natural constitution of the subject permits the identification of limitations and distortions which engender systematic errors (Bacon's 'Idols of the Tribe') in our representations of the world. And recognition facilitates corrections, e.g., the construction of instruments which compensate for deficiencies of the sense organs.

Likewise, a dialectical naturalism can assess and correct the commonsense picture of the world, which is the starting point of practical activity, including the fabrication and manipulation of instruments used in scientific investigations. A non-commonsensical outcome of a scientific investigation need not, therefore, be regarded as a sign of an inconsistency in reasoning. In this respect, a dialectical formulation of naturalistic epistemology is opposed to the philosophy of Lorenzen and his school, which rejects relativistic space-time structure because of its discrepancy with the spatial and temporal assumptions implicit in practical activity.

*Objection 4*

This objection rests essentially upon a factual premiss: that the historical and cultural components in human cognition dominate the biological components in all epistemologically significant respects. How difficult it is to assess the factual correctness of this premiss is indicated by the extensiveness of the anthropological literature dedicated to cultural universals and variables. However, the generic claims of naturalistic epistemology (i.e., theses (a) and (b) above) do not presuppose that the anthropological disputes concerning 'nature versus nurture' be settled unequivocally for the former. For naturalistic epistemology to be a viable enterprise, it suffices that there be some non-trivial biological constraints upon the cultural plasticity of human cognitive equipment.

Naturalistic epistemologists can argue, moreover, that the problem of relativism, which haunts the philosophers who insist upon the dominance of cultural and historical factors in human cognition, is to a large extent a corollary of their anti-naturalism. If epistemic principles are historically and culturally determined, then doubt is cast upon the possibility of a critique which will liberate intellectuals and populations from the grip of ideologies. Might it not be that the efforts of some critical theorists, like Habermas, to escape from relativism, would be greatly expedited by careful attention to the evidence concerning the biologically inherited cognitive faculties of children?

A critic might reply that resorting to the sciences in epistemological investigations neglects the fact that the articulation of the scientific method and the flowering of the sciences are themselves cultural achievements, fully realized only in a few of the innumerable human cultures. Consequently, even if the class of cultural universals is not empty, the scientific method cannot be understood in terms of universals. This objection is penetrating and cannot be answered adequately without an elaborate analysis. A crucial part of the answer is the distinction between a set of principles imposed culturally and a set of principles arrived at by inquiry when the culture permits and stimulates free inquiry. Whenever free inquiry occurs, it is undeniably a cultural contingency, but the *content* of the principles arrived at by inquiry is not dictated by the culture which permits the inquiry to proceed. This characterization of free inquiry resembles Habermas's thesis that beliefs are legitimized only if they are ideally arrived at by a process of free discussion in which all members of the society equally take part. A naturalistic epistemologist, however, would depart from Habermas and follow Peirce in trying to show both analytically and historically that there is a close connection between free inquiry and the scientific method. This divergence between the two philosophers is partly due to their preoccupations with different sets of problems, but certainly not entirely. It must be left to free inquiry to decide which route free inquiry should take.

## IV. THE NATURE OF THIS VOLUME

All the contributors to this volume have been associated in one way or another with the Philosophy Department of Boston University — as faculty members, graduate students, or regular visitors. During the last two decades, naturalistic epistemology has been one of the subjects of greatest interest among the Boston University philosophers. There have been many courses and seminars, many lectures at the Boston Colloquium for the Philosophy of Science, and several doctoral dissertations devoted to the subject. In addition, there were numerous discussions in our corridors of powerlessness. As a result, most of the contributors were aware of the ideas of others of the contributors and took them into account. There is nothing like a manifesto in this volume, because of the diversity of opinions of the authors. Nor is it a systematic introduction to the current state of opinion on naturalistic epistemology.[2] However, this volume is more than a collection of essays by philosophers contingently associated with one institution. It has a kind of loose integration, due to the intellectual exchanges among the authors. We were referring to these exchanges when we chose the subtitle, *A Symposium of Two Decades*.

*Boston University*

## NOTES

[1] A warning against taking this expression too literally is found in the Lord Chancellors's statements in W. S. Gilbert's *Iolanthe*: "my difficulty is, that at present there is no evidence before the court that chorused Nature has interested itself in the matter .... Now, an affidavit from a thunderstorm or a few words on oath from a heavy shower would meet with all the attention they deserve."

[2] The book which comes closest to being a systematic introduction to current work on naturalistic epistemology is the anthology of Kornblith (1985),which contains a number of excellent essays, including two by Quine, and a very extensive bibliography.

## REFERENCES


Achinstein, Peter: 1962. 'The Circularity of a Self-Supporting Inductive Argument,' *Analysis* **22**, 138—141.

Black, Max: 1958. 'Self-supporting Inductive Arguments,' *Journal of Philosophy* **55**, 718—725.

Brunswik, Egon: 1955. '"Ratiomorphic" Models of Perception and Thinking,' *Acta Psychologica* **11**, 108—109.

Descartes, René: 1951. *Meditations*. Tr. L. Lafleur. New York: Liberal Arts Press.

Glymour, Clark: 1980. *Theory and Evidence*. Princeton: Princeton University Press.

Geuss, Raymond: 1981. *The Idea of a Critical Theory*. Cambridge: Cambridge University Press.

Habermas, Jürgen: 1975. *Legitimation Crisis*. Tr. T. McCarthy. Boston: Beacon Press.

Hegel, G. W. F.: 1892. *The Logic of Hegel*. Tr. W. Wallace. London: Oxford University Press.

Kant, Immanuel: 1950. *Critique of Pure Reason*. Tr. N. Kemp Smith. New York: Humanities Press.

Kornblith, Hilary (ed.): 1985. *Naturalizing Epistemology*. Cambridge, Mass.: MIT Press.

Lorenz, Konrad: 1941. 'Kants Lehre vom Apriorischen im Lichte gegenwärtiger Biologie,' *Blätter für Deutsche Philosophie* **15**, 94—125. English translation, 'Kant's Doctrine of the A Priori in the Light of Contemporary Biology,' reprinted in H. C. Plotkin (ed.), *Learning, Development, and Culture: Essays in Evolutionary Epistemology*. Chichester and New York: Wiley & Sons, pp. 121—143.

Lorenzen, Paul: 1968. *Methodisches Denken*. Frankfurt: Suhrkamp.

Mannheim, Karl: 1936. *Ideology and Utopia*. New York: Harcourt, Brace, and World.

Nails, Debra: 1985. 'Plato's Naturalistic Epistemology,' paper presented to Boston Colloquium for the Philosophy of Science, November 26, 1985.

Peirce, Charles S.: 1934. 'The Fixation of Belief,' *Collected Papers*, Vol. 5. Ed. by C. Hartshorne and P. Weiss. Cambridge, Mass.: Harvard University Press.

Plato: 1935. *Plato's Theory of Knowledge; The* Theaetetus *and the* Sophist *of Plato*. Tr. Francis M. Cornford. London: Routledge and Kegan Paul.

Plato: 1945. *The Republic of Plato*. Tr. Francis M. Cornford. London: Oxford University Press.

Popper, Karl: 1972. *Objective Knowledge*. London: Oxford University Press.

Quine, W. V. O.: 1969. 'Epistemology Naturalized.' In *Ontological Relativity and Other Essays*. New York: Columbia University Press, pp. 69—90. (Reprinted in Kornblith, 1985.)

Shimony, Abner: 1970. 'Scientific Inference.' In R. G. Colodny (ed.), *The Nature and Function of Scientific Theories*. Pittsburgh: University of Pittsburgh Press, pp. 79—172.

Sosa, Ernest: 1985. 'The Coherence of Virtue and the Virtue of Coherence: Justification in Epistemology,' *Synthese* **64**, 3—28.

Whitehead, Alfred N.: 1933. *Adventures of Ideas*. New York: Macmillan.

# PART I

# HISTORICAL FIGURES

JUDSON WEBB

# IMMANUEL KANT AND THE GREATER GLORY OF GEOMETRY

One of the objections most often raised against Kant's treatment of the nature and foundations of geometry is that it cannot accommodate the rise of non-euclidean geometry, much less its success in providing a more accurate description of the physical world. Russell, however, argued that it was really the discovery by Hilbert and others of a complete axiomatization of euclidean geometry more than anything else that revealed the irrelevance of Kantian intuition for its foundations. The scholar G. Martin, on the other hand, has claimed that Kant was really the first and true champion of the axiomatic method, and indeed in such a way that by Kant's lights non-euclidean geometry was not only an inevitable logical possibility, but also not constructible in intuition. To evaluate all these claims it is imperative to determine the level of axiomatic consciousness of geometry actually obtained by Kant, for as Hilbert (1922) himself has well remarked,

> to proceed axiomatically . . . is simply to think with consciousness of what one is doing. In earlier times, when they did not use the axiomatic method, men believed in various connections with naive dogmatism. Axiomatics removes the naivete, but still leaves us with all the advantages of belief. (p. 161)

To this end I shall consider individually Euclid's principal axioms and definitions as they were formulated in Kant's time, examining how he understood or misunderstood the contents as well as the problems raised by each of them. Keeping track of differences between the fragmentary set of axioms bequeathed to Kant by Euclid and the complete set of Hilbert will help to clarify the force of Russell's claim that it was only these that Kant made up with intuition. This also facilitates the analysis and comparisons of his various arguments for the transcendental ideality of space, since several of them turn on properties of space expressed in different groups of axioms. But in particular, only one depends on the parallel postulate. This helps to clarify the dependence of Kant's ideality of space on its euclidean structure, and to show that some of his metaphysical expositions of it depend more on geometry than they seem. His definition of geometry as "the science of



*Abner Shimony and Debra Nails (eds.), Naturalistic Epistemology*, 17—70.

space" obliges any analysis of his foundations for it to consider seriously these arguments for the transcendental status of its object.

Section 1 examines Kant's attempt to ground Euclid's claim "to draw a straight line through any two points" on his transcendental subject whose 'I draw' must accompany any geometric proof. The problem is to see how, without any usable definition of 'straight' the subject can *a priori* draw such a line 'in thought' at all, let alone the unique plane through any three points. Section 2 looks at Kant's attempt to derive Euclid's postulate "to describe a circle around any point at any distance" as a corollary of the definition of a circle. I argue that his principle that in geometry "the *definition* is at once the construction of the concept" neither applies to the other construction postulates nor adequately meets Newton's claim that geometry must 'borrow' such postulates from mechanics. Indeed, a general aim of the apriorism implicit in Kant's view of geometry as the science of space is to free it from physical concepts and make certain claims about space independent of physical laws. Already in his 1768 paper on distinguishing different regions of space Euler's attempt to refute the relational theory of space is said to show only that it has difficulty in "giving a definite significance to the most general laws of motion." Euler had thus only put such a refutation "in the hands of the engineers," whereas Kant would invoke incongruous counterparts to put it "in the hands of the geometers themselves." A consideration of this argument in Section 3 leads also to Kant's defense of Euclid's license "to produce any straight line indefinitely." Since this postulate presupposes the infinity of space itself it is crucial to Kant's claim that geometry deals with space even though none of Euclid's axioms refer to space. Like incongruous counterparts themselves this postulate could not be adequately expressed until suitable order axioms were formulated. I argue that this axiomatic gap in the topology of euclidean space made geometric proofs dependent on the intuition of figures. An analysis of various uses made by Kant of congruous counterparts suggests that intuition functions primarily for him to break symmetries and restrict the scope of Leibniz's identity of indiscernibles. But in their preoccupation with Kant's use of these counterparts his commentators have tended to neglect the weakness of his scant treatment of congruence itself. In fact, it is in connection with Euclid's axiom that "figures which coincide are equal" that Kant encounters difficulties inherent in any attempt to ground geometry independently of experience. This axiom presupposes the rigid mobility of figures, but Kant (1965) is emphatic that "in space, considered in itself, there is nothing movable; consequently, the movable must be something that is found *in space only through experience*, and must therefore be an empirical datum" (A41). In Section 4 I examine Kant's attempt to evade this apparent contradiction by distinguishing between the motion of an object in space and motion as the act of his subject in 'describing space', and argue that this move cannot sustain the *a priori* necessity of the congruence theorems. Though awakened by Hume from dogmatic slumber about causality, Kant's ignorance of equally trenchant critiques by Hume and others of 'congruity' left him to slumber on in the naive belief that the motion required by these theorems was no more than such descriptive acts. On the other hand, congruence is an equivalence relation to which Leibniz applied his identity of indiscernibles, and we find in Kant's critique of this application further evidence of the role of intuition in breaking symmetries. Finally, Section 5 examines Kant's justification of the euclidean structure of space, including his claim to prove the parallel postulate "philosophically from concepts." I explore the common presumption that Kant knew of the possibility of non-euclidean geometry from the work of his correspondent J. H. Lambert, but am afraid that while the latter did indeed in an important sense 'discover' this geometry, Kant missed a golden opportunity to be in on it when he failed to accept Lambert's offer that they share their new ideas with each other. Closely related to this is Kant's misunderstanding of Leibniz's 'philosophical theory' of geometric similarity as the indiscernibility of two figures when viewed separately, which actually anticipates his final argument for the ideality of space found in his late reflections and lectures on metaphysics as well as the *Opus Postumum*. I argue that it is his most plausible such argument inasmuch as it 'idealizes' in his sense the most basic of the primary spatial qualities of objects, namely their size; but also that it leaves him unable to make such an argument for their form.

## 1. LINES, PLANES, AND THE DRAWING EGO

Kant bases mathematics on a 'manifold of intuition' which, in relation to sensibility is subject to the 'formal conditions of space and time', while in relation to the understanding is subject to the synthetic unity of apperception. That is, space and time are 'forms of intuition' and any representation therein must be capable of being accompanied by an 'I think', even though intuitions are just those representations "which can be given prior to all thought" (Kant 1965, B132). The manifold thus "has a necessary relation to the 'I think' in the same subject in which the manifold is found" (*ibid.*). The space of which geometry is the science is the mere form of outer intuition, but this form is only a manifold for *possible* knowledge: for

> To know anything in space (for instance, a line) I must *draw* it, and thus synthetically bring into being a determinate combination of the given manifold, so that the unity of this act is at the same time the unity of consciousness (as in the concept of a line); and it is through this unity of consciousness that an object (a determinate space) is first known. (B138)

In fact, we cannot even *think* of a line "without *drawing* it in thought, or a circle without *describing* it" (B154). So for Kant geometrical knowledge requires an "I draw" to accompany our spatial representations, a "figurative synthesis" of the productive imagination, and geometry is indeed

> . . . based upon this successive synthesis of the productive imagination in the generation of figures. This is the basis of the axioms which formulate the conditions of sensible *a priori* intuition under which alone the schema of a pure concept of outer appearance can arise — for instance, that between two points only one straight line is possible, . . . .(B204)

But the 'basis' provided here for this axiom by the subject is never explained in the first *Critique.* Setting aside the requirement that a straight line pass through any two points, however far apart, how do we know that it can draw a *straight* line at all? In *thought*, that is, prior to any experience and without any ruler to steady its hand? Kant says that this synthesis of a line is capable of "generating from a point all its parts one after another" (A163). But this does not tell us how it 'generates' a *straight* line. How then can the claim that the ego can do so be anything more than the euclidean postulate it supposedly justifies, namely, an assumption? Kant finally faced this question during his controversy with Eberhard, who enlisted A. G. Kästner to defend the Leibniz-Wolff tradition in mathematical philosophy. Kästner (1790a) stressed that Euclid's construction postulates for the straight line and circle were merely assumed and not proved. Kant (Akademie, V. 20, p. 411) replied that they are not proved "mediately through inference," but rather "immediatley through *description*": "for description, which occurs *a priori* in the imagination according to a rule, and is called 'construction', is itself the proof of the possibility of the object." To those who doubt that we can base more complex axioms on such description Kant answers

> That the possibility of a straight line and a circle cannot be proved mediately through inferences but only immediately by the construction of these concepts, which is not at all empirical, is only possible because among all constructions, that is, among all representation according to a rule, some must be *first*, such as the *drawing* or describing (in thought) of a straight line and the *rotation* of such a line around a fixed point, where neither the latter can be derived from the former, nor can either be derived from any other construction of the concept of magnitude. (*ibid.*)

But this still does not explain *how* Euclid's postulates can be proved by 'description', or any kind of non-empirical 'rule' for drawing straight lines in thought. We shall see in the next section that Kant ultimately rests his case for such postulates on appropriate definitions of the geometrical concepts, an approach, however, which forces on him a recognition of a basic difference between the straight line and circle postulates. Let us meanwhile consider the prospects for a Kantian synthesis of other purely projective postulates.

In the Transcendental Doctrine of Method Kant says that axioms occur only in mathematics, since only here can one construct concepts in intuition, "as, for instance, in the proposition that three points always lie in a plane" (A733 = B761). Now planes are also among the basic qualitative spatial objects left undefined by Euclid that Kant would synthesize figuratively by "drawing it in thought." But how can we draw a *plane* in thought? Even to draw one literally on paper requires all the *conventions of perspective*. In fact, Kant had continued the passage quoted above on the need to respectively draw and describe lines and circles in thought by insisting that "we cannot represent the three dimensions of space save by *setting* three lines at right angles to one another from the same point" (B154). But how can we 'set' them so in thought without a prior knowledge of perspective? One wonders whether Kant's figurative synthesis may not have drawn some inspiration here from Lambert who, in a section on the 'drawing of appearances' in his *Neues Organon* speaks of extending his newly introduced 'phenomenology' to a kind of 'transcendental optics' comprising both perspective and descriptive geometry. Indeed, he even envisaged a 'transcendental perspective' of all possible appearances. In any case, despite the attempt to base geometry on spatial intuition, Kant's transcendental synthesis of its figures wanes when we turn from the lines of plane geometry to the simplest objects of space, namely planes. It is hard to see how his ego can synthesize planes at all, let alone establish their unique determination by three points, without some kind of perspective. One thing is clear: Kant knew no 'rule' for constructing planes.

Kant's desire to guarantee *a priori* the applicability to experience of Euclid's geometry, in which lines and planes are effectively primitive concepts, thus limits his options for giving any meaning to these concepts. He could turn to nature itself by asserting, e.g. that light travels in straight lines, but this would make geometry depend on physical hypotheses and compromise its apriority. Geometry can have no 'naturalistic' basis by Kant's lights, only an *a priori* one. Indeed, this is just the view behind the 'pragmatic idealism' of Dingler recently developed by Lorenzen into a 'protophysics': geometry is a system of spatial measurement based on ideal norms prior to physics proper. As Lorenzen says, this idealism does "follow the line of Kantian philosophy," but in place of intuition it rather "starts with the simple truth that a brick is not a natural object but an artifact. We have to *make* its sides plane and we have to *make* it vertical" (Lorenzen 1969, p. 56). The most striking advantage of this pragmatic idealism over Kant's is precisely that one does not begin with lines and planes as primitives but rather can *define* them as qualities of physical objects produced according to norms such as the '*Dreiplattenverfahren*' which produces plane surfaces by rubbing three given material surfaces together until they pairwise 'fit' each other. Of course, one still needs a transitivity assumption for this relation, as becomes clear in Lorenzen (1984). This brings us to the one case where Kant's view seems plausible.

## 2. CIRCLES, DEFINITIONS, AND CONSTRUCTIONS

Only when we look at Kant's theory of definition can we see how he thought that geometrical postulates could be proved "immediately by description," or even how mathematical concepts could be "constructed in intuition" at all. For such concepts have definitions, and Kant thinks that "mathematical definitions are constructions of concepts, originally framed by the mind itself" (B758). They "*make* their concepts" whereas in philosophical definitions concepts are "only *explained*" (*ibid.*). Definitions thus enable the mathematician to meet Kant's demand that "bare concepts must be *made sensible* . . . by the construction of a figure which, although produced *a priori*, is an appearance present to the senses." When pressed to apply this to Euclid's postulate "to describe a circle around any point" Kant replied in a letter to Reinhold of 1789 only with a question:

> If a circle is defined as a curve all of whose points are equidistant from a midpoint, is not this concept given in intuition? And this even though the practical corollary that follows, viz. *to describe a circle* (as a straight line is rotated uniformly around a point) is not at all considered. (1967, p. 146)

The claim that Euclid's postulate actually *follows from his definition of a circle* occurs explicitly a month later in a letter to Herz where he calls it "a practical corollary of the definition" which could not be demanded "if the possibility of the circle were not already given in the definition" (*ibid.*). But this, Kant claims, just *means* that "it is exhibited in intuition, not actually on paper, but in the imagination *a priori*" (*ibid.*). In the *Critique* the line and circle postulates were both said to be "practical" in the sense of containing nothing but the synthesis by which we generate these concepts:

> Such a proposition cannot be proved, since the procedure it demands is exactly that through which we first generate the concept of such a figure. (A234 = B287)

In the case of the circle, of course, no 'proof' is needed: its synthesis is an immediate 'corollary' of Euclid's definition, which is the basis for the unity of our rule for generating circles. But in the absence of any comparable definition for 'straight' Kant was unable to derive the line postulate by means of his principle that "in establishing the objective reality of the concept, here as always in geometry, the definition is at once the construction of the concept" (1967, p. 145).

But as Kant will be forced to recognize when he tries to prove the parallel postulate, a purely mathematical definition for a geometrical concept does not itself guarantee its constructibility. What encourages this idea in the case of the circle is the metrical aspect of its definition: given a straight line we need only fix one end and move the other around it. However, as Leibniz pointed out a century earlier, the curve 'described' by the moving endpoint will be a circle only on the further assumption that our line moves *rigidly*, that is, without altering its length. Otherwise its 'synthesized' points will not be equidistant from a center. So Kant's attempt to derive the existence of circles immediately as a 'practical corollary' of their definition not only presupposes the kind of motion of something in space which according to the transcendental aesthetic can only be known from experience, but also the rigid bodies cited in his 1762 lectures on mathematics as the sole presupposition of mechanics. This brings us to Newton's thesis of the dependence of geometry on mechanics.

Newton challenged the traditional distinction of geometrical constructions from mechanical ones on the basis of their greater accuracy by arguing that

> he that works with less accuracy is an imperfect mechanic, and if he could work with perfect accuracy, he would be the most perfect mechanic of all, for the description of right lines and circles, upon which geometry is founded, belongs to mechanics. (Newton 1962, p. xvii)

Newton's point is that while mechanics is just as free as geometry to postulate the accuracy if its constructions, geometry requires the mechanical assumption of *movable* instruments for the description of its objects. Indeed, "it is the glory of geometry that from these few principles, brought from without, it is able to produce so many things" (*ibid.*). Newton concludes that

> . . . geometry is founded in mechanical practice, and is nothing but that part of universal mechanics which accurately proposes and demonstrates the art of measuring. But since the manual arts are chiefly employed in the moving of bodies, it happens that geometry is commonly referred to their magnitude, and mechanics to their motion. (*ibid.*)

A century later Kant acknowledges Newton's view that "geometry requires only two mechanical actions, which it postulates, namely to describe a straight line and a circle" (Kant 1970, p. 17). But only in the first introduction to the third *Critique* does he try to answer this view of geometry:

> This pure and sublime science appears to lose some of its honor when it admits that in elementary geometry it requires instruments, even if only two, for the construction of its concepts, namely the compass and ruler, which constructions alone it calls geometrical and those of higher geometry mechanical, since for the construction of the concepts of the latter more complicated machines are needed. But one understands by the former not the actual instruments . . . which could never produce those figures with mathematical precision, they ought rather to signify *a priori* only the simplest kinds of representation of the imagination, that no instrument can produce as well. (Akademie, V. 23, p. 198)

The geometrical constructions Newton found rooted in mechanical practice grow for Kant from the productive imagination. It is unclear whether Kant *consciously* believed that the imagination can anticipate *a priori* the possibility of movable rigid bodies, but he was clearly begging the question against Newton in claiming a degree of precision for constructions of the imagination that he denied for mechanical practice. Newton had indeed already contested this in an unpublished elaboration of his view:

> Geometry does not posit modes of description: we are free to describe them by moving rulers around, using optical rays, taut threads, compasses, . . . the unfettered motion of a careful hand, or finally any mechanical means whatever. Geometry makes the unique demand that they be described exactly. However, it has come to be usual to regard as geometrical everything which is exact, and as mechanical all that proves not to be, as though nothing could possibly be exact and mechanical at the same time. But this common belief is a stupid one, and has its origin in nothing else than that geometry postulates an exact mechanical practice in the description of a straight line and a circle. (Newton 1976, p. 289)

But he insists that geometry only postulates these "because it knows not how to teach the mode of effection. It does not construct a postulate: mechanics alone is capable of doing this" (*ibid.*). The glory of geometry is now counterbalanced by what Newton calls its 'disgrace' (*ibid.*) in being unable to construct its own postulates. Since Kant never read these words, his remarks above about geometry's apparent loss of honor in having to borrow its construction from mechanics may indicate that such talk was in the air. His latest and most careful attempt to save geometry from such disgrace reads as follows:

> In the most general sense one can call construction all exhibition of a concept through the (spontaneous) production of a corresponding intuition. If it occurs through the mere imagination in accordance with an *a priori* concept, it is called pure construction. (These are constructions which the mathematician must make use of in all his demonstrations.) Hence he can demonstrate by means of a circle which he draws with his stick in the sand, no matter how irregular it may be, the attributes of a circle in general, as perfectly as if it had been etched in a copper plate by the greatest artist. (Kant 1973, p. 111)

Thus we see that for Kant the productive imagination not only bestows on geometrical construction an exactness unattainable by mechanical instruments, but also insures the cogency of geometrical proofs even when accompanied by inexact figures. This latter guarantee — the 'hence' in Kant's explanation — is now believed to rest rather on

Hilbert's complete axiomatization from which all theorems follow purely logically. However, the conflict between Kant and Newton is not logical but epistemological. The glory they both saw in geometry was its ability to prove so much from so little, while the disgrace they disagreed over was its need to borrow the contents of it postulates from another more empirical science.

This brings us to Hintikka's ingenious reconstruction of Kant's theory of mathematical knowledge. He challenges the common view of the 'syntheticity' of a mathematical proposition which takes it as ultimately an inheritance of the *axioms*, claiming that it is rather a matter of *the way it is proved*. He agrees that one can find "the roots of Kant's thoughts in Euclid," but in the methodology of his geometry rather than its content. Specifically, he finds Kant's notions of intuition and construction both rooted in his recognition of the need for 'auxiliary constructions' in Euclid's proofs. According to Hintikka, a Kantian intuition is just "a representation of a particular entity in the human mind," while by 'construction' he merely meant "the introduction of such a particular to instantiate a general concept" (Hintikka 1982, p. 201). He proposes that for Kant "a mathematical argument is synthetic if it involves the use of 'auxiliary constructions,' i.e. the introduction of new particulars over and above those given in the conditions of the argument" (*ibid.*), and emphasizes the role of the postulates discussed above in guaranteeing such constructions. In fact, "the syntheticity of a geometrical theorem is on my account recognized from the use of auxiliary constructions in its proof" (*ibid.*, p. 213). Hintikka believes his interpretation of Kant enables us, moreover, to tie his difficult notions of intuition and construction to "procedures which are closely related to the instantiation rules of modern logic" (*ibid.*, p. 207). Having located the syntheticity of theorems "squarely within the framework of the Euclidean axiomatic and deductive treatment of geometry" (*ibid.*), we need not "pursue the syntheticity back to the axioms" (*ibid.*, p. 213).

But what about the axioms and postulates? It is clear from the above that Kant regarded postulates as synthetic *a priori* propositions, differing from theorems only in being "provable by description" rather than "mediately through inferences." Axioms, according to Kant, "in so far as they are immediately certain, are synthetic *a priori* principles". (A732 = B760) Hintikka has indeed isolated a 'synthetic' component due to the *use* of existential postulates, but it hardly applies to Kant's attempt to derive them as 'practical corollaries' of definitions. The syntheticity Kant found in the circle postulate is not the use of instantial variables but rather the synthesis in imagination of the circle it postulates immediately from the definition of the circle. It deals with the content of euclidean geometry, not its methodology. Nevertheless, Hintikka maintains that the essential property of Kantian intuitions is the purely logical one of their *singularity* rather than the epistemological one of their *immediacy* or their relation to space and time. On this view, the syntheticity of geometry has nothing to do with its being 'the science of space'. Indeed, Kant's proof of the ideality of space, on the ground that synthetic *a priori* knowledge of objects in space would be impossible if they were things in themselves, would clearly collapse. However unlikely it seems that Kant would have rested such an argument on the mere use of auxiliary construction in geometry rather than its content as 'the science of space', Hintikka's analysis is confirmed by many Kantian texts and will be seen later as a part of the explanation for a crucial weakness in Kant's conception of geometry.

## 3. ORDER, DIRECTION, AND INTUITION

In his fifth letter to Clarke, Leibniz argued that the conception of "a finite material universe, the whole of which moves around in an infinite empty space" was meaningless since "no change which could be observed by anyone would be occurring" (Leibniz 1956, p. 226). For if the universe moves to a different region in space it should be in a different state, but since no observation could discern any difference, it violates Leibniz's identity of indiscernibles which says that two things cannot differ in respect of place and time alone. It is impossible to objectively define any 'place' in Newton's space. But Leibniz thinks we can define the locution that "body B now occupies the *same place* as had body *A* before." He supposes a sufficient number of 'fixed existents' *C, E, G, F*, . . . which have undergone no mutual changes and defines A and B to have occupied the same place "when the relation of coexistence of *B* with *C, G, E, F*, etc. agrees perfectly with the relation of coexistence which *A* previously had with these bodies, supposing there has been no cause of change in *C, E, G, F*, etc." (*ibid*., p. 231). Leibniz expresses his belief that *only* such definitions are possible for spatial concepts by saying that space is only "an order of coexistence of things." He admits that "the mind, not content with agreement, seeks an identity, a thing which is truly the same, and conceives it as outside those subjects, and this is what is called *place* and space" (*ibid.*, p. 232). But prudence stops short of such identities and Leibniz says that he has followed Euclid himself "who, not being able to make it understood what ratio is absolutely, in the geometrician's sense, properly defined what is meant by *same* ratio. In the same way, in order to explain what place is, I have tried to define the same place" (*ibid.*). He anticipated here the importance of equivalence relations for defining scientific concepts. In fact, we are often not content with them, but replace them with abstract identities between classes conceived "outside those subjects."

Euler (1748) argued that this reduction of Newton's space to relations between bodies left his laws of motion without any adequate interpretation, and that even the concept of direction contained in the law of inertia "would not be absolutely explained by the relation or order of coexisting bodies". He pointed out that he was rejecting the identity of indiscernibles only for positions in space, however, not for bodies and intellectual things. Kant seems to have taken both points by the time of his 1768 paper on incongruous counterparts in which he seeks in the 'intuitive judgments' of geometry a proof that "*absolute space has its own reality independently of the existence of all matter*" (Kant 1968, p. 37). A left and right hand are so similar, says Kant, that "a complete description of the one must apply, in all respects, to the other" (*ibid.*, p. 41). Ant yet an 'internal difference' remains, namely, that they cannot occupy the same region of space. Since they are 'completely alike' in the relations between their parts the ground of their distinction must lie beyond them. If the first created object were a hand it would have to be a left or right hand, and on the relational view of space as the order of coexistence of bodies Kant thinks that no further space than that occupied by our original hand would exist. But then, he argues, it would have to fit equally well onto *either* side of a human body, which is absurd. He concludes that the difference between a left and right hand is "connected with absolute and original space" (*ibid.*, p. 49), that only their relations to such space can distinguish such space. In short, they *would* be identical were it not for absolute space. But if this argument is in fact directed against Leibniz's view sketched above, it is clear that Kant missed his mark. Indeed, from the mere fact that one can exhibit two bodies which cannot occupy exactly the same place in space, however similar otherwise, it hardly follows that one cannot define in the manner of Leibniz what it would mean to say that two bodies *could* do so at different times, much less that we could define place without reference to bodies.[1]

In later works Kant uses incongruous counterparts to prove rather the ideality of space as well as independence of sensible intuition from the understanding. Thus in the *Dissertation* he claims that between such counterparts

> there is a diversity which makes it impossible for their boundaries to coincide although they could be substituted for one another as far as concerns all things which may be expressed in marks intelligible to the mind in speech. And so it is clear that in these cases the diversity, I mean the discongruity, can only be noticed by a certain act of pure intuition. (Kant 1968, p. 69)

Clearly the intuition invoked here is not of a *singular object*, but rather of a *difference between two objects*. We intuit a 'discongruity' that cannot be defined in "marks intelligible to the mind." In the *Prolegomena* the same conclusion is reached: such a difference cannot be made "intelligible by any single concept, but only by their relation to the left and right hand, which bears immediately on intuition" (Kant 1953, p. 43). But Kant now qualifies his formulation of the 'complete equality' of such objects: in incongruous spherical triangles "nothing is found in either, when it is described *alone* and completely, which does not also appear in the other" (*ibid*., p. 42; italics mine). The same qualification occurred for the same reason in Leibniz's definition of *similar* figures as those "which cannot be distinguished when observed in isolation from each other": to insure in each case that the figures are *geometrically indiscernible*. In fact, everthing Kant says above about his incongruities applies equally well to Euclid's similar figures of different size: they cannot be made to coincide, but as Leibniz had emphasized in his *Analysis Situs* by means of an explicit indiscernibility axiom, everything that can be expressed in the language of geometry and proved about one of them holds equally well of the other. Of course, Kant thinks his incongruous counterparts only pose a *paradox* to those "who cannot get away from the concept that space and time are real quantities attached to things in themselves" (*ibid*., p. 41), but later in his metaphysical reflections and *Opus Postumum* he will implicitly recognize the paradoxical nature of the similarity principle also by basing another ideality proof for space on it. The relation between these ideality proofs (both are based on automorphisms) was obscured in Kant's time by the historical circumstance that while similarity was already defined in Greek geometry, Kant's counterparts have properties of direction and orientation which are undefinable in geometry as long as its language contains no betweenness relation or order axioms.

The first to include such a relation among his primitives was Leibniz who called it 'intersitum', but did not propose any axioms for it. His follower Kästner, attempting to fill gaps in Euclid's proofs concerning the intersections of auxiliary lines, was the first to formulate explicit axioms relating such intersections to the notions of the 'inside' and 'outside' of figures as well as the 'opposite sides' of a line.[2] A new axiom related to Kant's incongruities says that "an unlimited straight line divides the unlimited plane in which it is drawn into two parts which lie on opposite sides of it", thus enabling a geometer for the first time to make logical use of such words as 'sides' in proofs. But only much later will Pasch formulate the axioms for Leibniz's betweenness relation making possible the *definitions* of such intuitive notions as the 'sides' and 'orientations' of figures by means of suitable equivalence relations. In particular, one can then define two orientations for symmetrical objects in space and prove that those of opposite orientation can be carried into each other by reflection but not by proper motions.

Though we can thus define the intuitive quality of orientation conceptually "in marks intelligible to the mind" after all, Kant would still point out that neither 'right' nor 'left' can be so defined: *empirical* intuition or feeling is needed to identify particular directions or orientations. In Kant's last discussion of incongruities of 1786 this conclusion is strengthened by the claim that *nature itself* recognizes no internal differences between them. For, "in things in themselves ... there can be no conceivable difference in the internal consequences" (Kant 1970, p. 23). And yet

> ... since this difference admits indeed of being given in intuition, but does not admit to being brought to clear concepts and therefore of being intelligibly explicated (*dari, non intelligi*), it affords a good confirmative proof ... that space in general does not belong to the properties or relations of things in themselves, which would necessarily have to admit of reduction to objective concepts, but belongs merely to the subjective form of our sensible intuition of things or relations, which must remain wholly unknown to us as regards what they may be in themselves. (*ibid.*)

But since this proof involves differences in only *one* spatial property that can be "given in intuition, but not understood conceptually," namely, orientation, the conclusion ought rather to be that *orientation is not an objective property of things in themselves*. But these are the very words (*dari, non intelligi*) used by Lambert to describe the difference in size between similar figures, so we should not be surprised to find Kant later formulating a *euclidean* argument for ideality turning on the question of the objective magnitudes of things. Meanwhile, however, intuition was also essential for Kant in the conduct of geometrical *proofs* and it is this function which the order axioms are often seen as usurping and hence also as undermining his synthetic *a priori*. To examine this we look at how the issue came up in his controversy with Eberhard, which indeed provoked what M. Gram has called a "crisis of syntheticity" for Kant.

The discussion centered on Euclid's triangle-inequality which expresses the essential content of Kant's paradigmatic synthetic *a priori* proposition that the straight line is the shortest between two points. J. Schwab argued that if Euclid's proof is based on a suitable definition of a triangle it can establish analytically from concepts that, given triangle $ABC$, the sum of the sides $AB$ and $AC$ is greater than $BC$. Euclid extended $AB$ to a point $D$ such that $AC = AD$, so by Thales' theorem $\angle ADC = \angle ACD$. Also $\angle BCD > \angle ACD$ by Euclid's axiom that the whole is greater than the part, hence $\angle BCD > \angle ACD$. Since the greater angle subtends the greater side, we have $BD > BC$, and Euclid's inequality follows. For Schwab the threefold crux of the matter was (i) to derive Euclid's postulate "to extend a straight line indefinitely" from the definition of a triangle as "a plane bounded by three straight lines" (Schwab 1791, p. 399), which implies analytically, he thinks, that it has a finite area and hence finite sides. Since it is also

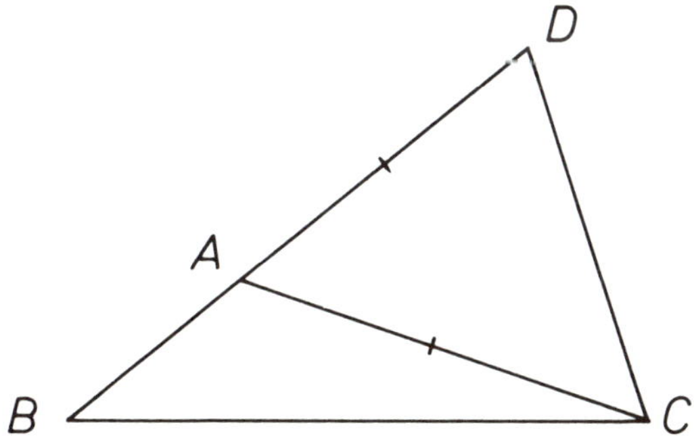


Fig. 1. Schwab's figure.

analytic that "any finite magnitude can increase" (*ibid.*, p. 400), Euclid's postulate follows. (ii) To prove Thales' theorem analytically from concepts: in addition to the extension postulate, Euclid's proof of it uses only the axiom that "two straight lines cannot enclose a space" and the congruence theorem for triangles. (iii) The obvious analyticity of the part-whole axiom. A. W. Rehberg objected on Kant's behalf to Schwab's *use of the figure*: for "a proof from concepts must be conducted entirely with mere words; for everything contained in a concept can be unfolded in words" (Rehberg 1972, p. 449). He doubted that the extension postulate could be proved with mere words without any figure. He admitted, however, that Schwab had revealed more analytic steps in the proof than met the eye. In fact, the only proposition used in it that could not possibly be proved analytically, according to Rehberg, was the axiom that two straight lines cannot enclose a space, for as Kant himself had remarked, "there is no contradiction in the concept of a figure which is enclosed within two straight lines, since the concepts of two straight lines and of their coming together contain no negation of a figure" ($A221 = B268$). Rehberg granted that both Euclid's congruence principle and his part-whole axiom were analytic, again following Kant with respect to the latter, though we shall see that Kant held steadfastly to synthetic apriority of the former. In fact, a purely conceptual proof of the triangle-inequality can use, by Kant's lights, neither figures, extension, nor congruence.

The essential reason why figures are needed in the proof, however, was obscured from Kant and his followers by the very analyticity of the part-whole axiom. In fact, without any order axioms a figure is needed to *apply* this axiom, which is invoked to show that $\angle BCD > \angle ACD$. But nothing in the words of the proof tells us *which of these angles is the part and which is the whole*. In the absence of order axioms only a figure can show us that $\angle ACD$ is part of $\angle BCD$. My concept of whole tells me that the angle *BCD* is greater than the angle *ACD if it is* the whole, but only my intuition of the figure shows me that it is. Here is another Kantian role of intuition in addition to distinguishing symmetries in space. Instead of intuiting a distinction we cannot define, we intuit an inclusion relation that does not follow logically from Euclid's own axioms. The order axioms do not completely usurp the first function: though they enable us to define and prove the existence of opposite handed objects, they cannot tell us which is the right one or the left. They do, however, obviate the need for intuition in applying the part-whole axiom, for once formulated they *will* tell us which is the part and which is the whole in the above proof. Kant might still claim that the order axioms are themselves based on spatial intuition; but since they are not construction postulates it would not be an intuition *of* the synthesis of geometric objects, but rather the intuition *that* certain relations hold.

Finally, betweenness is also needed to formulate the intention of the extension postulate, for without an axiom stating that exactly one of any three collinear points is between the other two we can go on 'extending' a straight line forever without insuring that we must always obtain *new* points. This is analogous to the problem of formulating the Peano axioms that insure that our intuition of counting will always produce new numbers. But in responding to Kästner's objections to the use of infinity in geometry, Kant (Akademie, V. 20) sharply distinguished the extension postulate from the counting principle in arithmetic:

> When the geometer says that a line, however far one has produced it, can always be extended further, that does not mean the same thing as when it is said in arithmetic of number that one can always increase it without end through the addition of other units or numbers (for these additional numbers and the magnitudes they express are *in themselves* possible, without having to be admitted with the previous ones only as parts of one magnitude), rather to say that a straight line can be extended into the infinite means no more than: the space in which I describe the line is greater than every space which I may describe in it, and so the geometer bases his task, of extending *a* space (of which there are many) into the infinite, on the original idea of a unique, infinite subjectively given space. (p. 420)

This nicely anticipates the constructivist claim in arithmetic that one can generate an infinity of integers without presupposing the existence of infinite sets in set-theory. Kant's geometer, however, cannot be assured of extending his lines as he pleases without a prior intuition of an infinite given space. This implies for Kant that the extension postulate "could never be inferred from concepts" (1953, p. 41), though he never suggests that it might be derived as a 'practical corollary' of a definition of straight line: its content is tied to a global property of space itself. Nevertheless, the geometer, says Kant, can leave the problem of "the subjective ground of the possibility of space, that is, its ideality" entirely to transcendental metaphysics. One strand of the proof of Kant's favorite synthetic *a priori* proposition thus leads directly back to his ideality of space, this time proved by analyzing the only sense in which he thinks it can coherently be regarded as 'infinite'. The use of the extension postulate in this proof also illustrates Hintikka's idea that the syntheticity of a theorem is tied to the introduction of new points in its proof rather than inherited from the axioms. However, there are proofs of the triangle inequality which do not use this postulate, so this idea can be further tested by examining the remaining loose ends in our analysis of its proof, namely, Thales' theorem and the congruence theorem for triangles.

## 4. CONGRUENCE, MOTION, AND PROOF

The Greek axiomatization of geometry marked for Kant the revolutionary beginning of science. The Egyptians, he admitted, had practical knowledge of geometry, but it consisted of "mere empirical surveying, which enabled them to measure according to certain rules without any proof" (Kant, Akademie, V. 23, p. 894). The Greeks raised such knowledge to the level of a science by discovering the possibility of proving it. In fact, they really discovered, Kant thought, without being able to quite say so, the synthetic *a priori* nature of geometrical propositions, such as the theorem that the sum of the angles of *any* triangle must be 2R. The Egyptians may have established this empirically by measuring the angles of various triangles and adding them up, he admits, but this could never establish their *necessity.* This possibility occurred only when

> A new light flashed upon the mind of the first man (be he Thales or some other) who demonstrated the properties of the isosceles triangle. The true method, so he found, was not to inspect what he discerned either in the figure, or in the bare concept of it, and from this, as it were, to read off its properties; but to bring out what was necessarily implied in the concepts that he had himself formed *a priori*, and had put into the figure in the construction by which he presented it to himself. If he is to know anything with *a priori* certainty he must not ascribe to the figure anything save what necessarily follows from what he has himself set into it in accordance with his concept. (Bxi—xii)

Whatever its difficulties, Kant's formulation does have an advantage over modern ones of applying equally well to proofs which, presumably like Thales', were not based on *any* explicitly stated axioms. Exactly what Thales' concept of isosceles triangle was and how he constructed it we shall probably never know. But we can be reasonably certain that in proving the equality of its base angles he used some congruence principle like 'side-angle-side'. Moreover, Kant's formulation implies that such a principle is "necessarily implied in the concept" of a triangle: unless he can justify this, he cannot explain what the Greeks knew about isosceles triangles that the Egyptians didn't.

Euclid tried to prove this congruence principle by means of a new axiom stating that "figures which coincide with one another are equal." But when he 'applies' the triangle *ABC* to triangle *DEF* (see Figure 2),

Fig. 2.

given the indicated equalities, he also presupposes the converse of this axiom, that is, that equal figures *can be made to coincide by moving them*. The medieval philosopher Adelard of Bath pointed out that this proof also assumes that this 'application' produces no change in the triangle moved, that it is a '*triangulis indissolutis*'. Such proofs were often criticized and Euclid himself seems to have taken pains to avoid them. Indeed, anyone answering sceptical attacks on geometry with Plato's thesis that its figures really stand for abstract *forms* was hard pressed to explain how such objects can move.

The revival of Greek learning in the Renaissance also revived this scepticism and Isaac Barrow defended attacks on the congruence principle by observing that those "who despise and reject it in mathematical Demonstrations as favoring too much of mechanical Bungling do endeavor to overthrow the very Basis of Geometry, but without either Wisdom or success" (Barrow 1734, p. 187). But to clear geometry of 'mechanical bungling' he invoked a mentalistic formulation: geometers "do not perform their *Congruity* by the Hand, but by Thought, not by the sense of the eye, but by the Judgment of the Mind. They suppose an accurate and perfect *Congruence* which no hand can perform, nor any sense discern, and from that Supposition draw just and logical Consequences" (*ibid.*). Not that hands are useless to the geometer, for "what the Mind demands to be understood, the Hand can execute in part, and the Praxis can in some measure emulate the Theory (*ibid.*, p. 188). But the theory and praxis of geometers both depend on *mechanics*, and "they who asperse the knowledge of Mechanics in those Demonstrations where Congruity is applied, do rather accuse them of the greatest Perfection, than of any defect or Blemish" (*ibid.*). The perfection that geometry takes from mechanics emerges clearly when, having defined congruence as the possibility of occupying the same place, Barrow says that "an *Occupation of Place* is not more extrinsically, nor less intrinsically conjoined with Magnitude than Motion . . . and as Magnitudes wheresoever placed are conceived equal, so Figures howsoever quiescent and unknown do obtain and preserve the same Nature" (*ibid.*, p. 224). Barrow's peculiarly clear formulation of the dependence of congruence on the mechanical assumption of rigidly movable magnitudes presumably had some influence on the views of Newton examined in section 2 above. It also influenced Hume, but failed to stem his sceptical attack on geometry in the *Treatise*.

To Hume it was "evident that the eye, or rather the mind, is often able at one view to determine the proportions of bodies, and pronounce them equal, greater, or less than each other, without examining or comparing the number of their minute parts" (Hume 1978, p. 47). Such judgments may be mistaken, "but we often discover our error by a juxtaposition of the objects, or . . . by use of some common and invariable measure which, being successively applied to them, informs us of their different proportions" (*ibid.*). The movable rigid body, which Hume implicitly sees to be a necessary condition for the transitivity of congruence required for indirect measurement, thus enters his account as a *corrective to judgments of perceptual equality*, leading to "a mixed notion of equality." But as sound reason convinces us that "there are bodies *vastly* more minute than those which appear to the senses" we realize that *no* instruments or art can prevent error and uncertainty in our measurements. Nevertheless, we imagine that the equality of two figures cannot survive the removal of such minute parts from one of them, and "we therefore suppose some imaginary standard of equality by which the appearances and measuring are exactly corrected" (*ibid.*). But as our very idea of equality is the mixed one of appearances corrected by "invariable" rulers, "the notion of any correction beyond what we have any instruments and art to make, is a mere fiction of the mind" (*ibid.*). But since no exact standard of equality exists, Hume concludes that we have no exact standard for *straight lines* either. Indeed, Hume challenges the geometer to prove to him that two straight lines cannot have a common segment, for "supposing these two lines to approach each other at the rate of one inch in twenty leagues, I perceive no absurdity in asserting that upon contact they become one" (*ibid.*, p. 52). Alas, Saccheri had just published unbeknownst to Hume a milestone in geometry in which he tried unsuccessfully to prove the parallel postulate by reducing to absurdity a similar notion of asymptotic straight lines. Hume's main conclusion is that geometry is a fallible science based on a mixed notion of 'congruity' derived from perceptual and mechanical experience of measuring.

Whereas Barrow and Hume analyzed the mechanical and experiential basis for congruence, Leibniz also analyzed its logical role in geometry by basing various geometrical calculi on it. In the *Calculus Situs* he takes congruence as the sole primitive relation and shows it to be an equivalence relation which 'represents identity' in the calculus. Thus congruent figures can be identified in the calculus, since it abstracts from their place, which it cannot express. This is what Leibniz means when he says that geometrical truths can be reduced to a principle of identity: they can sometimes be proved by means of a congruence relation sharing the main logical properties of the identity relation. A classic case of this was Pappus' proof by reflection of Thales' theorem without any auxiliary constructions. But while congruent figures are only formally indiscernible in his calculus, Leibniz also held a *material* version of the indiscernibility principle claiming that

(I) Two bodies cannot differ in respect of place and time alone (some other 'internal difference' is always necessary),

which expresses for Leibniz a fact about real things in nature which he says "overthrows the whole of corpuscularian philosophy" which postulated congruent atoms individuated only by their places in space. It follows that such notions "are mathematical, that is, they are abstract and not real" (Leibniz 1973, p. 133). One must distinguish then between the indiscernibility principles which govern Leibniz's abstract calculi for congruence and similarity and (I) allegedly holding for real things in nature. Only the latter conflicts with transcendental idealism, since according to Kant "a location *b* can contain a thing which is completely similar and equal to another in location *a*, just as easily as if the things were inwardly ever so different. Difference of locations, without any further conditions, makes the plurality and distinction of objects, *as appearances*, not only possible but also necessary" (B328, italics mine). So a mere difference in their location can individuate two cubic feet which are conceptually indistinguishable since "these locations are conditions of the intuition wherein the object of this concept [of a cubic foot of space] is given; they do not, however, belong to the concept but entirely to sensibility" (B338). Again, Kant invokes empirical intuition to distinguish otherwise 'completely similar and equal' things, this time by their different locations in space. But this depends on his ideality of space and we cannot but wonder why Kant doesn't invoke also his incongruous counterparts against Leibniz here, or anywhere in his masterpiece on transcendental idealism. He used them three times from 1770 to 1786 to establish intuition and ideality of space, presumably against Leibniz and his indiscernibility principles. But here in the *Critique* itself, where he does explicitly confront both together in the Amphibolies of Reflection, the incongruities are conspicuous by their absence. Does Kant now see on reflection that *congruent* things will serve even better as 'completely similar and equal' objects that can only be distinguished by the conditions of sensibility? In fact, Kant's remarks here really comprise yet another ideality proof: since empirical intuition can distinguish positions in space which the understanding cannot, position must be a property of appearances and not of things in themselves. If space were a property of the latter, we'd have to identify congruent things. Again, it seems that Kant has made, explicitly or only implicitly, an ideality argument for each of the three automorphism groups for euclidean space, in which the properties of position, size, and orientation are correspondingly 'idealized' in his sense. But as Leibniz knew only too well, it is a difficult matter at best to free congruence from its presuppositions of rigidity and motion, which brings us to the question of how this can be managed in transcendental idealism.

In the first edition of the *Critique* Kant said that all concepts involving *both* space and time "presuppose something empirical." In particular,

> Motion presupposes the perception of something movable. But in space, considered in itself, there is nothing movable; consequently, the movable must be something that is found *in space only through experience*, and must therefore be an empirical datum. (A41 = B58)

But how then, asked his follower C. G. Schütz, can geometry be the *a priori* science of space if to know anything in space I must *draw* it? "Drawing is however a kind of motion; motion is an empirical concept; therefore it appears that even lines, consequently figures, and also the conic sections require empirical support in order to be represented" (Schütz 1785, p. 43). Kant replied directly to this question in the second edition of the *Critique* by distinguishing between the motion of an *object* in space and motion as an *act* of describing space:

> Motion of an object in space does not belong to a pure science, and consequently not to geometry. For the fact that something is movable cannot be known *a priori*, but only through experience. Motion, however, considered as the describing of a space, is a pure act of the successive synthesis of the manifold in outer intuition in general by means of the productive imagination, and belongs not only to geometry, but even to transcendental philosophy. (B155a)

These are just the descriptive acts we have seen Kant invoking in correspondence and in reply to Kästner to derive construction postulates as 'practical corollaries' of definitions. But can the *geometer* prove anything by them? The circle postulate is certainly such a corollary for a glazier who is practically certain that he can describe a perfect circle by turning the cutting wheel of his glass cutter around its fixed point; but only because he believes his instrument to be *rigid*, something he could not imagine *a priori*. Even if straight lines can be synthesized by acts of imagination, it is far from clear that even the further ability to move *them* in thought can secure the circle axiom. I can, of course, imagine them to turn rigidly, but I can *just as easily imagine them not to do so*, and with this admission we lose the universality and *necessity* of our axiom. In short, the synthesis of a circle requires much more than that of a straight line: the latter requires only a moving point, the former a property of the movable *object* so synthesized. In the case of the circle we seem to have some control over this motion inasmuch as we keep one end of the line fixed. But in the more important and general case of Euclid's congruence theorem we lose even this, for now we must move a *figure*: after synthesizing it we must, so to speak, set it free to move on its own. Nevertheless, Kant claims in the *Prolegomena* without any hesitation that

> All proofs of the complete congruence of two given figures . . . finally come to this, that they coincide with each other; which is obviously nothing other than a synthetic proposition resting on immediate intuition. This intuition must be given pure and *a priori*, otherwise the proposition could not hold as apodictically certain, but would have only empirical certainty. (Kant 1953, p. 40)

But why *must* they coincide after being moved? How can their use in 'describing space' insure this? And is this description the 'construction' of the proof? Unfortunately, Kant never answers these questions. However, his follower A. W. Mellin tried to justify Kant's position here in his *Dictionary of Critical Philosophy* by basing Euclid's proof of the congruence theorem on 'pure transcendental motion', even laying out the proof in the euclidean format that Hintikka has shown to be important for Kant. But for its 'construction', which begins when one "lays in thought the one triangle on the other," Mellin must formulate a new postulate which is lacking in Euclid, namely, "to lay one figure arbitrarily on another, in thought." But this is hardly a postulate in the usual sense, for *no new points or lines beyond the given data are produced.* Nor does he explain how we know that such movement does not *change* the given relations. He meets the familiar charge of Newton that geometry depends on mechanics here by claiming that his construction is not "a mechanical application of the triangle in experience," but rather occurs "like any other construction, merely in thought (in the pure or non-empirical imagination)" (Mellin 1801, p. 798). Yet Mellin insists that "the demonstration shows that the coincidence, under given conditions, *must* follow, that is, necessarily (apodictically), and consequently also for *all* definite triangles . . . must hold, because empirical determinations have no influence here" (*ibid.*, p. 799). But this does not *logically* follow even with Mellin's new postulate, for we can imagine the motion disrupting the given conditions. Neither Kant nor Mellin explain why it is that, in Barrow's (1734) words, "as Magnitudes wheresoever placed are conceived equal, so Figures howsoever quiescent and unknown do obtain and preserve the same Nature". Barrow and Newton insist that this is borrowed from mechanics. For Kant it must be necessarily implied in our concept of triangle and be 'put into the figure in the construction by which we present it'. But here there are no *auxiliary constructions* of the kind to which Kant could point in his famous description of how the geometer 'is guided throughout by intuition' in his proof that Euclid's angle sum holds for all possible triangles. Here, as we might echo what he said of the philosopher's vain efforts to derive that theorem from his concept of triangle, the geometer has only his triangle with given relations to another, and he must somehow move it in thought on the other one without changing it in any way, without any bungling.

How strong this assumption really is emerges from a theorem that Saccheri was able to base on congruence, namely

($S$) If even *one* triangle has an angle sum of $2R$, then they *all* do.

The 'universality' problem for Euclid's angle sum is thus concentrated entirely in the congruence principle itself: there is no problem of the kind imagined by Kant in proving it for all triangles from a particular triangle, *if* one know the rigid mobility of one's figures. In short, ($S$) shows that the induction problem for geometry centers on congruence, which implies as we now say, the 'constant curvature' of space. But what, as Hume might have asked Kant, is the necessary connection between my concept of a triangle and its rigid mobility? The congruence theorem basically expresses the 'causal inertness' of space: if a body changes in motion we look for a cause in other bodies and not in space itself. But in the *Treatise* Hume had even questioned whether every effect need have a cause, and the same reasoning he applied to causality and induction applies at least as well to congruence: I can just as easily *imagine* a triangle not surviving transport intact as I can the sun not rising tomorrow. It is also worth noting here that, even before Helmholtz and Einstein, and independently of non-euclidean geometry, classical physics posed a dilemma for geometry by pulling out from under it the very rigid bodies on which its congruence principles seemed to some to depend, and we conclude this section by reviewing how Kant's idealism founders over it.

Kant's dilemma over congruence begins with his infamous proof of the *apriority* of space on the grounds that "We can never represent to ourselves the absence of space, though we can quite well think it as empty of objects" (A24 = B39). It follows that our representation of space is independent of any objects at all, let alone rigidly movable ones, and hence that the existence of such bodies is not necessarily implied in our concept of space, but must be, as Kant says, an empirical datum of experience. But geometry, he says, is "a science which determines the properties of space synthetically, and yet *a priori*" (B40), and hence its congruence theorems must express a necessary property of space. However, their natural interpretation presupposes rigidly movable figures, otherwise Euclid's proof collapses. Since Kant could not give up the apriority of space and its consequent independence from movable bodies, he was forced to find another less natural and certainly inadequate basis for congruence in his subjective acts of 'describing space'. These play the role, so to speak, of his missing 'anticipations of motion'. In sharp contrast to this, his contemporary Boscovitch tried to compensate for the lack of ideally rigid bodies in nature which classical physics had begun to reveal, and the consequent failure of the transitivity law for physical congruence, by formulating new laws of attraction (for an atomic theory) which could still provide an acceptable criterion for "the equality of measured intervals" even though under translation "the standard of comparison will not remain identically the same" (Boscovitch 1966, p. 204). As Einstein (1961) later observed, "the subtlety of the concept of space was enhanced by the discovery that there exist no completely rigid bodies . . . . The structure, whose possible congruences are to be described by euclidean geometry cannot therefore be represented apart from physical concepts" (pp. 142—143). By adapting another well known remark of Einstein, we can summarize Kant's dilemma like this: insofar as geometry is the science of space it is not certain and insofar as it is certain it cannot be the science of space. Finally, Kant's inadequate treatment of congruence left him unable to clarify just what Thales could have known about isosceles triangles that the Egyptians didn't, or what advantage the geometer has over the philosopher in trying to prove congruence theorems. However, the decisive advance of the Greeks lay in their theory of parallels and similarity by which they began to measure the entire world and the heavens, and here is where some of Kant's ideas come into their own.

## 5. PARALLELS, FORMS, AND NUTSHELLS

Kant distinguished mathematical knowledge based on the construction of concepts very sharply from philosophical knowledge based on their analysis. The latter can only bring to light what is already contained in the concept while mathematical construction adds something new to it. While the philosopher would vainly analyze his concept of triangle for its angle sum, the geometer succeeds by taking a more constructive approach:

> He at once begins by constructing a triangle. Since he knows that the sum of two right angles is exactly equal to the sum of all the adjacent angles. . . . He then divides the

external angle by drawing a line parallel to the opposite side of the triangle, and observes that he has thus obtained an external adjacent angle which is equal to an internal angle. . . . In this fashion, through a chain of inferences guided throughout by intuition, he arrives at a fully evident and universally valid solution of the problem. (A716 = B744)

The example nicely illustrates Kant's distinction though as we have seen, if the geometer fills in this proof there will be steps requiring congruence that he can take only on the advice of mechanics or transcendental philosophy. In fact, such steps are required to *draw a parallel.* Moreover, for the proof to work, the geometer must know that the parallel he draws is the *only* one he could possibly draw, which knowledge can hardly be itself contained in the construction. An infinity of different constructions can be shown by congruence to produce a parallel, and the parallel postulate says they all lead to the same one. Clearly no one of them can speak for all the others. Such uniqueness problems confirm, I believe, an important observation of C. Parsons (1980), who points out that, though playing no official role in the *Critique*, Kant needs propositional *intuitions-that* as much as singular *intuitions-of* objects. Thus, in the case of the parallel, his geometer must not only construct an intuition, i.e. *of* a single parallel, he must intuit *that* no other construction can lead to a different such intuition. And so with uniqueness generally: one must intuit-that various intuitions-of are of the same object. Kant's intuitivity thesis is thus more complicated than meets the eye, and the question raised by Hintikka's analysis is whether it can be simplified by reducing intuitions-that to intuitions-of. But how conscious was Kant of all that his thesis involves, particularly in connection with the problem of parallels? Some light on this is shed by some investigations into parallels undertaken by Kant himself shortly after he wrote the advertisement above for the geometer's method.

Interestingly enough, Kant was more consciously concerned about the *existence* than he was the uniqueness of parallels, which is understandable if we recall the role he gave to definitions. When the mathematician 'constructs a concept in intuition', he does so from a definition of which, if appropriate, as we saw in the case of the circle, the construct is a 'practical corollary'. Euclid's definition of parallel lines, however, is not so obliging: they are "straight lines which, being in the same plane and being produced indefinitely in both directions, do not meet one another in either direction", which offers neither a 'rule' for constructing parallels, nor suggests any 'synthesis' of them. As Kant puts it in his notes on parallels: "I think that from a definition which does not at the same time contain the construction of the concept itself, nothing follows that would be a synthetic predicate" (Kant, Akademie V. 14, p. 31). He thus replaces it with Wolff's definition of parallels as "straight lines which are everywhere equidistant," hereafter "Wolff-parallels." But Kant finds that these are really no easier to construct *a priori* that Euclid's. In fact, the latter's postulate is equivalent to the *existence* of Wolff-parallels.[3]

Kant's interest in the problem of parallels was also stimulated by the theory of parallels developed by his friend and commentator J. Schultz, which is based on the idea that "the magnitude of a plane angle is determined by the amount of the plane that lies between its sides prolonged without end" (Kant, Akademie, V. 13, p. 132). Suppose

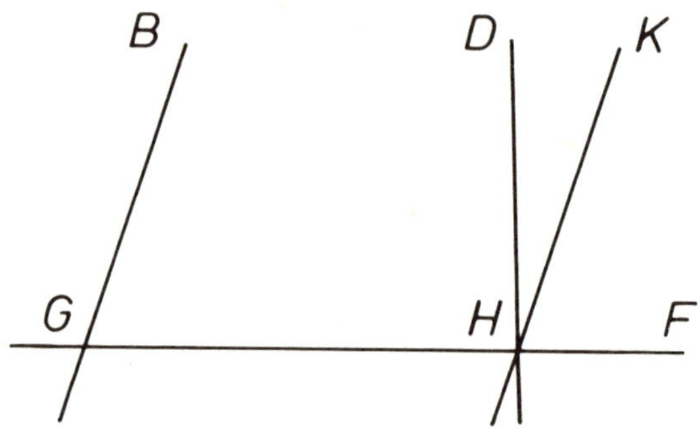


Fig. 3. Schultz's figure.

angles *BGH* and *KHF* are equal and *DH* is perpendicular to *GH*. To prove that *BG* must intersect *DH* if both are produced indefinitely, Schultz introduces two kinds of infinite objects based on this idea. The area contained in the angle *DHK* when its sides are prolonged without end is called an 'infinite area', which Schultz claims bears a definite ratio to the whole infinite plane, the ratio of its angle to 360°. The infinite area, on the other hand, contained by *GH* and the parallels *BG* and *HK* prolonged without end is called a 'strip'. Schultz argues that strips all bear vanishing ratios to infinite areas; unlike such areas, no finite number of strips covers the whole plane. The strip *BGHK* has such a ratio to the infinite area *KHF* because the latter is equal to the infinite area *BGH*. But the infinite area *DHK* has a finite ratio to that of *KHF*, hence the strip *BGHK* has a vanishing ratio to the infinite area *DHK*. It follows according to Schultz, that *BG* must eventually meet *DH*, otherwise the infinite area would be *part* of the strip. However, his proof that the infinite areas *BGH* and *KHF* are equal presupposes that *BG* and *HK* are not just Euclid-parallels, but also Wolff-parallels. In short, he presupposed something equivalent to what he was trying to prove. But no one was aware of this for some time.[4] Kant himself wrote to Schultz in 1784 that "I have no doubt that your ... ingenious theory of parallel lines will broaden and extend human knowledge and contribute to your deserved fame" (Kant 1967, p. 113). However, it is clear from Kant's notes on parallels that he preferred a proof that did not use Schultz's infinite areas and strips; indeed, in such a proof the fact that these areas are not entirely given made it much less clear that the geometer was being 'guided throughout by intuition'. But Kant only claimed to find a 'philosophical proof' of Euclid's postulate (a claim repeated three times in the *Opus Postumum*), not one which was "mathematically represented". Specifically, his theorem that a line perpendicular to one of two Wolff-parallels is perpendicular to the other, is only proved conceptually, since according to Kant it follows from the concept of such parallels that they alone among pairs of lines 'have a determinate distance between them'. Briefly, it is based on concepts and not intuitions. Kant is thus forced to broaden his notion of proof in geometry, distinguishing *philosophical* proofs from concepts which, though foregoing any construction, "can still be proved with complete rigor but not euclideanly," *mathematical* proofs which use infinite objects such as Schultz's areas and strips, and purely *geometrical* proofs in which "the magnitudes whose relations are to be established can be entirely given" (Kant, Akademie V. 14, p. 48), indeed "*ohne Herbeiziehung unendlicher Flachen*". Since its magnitudes are finite, geometrical proofs are preferred, though mathematical ones are legitimate, even "better than merely philosophical ones" (*ibid.*). These distinctions mark Kant's growing concern with the problematic nature of his metaphysical exposition of space as 'an infinite given magnitude', no doubt heightened by Schultz's own insistence that his infinite areas are 'given in pure intuition' in Kant's sense. In fact, Schultz's theory of parallels was the target of some sharp criticism in Kant's controversy with Eberhard. Kästner (1790b) argued against it that only the potential, not the actual infinite was admissible in geometry, and Kant's reply to him seems to tacitly concede the point.[5]

This brings us to the question of the relation of Kant's epistemology to euclidean geometry. As it happened, Kant's own approach to parallels did not lead to an equivalent to Euclid's postulate directly related to his own philosophical notions. A little more 'axiomatic consciousness' of Wolff-parallels in Hilbert's sense would have shown him that all he needed construct in intuition was a *single rectangle* to prove this postulate. However, by an 'intuition' of a rectangle, he clearly could not have meant anything like an instantial variable introduced logically from such an existential axiom, e.g. 'there exists a rectangle'. It would rather have to be something that established the axiom itself *a priori*. But as rectangles bear no obvious relation to the main ideas of Kant's idealism, it will be useful to turn to another approach to parallels which does bear such a relation, namely, the theory of similarity and forms of Leibniz and Lambert.

Leibniz seemed to vacillate between basing geometry on congruence and basing it on similarity: As he says, "*geometria duplex est, una quae utitur solo principio congruentiae, altera quae utitur et principio similitudinis*" (Leibniz 1903, p. 525). Whereas congruence is a matter of quantity, many problems of geometry depend on only the similarity of figures, concerning which "Euclid himself was forced to assume certain obscure axioms without proof" (Leibniz 1976, p. 254), above all his axiom of parallels. But calling figures similar if they have the *same form* is useless without any independent definition of form, so Leibniz defines them as we saw as figures "which cannot be distinguished when observed in isolation from each other" (*ibid.*, p. 255). Lacking a general concept of similarity the ancient geometers defined such figures by the equality of their corresponding angles, but Leibniz saw that this special case failed to reveal the general nature of similarity. To illustrate the use of his definition in proving the general similarity of such equiangular figures, he formulates its indiscernibility axiom

(i) "Things which are indistinguishable by their determinants (data sufficient to define them) cannot be distinguished at all" (*ibid.*).

Given equiangular triangles *ABC* and *LMN* Leibniz claims that the base *BC* and angles *B* and *C* determine the former while base *MN* and angles *M* and *N* determine the latter. From these data (which are tacitly assumed to be) sufficient to determine each triangle it should follow

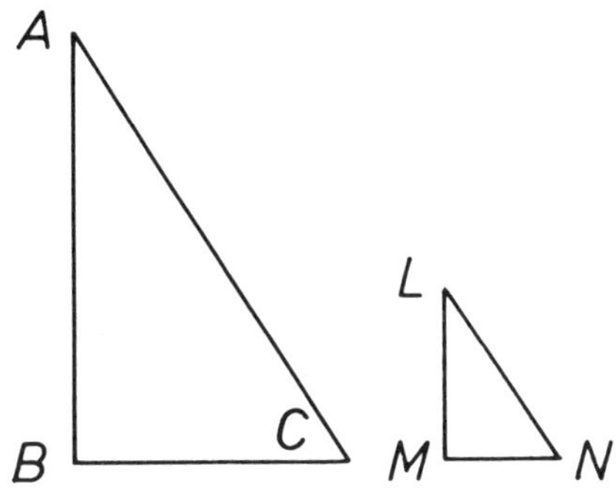


Fig. 4 Leibniz's figure.

from (i) that they cannot be distinguished at all if viewed in isolation from each other, for

> in each triangle, the base and two adjacent angles are given, but the base cannot be compared with the angles. Therefore there remains nothing which can be examined in each triangle viewed singly by its determinants, except the ratio of each given angle to a right angle . . . that is, the magnitude of the angle itself. And since these are found to be the same in both triangles, it follows necessarily that the triangles singly cannot be distinguished and therefore are similar (*ibid.*).

The triangles are distinguished by magnitude, but "this can be known only by observing together either both triangles at the same time or each with some other unit of measure" (*ibid.*). The axiom (i) also implies that triangles with proportional sides are similar in Leibniz's sense since "if the sides are given, the triangles are given, and it suffices (according to our axiom) that no basis for distinguishing can be found in the ratio of the sides, so that we may conclude that such a distinction can be found in no other circumstances in the triangles viewed singly" (*ibid.*, p. 257).

Leibniz's proofs went to the heart of euclidean geometry which, as Lambert will discover, is unique among geometries in maintaining the double standard of an absolute measure for angles but a merely relative one for length. Hence length and size can only be determined by the use of real standards imported from outside geometry, whereas quality and form can be defined intrinsically by pure numbers. Units of length involve a "truly arbitrary element," says Leibniz, "for since a straight line is uniform and indefinitely long nature does not indicate any boundaries on it" (Leibniz 1981, p. 303). The consequence that size can only be "given, not intelligibly conceived" will later be seen to play a crucial role in some of Kant's ideality arguments.

Kant apparently knew of Leibniz's theory of similarity only through Wolff, whose entry on similarity in his mathematical lexicon notes the distinction between angles and size, which "can only be given unclearly to the imagination, however cannot be explained in words or clearly conceived by the understanding" (1965, p. 1179), but misses Leibniz's point that angles *can* be so conceived. Wolff repeated Leibniz's claim that since ancient geometers had no general concept of similarity they had to assume unproved axioms (such as that of parallels which Wolff himself tried to prove with *his* parallels), which became theorems with a more general definition. Kant reacted to this theory in his 1763 paper on the clarity of natural theology in which the distinction between mathematical and philosophical knowledge emerged for the first time. Philosophers who thought they had learned rigor from geometry had failed to notice, according to Kant, that "geometers acquire concepts by means of *synthesis*, while philosophers can only do so by means of *analysis*" (Kant, 1968). Mathematicians had indeed sometimes tried to frame analytic definitions, but it was a mistake:

> It was in this way that Wolff considered similarity in geometry with the eye of a philosopher, intending to subsume the geometrical concept of similarity under the general notion of similarity. He could, however, have well abandoned the task, for when I think of geometrical figures in which the angles included by the lines of the outline are reciprocally equal, and the sides which they include have one and the same relation, then this can always be regarded as the definition of the similarity of geometrical figures; and similarly for the remaining spatial similarities. The general definition of similarity is of no consequence to the geometer. (*ibid*., p. 7)

Clearly Kant did not appreciate the need of a more general definition like Leibniz's to cover the similarity of the arbitrary curvilinear figures for which he had discovered a differential calculus. But neither does Kant seem to have aware of the fact that the *existence* of similar figures is neither a consequence of Euclid's definition nor the Leibnizian one. It was still only the former who had proved their existence, precisely by means of his parallel postulate. J. Wallis (1663), on the other hand, had shown that this postulate is provable if we postulate the existence of triangles of arbitrary size similar to a given one, demonstrating the important equivalence of Euclid's postulate with form independent of size implicit in the similarity postulate. But the latter was more complicated than Euclid's postulate, which was already regarded by many as too complicated for a postulate. A century later G. Klügel (1763) examined at Kästner's instigation twenty-eight more attempts published by geometers to prove Euclid's postulate and found all of them full of fallacies and paralogisms, which of course only stimulated even more such attempts.

So just as geometers were becoming *increasingly uncertain as to whether the parallel postulate was provable*, Kant was basing an explanation of the supposed certainty of their knowledge on their use of concrete symbols, "for since mathematical signs are sensuous epistemological tools, it is possible to know that no concept has been neglected . . . it is possible to know this with the same confidence with which one is assured of what one sees with one's eyes" (*ibid.*, p. 24). We have seen, however, that the concept of betweenness and order was long neglected until Kästner's work precisely because of the reliance on concrete figures. While Kant then was claiming in a Prize Essay that only geometry's disdain for analytic definitions and its use of such symbols spared it "exactly the same unfortunate discord" that plagued philosophy, Kästner was quietly urging that without a closer analysis of its neglected concepts of order and congruence, geometry may never resolve the growing discord over the status of its basic postulate. We can only wonder whether Kant's own belated awareness of this discord many years later motivated his own interest in finding out what "the problem of parallels" was all about, indeed how such a problem was possible at all given his assurances of 1763. At that time he was also convinced that geometers had finally resolved the old puzzle of the infinite divisibility of space by means of their symbols; for

> If the geometer, for example, wishes to demonstrate that space is infinitely divisible, he takes, say, a straight line, standing vertically between parallel lines, and projects other lines from a point on one of the parallel lines, which intersect such lines. He recognizes with greatest certainty, from this symbol, that the division must go on endlessly (*ibid.*, p. 7).

This raises anew a question raised in connection with Kant's treatment of the extension postulate: how can geometry be the science of space if none of its axioms ever mention space? In particular, what has the parallel postulate to do with space? The answers to these questions, as well as Kant's possible knowledge of non-euclidean geometry, are bound up with his relation to J. H. Lambert, to whom he had initially

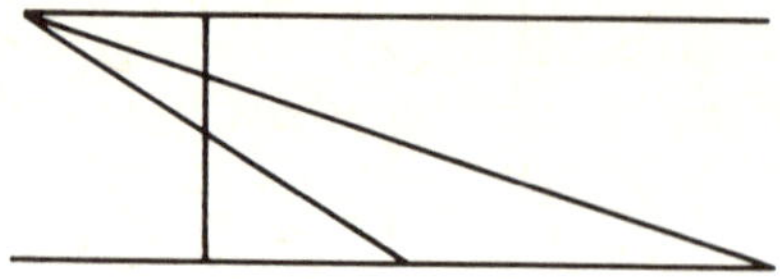

Fig. 5. Kant's figure. (See Akademie, V. 1, p. 478. Kant's later critical reflection (Akademie V. 18, p. 379) relates this infinite divisibility to an "absolute magnitude" for space, but not in the precise sense that we shall presently see Lambert giving to this idea.)

intended to dedicate the *Critique of Pure Reason* before his untimely death.

In his *Architektonik* Lambert proposed that philosophy, after developing a general theory of similarity, should investigate *form*: the form of things, the form of appearances, and the form of knowledge. This would make form 'abstract and transcendent'. To approach space and time in this way, he explains the intrinsic measure of angles, remarking that nobody doubts that they are thereby "intelligible without having to be presented." But linear magnitude and size must be "immediately presented and perceived . . . not conceived," or, as he quotes Wolff's 'theorem' proved from his definition of quantity:

(*Q*) "*Quantitas dari sed non per se intelligi potest*" (cf. Wolff 1736, §196)

(which we recall are the words Kant himself uses later to characterize the difference between his incongruities). Lambert stresses, however, the case of angular quantities as showing that (*Q*) cannot hold, as Wolff thought, for all quantity. We can always assume that parts of space are larger or smaller as we please, since no such assumption is more intelligible than any other "if we consider them in themselves without any relation to other things in the world which have a definite magnitude" (Lambert 1771, p. 325). Geometry is also the 'science of space' for Lambert, but as he observed in a letter of 1766 to Kant, "Euclid does not derive his elements from either the definition of space or that of geometry, but begins instead with . . . the simplest elements of space" (Kant 1967, p. 53). Lambert thus tries to formulate the axioms of geometry as statements expressing properties of space itself, beginning with

(*A*) Space has no determinate unit (*bestimmte Einheit*) (Lambert 1771, p. 62),

which is Lambert's ingenious replacement for the parallel postulate, whose truth, he says, we can represent only "with more effort and less evidence" (*ibid.*, p. 305) than that of other axioms (see Note 7). In fact, stimulated by Klügel's dissertation, Lambert tried in his *Theory of Parallel Lines* (1766) to prove Euclid's postulate himself, but 'discovered' non-euclidean geometry instead.

Clearly perceiving that Wolff's redefinition of parallels only made the problem worse, he followed Saccheri's approach of pursuing the deductive consequences of the two possible non-euclidean angle-sums for triangles until they were reduced to absurdity. But this reduction, he explains, like any correct deduction, must be carried out in an "absolutely symbolic way" which completely abstracts from "the representation of the matter" (*Vorstellung der Sache*). Such a deduction from Euclid's "other axioms" must thus treat them "as so many algebraic equations, that one already has as previously given, and that one must solve for *x, y, z*, . . . without looking back to the matter itself" (Lambert 1895, p. 162).[6] Lambert's main tool is the quadrilateral named after him having three right angles whose fourth angle, on the assumption that the angle-sum for triangles $< 2R$, is less than a right angle. The difference between the fourth angle and $R$ is the *defect* of the Lambert quadrilateral, and similarly for the difference between the angle-sum of a triangle and $2R$. Lambert shows that defects are additive in that they are the sums of the defects of the triangles into which a given figure can be decomposed. This enabled him to solve the difficult problem of *area* for non-euclidean figures by proving it to be proportional to their defects. From this follows

(*L*) If the angle-sum of triangles $< 2R$, then there is an absolute measure for every line, area, and volume in space.

Lambert's proof uses his area theorem to establish a one-to-one correspondence between angles and Lambert quadrilaterals. As the angle $G$, e.g. goes from 90° to 0, the area of ADGB goes from 0 to infinity, and similarly for the angle $J < G$ and *ACJE*. Thus $G$ is the 'absolute measure' of *ADGB* in the sense that, like all angles, it is intrinsically definable in geometry itself and uniquely correlated with

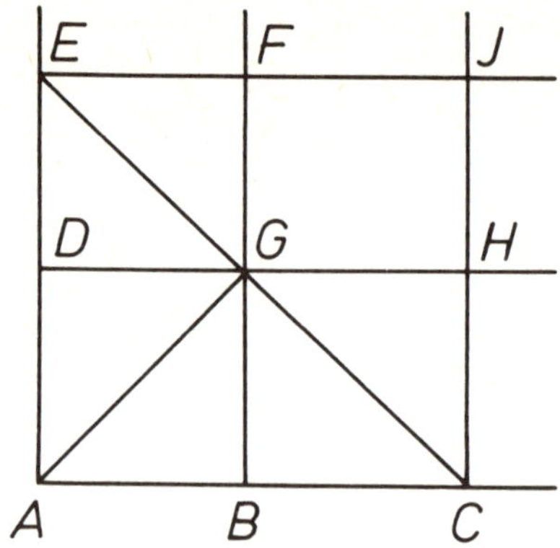


Fig. 6. Lambert's figure.

this quadrilateral. If a Paris foot, says Lambert, happens to correspond to 80°, we may increase *ABGD* until its acute angle is 80° and take its side *AB* = *AD* for the absolute measure of the Paris foot. Theoretically then, such a magnitude would not need to be exemplified by a physical object nursed under ideal conditions. But although (*L*) contradicts (*A*), a proposition, says Lambert, which "no one has ever doubted," it is nevertheless not absurd; and, in particular, *it cannot be refuted by using* (*Q*) *without begging the question*, namely, why (*Q*) should hold for linear but not angular magnitudes.[7] Indeed, there is, says Lambert, something so "alluring" about (*L*) that he would just as soon wish that the angle-sums of triangles *were* $< 2R$, a sentiment Gauss will share later when he writes to Gerling that "it seems paradoxical, that a constant as it were can be possible *a priori* but I find nothing in it contradictory. It would even be desirable that Euclid's [geometry] were not true because then we would have *a priori* an absolute measure" (Gauss 1900, p. 169). Lobachevsky was perhaps the most allured of all the great geometers by this prospect, but he admitted that "we are not in a position to conceive what kind of relation between things could hold in nature that would bind such different things as lines and angles" (Lobachevsky 1898, p. 24).

Among the "inconveniences" one had to tolerate as a price of the 'advantages' of (*L*), Lambert listed the absence of similar figures of different sizes and the consequence that "no figure could be represented in other than its absolute magnitude" (Lambert 1895, p. 201). The question is whether and why this would. incovenience Kant's transcendental basis for geometry; whether, that is, it is a consequence of his thesis that space is only the form of our outer intuition that the figures we synthesize in it could not possibly be represented in their absolute magnitude. Or whether, indeed, he was at all aware of such a possibility. To pursue these questions we examine next the role of form and figure in Kant's idealism, and then his treatment of magnitude.

In the Transcendental Aesthetic figures are introduced along with pure intuition as follows. The "*form* of appearance" is said to be "that which so determines the manifold of appearance that it allows of being ordered in certain relations" (*A*20 = *B*34). While the sensation or matter of appearance can only be given "its form must lie ready for the sensations *a priori* in the mind, and so must allow of being considered apart from all sensation" (*ibid.*). Kant calls such form "*pure intuition*". And now

> . . . if I take away from the representation of a body that which the understanding thinks in regard to it, substance, force, divisibility etc., and likewise what belongs to sensation, impenetrability, hardness, colour, etc., something still remains over from this empirical intuition, namely extension and figure [*Gestalt*]. These belong to pure intuition, which, even without any actual object of the senses or of sensation, exists in the mind *a priori* as a mere form of sensibility. (*A*21 = *B*35)

Here is Kant's famous equation: a pure intuition is the form of an empirical intuition, consisting of a shape or figure in extension. However, as Kant has not listed size or magnitude as one of the properties he has 'taken away' from a body to arrive at its figure, he is not yet implicated in euclidean form. Kant's pure intuitions are figures in an infinite space, indeed, "all figures are only possible as so many different modes of limiting infinite space" (*A*578 = *B*606). But since they are limitations, he claims, of a *unique* such space, Kant would like to call *it too* a pure intuition: space is the pure intuition whose parts are also pure intuitions, *viz.* the figures. Figures or shapes also arise when "A concept of space and time, as quanta, can be exhibited *a priori* in intuition, that is, constructed, either in respect of the quality (figure) of the quanta, or through number in their quantity only (the mere synthesis of the homogeneous manifold)" (*A*720 = *B*748).

In general, figure is "the outer form of an object" (cf. R4432), as well as an object of geometry, but I have found no place in Kant's published writings where he makes figure or shape independent of the size of the object as a matter of principle. He may have intended this however in the *Opus Postumum* where he says that "Quality is the determination of a thing insofar as it is not increased even though the thing itself is magnified [*vergrossert*], for example, figure" (Kant, Akademie V. 21, p. 458). Here we have euclidean figures, if by '*vergrossert*' Kant meant 'enlarged with preservation of shape'. That he just may have is suggested by a late reflection tying this notion directly to his idea that space and time themselves are mere "forms": indeed

($r$) If space and time were given in themselves as something infinite, they would have to be regarded as infinite magnitudes. Now they are nothing but forms, to be magnified (vergrossern) or reduced (verkleinern) without limit. (*R*6420)

To K. Mainzer (1980) it is "obvious" that ($r$) refers to Wallis's similarity principle. Of course, this only makes prima facie sense for space, not time, for as Kant says, our form of inner intuition "has to do neither with shape nor position" (*A*33 = *B*50). But then "just because this inner intuition yields no shape" we represent time "by a line progressing to infinity" (*ibid.*); and since Kant meanwhile had in the *Prolegomena* characterized the straight line as "one which is similar to itself in all its parts (equal as well as unequal)" (1953, p. 140), Mainzer may well be right about Kant's meaning in ($r$).[8] If so, Kant was indeed claiming that no figures in space have absolute magnitude; that is, to any given figure $f$, there exists an arbitrarily larger or smaller figure $f'$ similar to $f$. But can Kant *construct* $f'$ in intuition? Not unless he can so construct Wolff-parallels, but as we have seen he was only willing to claim a "philosophical proof" for their existence, and hence also for the parallel postulate. It should follow that the existence of $f'$ can only be proved philosophically, but I don't believe that Kant was aware of Wallis's result. In any case, it is clear that without such a proof he could not claim that the form of our intuition excludes Lambert's possibility of representing the given figure $f$ in its absolute magnitude.

In fact, the "inconvenience" Lambert saw in such a possibility comes only to this: that insofar as we imagine that we apprehend or can construct such an $f'$, we must also imagine both $f$ and $f'$ to be *very small, absolutely*. In short, we would have to admit that the geometer can construct only an absolutely small portion of space, but also that he can give a *geometrical* meaning to this claim. From the standpoint of a geometry based on Lambert's ($L$), the fact that the angle-sums of all triangles we can actually measure are not perceptibly different from $2R$ simply confirms this hypothesis. And however incovenient it would be for Lambert to admit that, because of the way it looks, his figure for proving ($L$) is really very small, his proof did in fact comply, perhaps more than any before its time, with Kant's 'true method' of geometry, namely "not to inspect what he discerned either in the figure . . . and from this, as it were, to read off its properties; but to bring out what was necessarily implied in the concepts that he had himself formed *a priori*, and had put into the figure in the construction by which he presented it to himself" (Bxii). Indeed, Lambert had formed *a priori* a new concept of quadrilateral and found that it necessarily implied the possibility that space contains *a priori* an absolute standard of magnitude. This brings us to Kant's later writings and reflections on the relativity of magnitude and his attempts to infer ideality from it.

In Kant's late reflections ideality arguments come in bunches:

> A proof of the ideality of space is that it is a magnitude which can always be represented as a part of one still larger (that is, infinite). A second is that, however small we may represent space to ourselves, I can imagine all things given to me in a still smaller space, without being able to distinguish the things in the one, considered inwardly, from those in the other. It is not therefore the embodiment of things in themselves. (*R*6338)

Kant would clinch the first one by arguing that if space were a property of things in themselves, it would have to be an actually infinite magnitude. The second argument, however, depends on the similarity principle, as Kant makes a little more clear in the next reflection:

> A quantity . . . which can only be considered part of every other, is infinitely small. That all extended things in the world can be brought into a drop of water, or even an infinitely smaller space, proves the ideality of space, if everything is considered only relatively, not absolutely large or small. (*R*6344)

However, this argument seems to depend only on infinite divisibility when he reflects further that "as large as the given world may be, it can still be contained in the space of a waterdrop or the head of a pin, because of its infinite divisibility. Therefore, it cannot be composed of things in themselves" (*R*6346). Here, as perhaps also in the reflection ($r$), I think Kant was conflating the continuity with the similarity principle, but in any case the issue in these arguments now is clearly one of the *size* of things. As yet another reflection begins: "All size is in appearances only relative, but in things in themselves it is absolute." (*R*6398) This is a crucial premise in the argument which follows: if we apprehended things in themselves in space, they would have to have some absolutely determined size.

Kant's fullest formulation of this ideality argument occurs in his 1795 lectures on Baumgarten's metaphysics, in which he reasons as follows. If only *some* things in the universe were shrunk to the size of a pea, such a change would be detected by comparison of magnitudes. However,

> if one assumes that the entire universe is changed, then no such comparison is possible, and hence no change of magnitude is conceivable; the smallest sphere would be the same size as the largest in respect of our subjective representation, since we could not compare them with any third object. (Kant, Akademie, V. 29, p. 997)

In short, "size or smallness can never be known absolutely, but only through relation" (*ibid.*). But now, Kant argues, "From the fact that we are not in a position to determine the magnitudes of things in and of themselves, it follows that we cannot represent them to ourselves as they are in and of themselves, but only as they appear to us" (*ibid.*). Hence ideality: space and the things in it, concludes Kant,

> are determined merely according to the laws of sensibility, by which they are brought into relations. Space and time therefore are not conceived by the understanding . . . but are merely the subjective form in which things appear to us. (*ibid.*, pp. 997—8)

This ideality argument depends on the euclidean property of scale-invariance, as expressed in Lambert's axiom ($A$), for if the geometry of 'space and things' were hyperbolic, the changes envisioned by Kant would not be indiscernible, but rather betrayed by changes in angles and shapes. Kant's smallest sphere would then be distinguished from his large one by the greater value of the ratio of its circumference to its diameter. Only in euclidean space can we imagine universes of different sizes which are nevertheless completely similar, and hence must *present the same appearances* to all observers in them. This is the nub of Kant's argument: we can only determine those properties of the universe that it would share with one similar to it, that is, its 'form' or the way it 'appears' to us. And since any universe is in space, we may call space "the subjective form in which things appear to us". This argument introduces the language of 'appearances' and 'form' more convincingly than Kant's other ideality arguments; indeed, it is hard to think of any other properties that two similar universes or configurations have in common other than their appearance or form. However, Kant has nowhere tried to prove that space, as a form of intuition, must satisfy Lambert's axiom, though this is possibly what he *meant* to say in reflection (*r*).[9] Kant himself hit upon a serious difficulty in his argument we he wondered later in the *Opus Postumum* "Whether it could be said that, if the entire universe of all bodies became a thousand times smaller, this would be a true change. Or whether this statement itself is not contradictory or meaningless" (Kant, Akademie V. 21, p. 4). He thus anticipated the positivistic critique of unverifiable propositions that grew out of discussion of the overnight doubling of the universe imagined by Poincaré.

Nevetheless, Kant's euclidean ideality argument was a strong one in the sense of admitting a natural extension to include the physical laws known in his time. Thus Laplace, noting that all laws of attraction then known, were like the law of gravitation, inverse square laws, argued as follows:

> The law of attraction, inversely as the square of the distance . . . appears to be the law of all forces, of which the action is sensible at a distance, as has been recognized in electrical and magnetic forces. Hence . . . this law should be regarded from its simplicity and generality, as rigorously true. One of its remarkable properties is, that if the dimension of all bodies in the universe, their mutual distances and velocities, increase or decrease proportionably, they would describe curves entirely similar to those which they at present describe; so that if the universe be succesively reduced to the smallest space imaginable, it would always present the same appearances to observers. These appearances are consequently independent of the dimensions of the universe. (Laplace 1830, pp. 321–2)

Here it is not just the geometry of space which excludes any 'determinate unit' in it, but nature's laws of attraction between bodies in space as well. Kant might have argued more generally that the independence of all physical phenomena in space from the dimensions of the universe showed that space comprised only the 'form of appearances'. The possibility that he considered such an extension of his argument is suggested by several passages in the *Opus Postumum*, one in fact in which he says outright that "the principle of the ideality of objects in space and time was first introduced through the world-system of Newtonian attraction" (Kant, Akademie V. 21, p. 55). And he delights in "the possibility according to Newton and de Luc that the entire universe could be conceptually represented in a nutshell. This proves precisely that all our sensible representations give us nothing but appearances whose form is space and time (subjective)" (*ibid.*, p. 501).

It is instructive to see how these euclidean and more physical arguments developed gradually out of the direct ideality argument of the first edition of the *Critique*, which is then sharpened by the transcendental exposition of space in the second edition.

The conclusions that space (i) "is nothing but the mere form of appearances of outer sense" which (ii) does not "represent any property of things in themselves" was first inferred in 1781 from its metaphysical exposition together with the premise that "no determinations, whether absolute or relative, can be intuited prior to the existence of the things to which they belong, and none therefore, can be intuited *a priori*" (*A*26 = *B*42). Thus (ii) but *not* (i) follows from the major premise 'Nothing that can be intuited *a priori* can belong to things in themselves' and the minor premise 'Space is an *a priori* intuition'. It is not clear what would entail (i) for the basic reason that it is unclear what (i) *means* (see Note 8). But the major premise was felt by rationalists to beg the question about objective realities, and the four metaphysical arguments for the minor premise were felt to be too psychological or obscure to have much force. Even granting that space is an *a priori* intuition does not by itself give us any of those 'determinations' of spatial objects which Kant says can be intuited prior to things. The only clue in the first edition of the *Critique* is the remark about a 'substantial phenomenon in space' that "its inner determinations are nothing but relations, and it itself is entirely made up of mere relations" (*A*265).

This gap is filled in the second edition by the 'transcendental' exposition of the concept of space, which tries to show that 'other *a priori* synthetic knowledge' of the kind we *have* in geometry, is only possible if space is an *a priori* intuition. Indeed, "Geometry is a science which determines the properties of space synthetically, and yet *a priori*" (*B*40). And though the new major premise of the main ideality argument that 'we could never prove such synthetic *a priori* propositions about things in themselves' still begs the question, it nevertheless leads Kant to be more explicit regarding just what kind of knowledge we have of our intuitions that does not qualify as knowledge of things in themselves. In fact, he now writes in 'confirmation' of his ideality theory, "everything in our knowledge which belongs to intuition . . . contains nothing but mere relations; namely, of locations in an intuition (extension), of change of location (motion), and of laws according to which this change is determined (moving forces)" (*B*67), whereas "a thing in itself cannot be known through mere relations" (*ibid.*). It can only be known, says Kant, through "inner properties" (*ibid.*). So the 'direct argument' for ideality, *as soon as Kant tries to clarify the claim that in space we are given* 'nothing but relations', anticipates the later argument based on euclidean space and possibly even that of Laplace appealing to physical laws. Specifically, the relations of 'location' in space are relative, as implied by the congruence principle of geometry, and so is, according to Kant in 1786, 'all motion that is an object of experience' (1970, p. 19). Above all, the merely relational character of spatial phenomena is shown now by the "laws according to which this change is determined (moving forces)," and the later argument, if we suppose it to be essentially the one formulated in part by Laplace, does indeed show that the force laws envisaged by Kant permit only the determination of relations between bodies that appear to us in phenomena, but not their sizes. To conclude anything about things in themselves we need the additional premise quoted before from Kant's reflections that "all magnitudes in appearances are relative, but in things in themselves are absolute". Then we may conclude that our knowledge of spatial phenomena, as specified by the laws known to Kant and Laplace, consists only of relations, not of an intrinsic property of things themselves. And that such knowledge does not, as Kant had already put it in the first edition of the *Critique*, enable us to get at "the absolutely inward [nature] of matter, as it would have to be conceived by pure understanding" (*A*277), to be known as a thing in itself.

The fact that each of Kant's ideality proofs only subjectivizes the particular spatial property on which it turns leads back to an old question often raised against Kant's idealism, notably by Herbart who asked "from whence the definite shapes of definite things?" Kant's exposition of the non-empirical character of our space concept isolates the form from the matter of perception: the latter comprises non-spatial, unordered, qualitative sensations given *a priori*. But why, asked Herbart, do I always represent this object as rectangular, that one as triangular, this one as round, that one oval? The reason clearly cannot lie in pure *a priori* intuition itself. Nor could hardly lie in those sensations either, for they are supposed to be purely qualitative and unordered in any spatial way. But if so, why can't I order and synthesize them into a spatial manifold as I wish? Or as Adickes put it: is the definite synthesis leading to definite shapes "a completely free construction or only the reconstruction of an already existing empirically perceivable enduring action"? In short, is the shape or form of an object a primary, secondary, or transcendental quality? Is it a quality of appearances or of things in themselves? These questions, which have divided many Kantians, came up inevitably in my analysis of Kant's ideality arguments, of which there are two kinds besides his direct argument.

First there are those that turn on alleged contradictions that result from supposing that things themselves could be infinite or infinitely divisible. The conclusion in such cases is that only appearances, for example, are infinitely divisible; that is, the law of continuity holds only for intuitions. Then there are those which turn on the basic relations or automorphisms of space: the rigid motions, reflections, and similarities. Thus the difference between congruent objects can be intuited but not conceived, hence position is a property of appearances, not of things in themselves. The difference between incongruities can be intuited but not conceived, hence orientation is a property of appearances, not of things in themselves. Finally, since the difference between similar non-congruent objects must be given in terms of a unit of measurement which must be given in empirical intuition, size is purely relative and hence again, a property of appearances, not of things in themselves. (Kant has focussed on the case where one object is the whole universe and the other a "transformation of it into a nutshell," and concluded that, since it makes no sense to attribute one size rather than another to the material universe in space, space itself must comprise only the form of its appearance.) But now we have run out of the basic automorphisms of euclidean space: the relativity of position, orientation, and size still leave us with a definite shape for every object in space. The difference between a circle and a square can not only be intuited, but *also* conceived. Thus only the direct argument can be tried for the ideality of shape or form: 'But how could I possibly intuit form *a priori* if it were a property of things in themselves?' To a realist, the answer seems obvious: because geometrical optics shows how at least very faithful *projections* of a thing itself are given to me, and through experience I can integrate them into its shape. Kant would perhaps argue that by the time these projections are processed for perception, they could have no more resemblance to the thing itself than does a word, or any other symbol. As he put it in the *Opus Postumum*, "form here constitutes the sight of objects, not conversely (otherwise empiricism would hold)". In any case, the ideality of form is one of Kant's basic presuppositions, hence not something for which he can easily argue. He can only reflect that "we can indeed have a concept of a (form) shape without something being there, but not of a reality; for this presupposes sensation and therefore empirical intuition." (Akademie, V. 18, p. 201)

From the standpoint of naturalistic epistemology, Kant's position seems backwards. Instead of allowing for a gradual accommodation of his categorical and sensible framework to the objective natural world, he produces arguments for saying that what is real in experience must conform to the conditions of sensibility. Nevertheless, the neurobiologist G. S. Stent finds in the discovery of straight-line and parallel detectors in the cortex a justification of Kant's euclidean *a priori*. The evidence shows that "a geometry based on straight, parallel lines . . . is the most immediately compatible with our mental equipment" (Stent 1978, p. 166), and hence that, at this stage of evolution, "Euclidean geometry and its nonintersecting coplanar parallel lines is the 'natural' geometry, at least for men" (*ibid.*). However, this is only a processing mechanism; it can hardly yield fine-tuned *knowledge of geometric axioms*. At the very most, it only helps us detect the existence, not the uniqueness of parallels. Long ago, Peirce suggested, as an answer to the question of the 'exact truth' of the parallel postulate, that

> if there is some influence in evolution tending to adapt the mind to nature, it would probably not be completed yet. And we find that other ideas require correction. Why not this [postulate] too? Thus, there is some reason to think this natural idea is *not* exact (Peirce 1958, p. 74).

When the postulate is interpreted physically, as part of the description of the natural world at large, there is even more reason to think this now. However, the geometry based on it remains obviously a very natural one for man.

But it is also clear that Kant's attempt to establish its truth prior to experience fails. In fact, his approach founders already with the congruence principles and, it seems, necessarily so. Nor giving him that, does it seem possible for him to establish that Lambert's figure is not really a small Lambert quadrilateral. However, we can see from his late ideality arguments that their motivation was perhaps more aesthetic and moral: to show that the apparently immense world we inhabit is really neither large nor small in itself, and hence that its sublimity — the fact that we apprehend it as 'absolutely great' — is an *aesthetic judgment*. The triumph of mind over nature already implicit in the ideality of space is thus completed by the ideality of sublimity, which "does not reside in any of the things of nature, but only in our own mind, insofar as we may become conscious of our superiority over nature within, and thus over nature without us (as exerting influence upon us)" (Kant 1928, p. 114).

*Boston University*

## NOTES

[1] See Buroker (1981) for an extensive analysis of Kant's incongruity arguments and their Leibnizian targets. This fascinating book raises acutely a question which has long puzzled me: if the congruities are actually the origin of Kant's idealism, and Buroker's case for this is stronger than I ever thought possible, then why of all places should they disappear from the *Critique* itself? My own suggestion, which occurs in the next section, is related to criticisms of the incongruity arguments made long ago by Adickes (1924, p. 240), who argued that the critical turn Kant gave them from 1770 on could just as well be applied to any other spatial relations at all. Adickes took this as a kind of *reductio ad absurdum* of the force of such arguments against either Newton *or* Leibniz. I believe, however, that although he was right in claiming that such arguments can only beg the question against Leibniz, Adickes overlooked the kernel of truth in the seeming absurdity to which he would reduce Kant's incongruity argument: such an argument cannot be applied to *any* arbitrary spatial relation, but only to those corresponding to the basic automorphisms of space, such as 'is congruent to,' 'is similar to,' and 'is a reflection of.' (See Mainzer (1978) for a useful modern formulation of this aspect of Kant's theory.) I will also argue that Kant himself actually *did* make such arguments, but that he failed to clearly appreciate that each of them depends on a Leibnizian indiscernibility axiom, undoubtedly because he was ignorant of the relevant Leibnizian texts. Thus Kant speculates at the beginning of his 1768 venture into incongruity that Leibniz's *Analysis Situs* "was probably never anything more than a thing of the imagination. I do not know how far the object, which I propose to examine here, is related to that which the great man had in mind" (1968, p. 36). Little did he know that Leibniz had not only communicated his Analysis to Huyghens, but based it on an indiscernibility axiom on which he would himself much later and unknowingly base the ideality argument for space in which he seemed to have the most confidence.

[2] See Goe (1962) for details of and references to Kästner's pioneering work in the formulation of new axioms. See also Friedman (1985) for a revealing discussion of the significance of the order axioms for some of Kant's arguments. He makes the point that some of them should be seen in light of the fact that Kant knew only a monadic logic and not the theory of relations required for the logical formulation of these axioms, which would have enabled me to sharpen my analysis of Kant's treatment of the extension postulate. It does lead in Friedman's hands to a sharper analysis of the metaphysical expositions of space, and I can only regret that his interesting paper was not available to me when I wrote the main body of this one.

[3] See Adickes (1924, pp. 16—17) for a useful summary of Kant's work on parallels and Kant (Akademie, Vol. 14, pp. 25 ff) for his extraordinarily helpful notes on Kant's

text. Kant's use of Wolff-parallels shows, incidentally, that he most probably did not read Lambert's 'Theorie der Parallellinien' before it was published posthumously in 1786, for Lambert clearly pointed out the circle this approach leads to: Euclid's postulate is now contained already in Wolff's definition.

[4] See Schubring (1982) for a full analysis of Schultz's theory of the infinite and his attempts to prove Euclid's postulate by means of it.

[5] We should recall, lest we think that Kästner's objection to Schultz's use of the infinite was the purely philosophical issue it is now, that the more famous such objection of his student Gauss was also directed against a fallacious use of infinite figures to prove the parallel postulate. One would never know from reading just Kant's response to Kästner's papers (Akademie, V. 20, pp. 411—23) that the man he addressed in a letter at the time as "the Nestor of all philosophical mathematicians in Germany" was questioning the validity of a specific kind of proof for the main axiom of geometry. But Kant was in a delicate position with Schultz: the man who had worked so tirelessly and unselfishly to expound and defend the critical philosophy was now, in the crisis raised by Eberhard, tying its reputation too closely to his proof of the parallel postulate for the peace of mind of Kant, who would have preferred to prove it "*ohne Herbeiziehung unendlicher Flachen*." On August 2, 1790 Kant wrote to Schultz, who was working on his reply to the attacks marshalled by Eberhard, suggesting "*unmassgeblich*" that Schultz not deal in his review with those places were Kästner appeared to allude to his own theory of the infinite, but save his self-defense instead for his commentary (Schultz 1792). Kant had also sent, however, a copy of his own response to Kästner, and flattered himself that it might enable Schultz to bring his theory "into agreement with what the *Critique* says in those parts of the Antinomy dealing with the infinite in space." Upon receiving Schultz's review he glanced over it and immediately dashed off a note on August 15 to Schultz praising him for a masterful job. But the very next day on August 16 he wrote again at greater length to Schultz, expressing a "*Bedenklichkeit*," again "*unmassgeblich*," that occurred to him on reading through the review more closely: "It concerns the passage on your theory of parallel lines." Kant was afraid that it would only help Eberhard by providing additional targets for the mathematicians he had enlisted as "borrowed plumage" to dignify his attacks on the *Critique*, thereby making it more difficult for future critics to justly appreciate Schultz's philosophical claims. Especially since, Kant claimed, the passage preceding it was sufficient to meet any objections to Schultz's theory based on misunderstanding; but he left the matter to Schultz's "wise judgment." Now in the passage Kant apparently thought would be sufficient, Schultz was responding to Kästner's claims that only the "potential," not the "actual" infinite was admissible in geometry, and that the reason the parallel postulate lacks the evidence of the others lies "not in *infinite space* in the new meaning [Schultz's] of these words, but rather that one has only a clear (*klaren*) concept of the straight line, not a distinct (*deutlichen*) one" (1790b, p. 414). Schultz replied that whereas Eberhard might easily see this as evidence that the apodictic certainty of geometry depended not on *a priori* intuition, but rather on the distinctness of its concepts, or in Eberhard's language, on the "supersensible" or "intelligible," precisely the opposite was true, since the geometer obviously could not dispense with infinite space. Now if Kant thought this, together with his remark already incorporated by Schultz to the effect that the geometer could leave the question of how this infinite is "given" to the metaphysician, was sufficient to answer the claim that the problem of parallels had weakened the apodictic

certainty of geometry, he was surely wrong. In any case, Schultz, after pointing out that the assumption that two coplanar lines perpendicular to a third one could meet would lead to the impossibility that they could enclose a *finite* area, proceeded to meet Kästner's claims head on: "If however one of these straight lines is perpendicular to the third and the other not, and one assumes that they do not meet, then are all parts of the infinite plane between them *unlimited*, so it is obviously clear that the impossibility of *this* assumption can not be proved as long as the ratio of this *unlimited* part of the plane in respect of the assumed angle is not known; and this leads then to a theory of parallel lines which is not only the true, but also the only one possible" (Akademie, V. 20, p. 422). Though it is difficult to know exactly what Kant thought about this, it is clear that there is nothing in either the *Critique* or his own answer to Kästner that even scratches the surface of the problems raised about geometry by Gauss' old teacher. On the other hand, had Kant by now known anything at all about Lambert's work on parallels, which had been published four years before, he surely would have called Schultz's attention, however "*unmassgeblich*," to at least the *possibility* of another "theory of parallel lines." I leave the reader to judge.

[6] It should be noted that this was really the first coherent *formulation* of the problem of parallels: that Euclid's postulate be deduced, as we now say, *purely logically*. To appreciate this consider Lambert's two celebrated "semantic" observations. He proved under his second hypothesis ($> 2R$) that the area of triangles would be proportional to their excess which, in view of the known formula $A = r^2 (\alpha + \beta + \gamma - \pi)$ for spherical triangles leads to his observation that "Here it seems to me remarkable that the second hypothesis is fulfilled (*statthat*) when instead of a plane triangle one takes a spherical one, because in this both the sum of the angles is greater than 180 degrees and also the excess is proportional to the area of the triangle" (Lambert 1895, p. 202). So despite having "easily refuted" this hypothesis — by means of a "purely symbolic" proof which used, however, the *extension postulate* — Lambert finds what we now call a *model* for it, a structure in which it is "fulfilled," though apparently (?) without full awareness that this postulate is not. And now, presumably noting that the above formula for the area of spherical triangles goes over into $A = r^2 (\pi - \alpha - \beta - \gamma)$ upon substituting '$\sqrt{-1}\, r$' for '$r$' — a special case of what he proved for plane triangles under his third hypothesis ($< 2R$) — Lambert *explains the extreme difficulty he was having refuting it* with the suggestion that "I should almost conclude from this that the third hypothesis takes place (*komme vor*) in an imaginary sphere. At least there must always be something (*Etwas*!), why it cannot be refuted nearly as easily in the plane as in the case of the second hypothesis" (*ibid.*, p. 203). Now just how deeply Lambert's first sentence here is bound up with his introduction of hyperbolic functions into mathematics, and to what extent he may be said to have "anticipated" Beltrami's pseudosphere, I cannot say. Expert opinion seems divided. However, I do believe that Lambert's second sentence anticipates our completeness principle for "purely logical" deductions and refutations. Indeed, it seems to me that he has suggested nothing less than: (c) if you cannot refute (i.e. derive a contradiction from) an hypothesis (together with "other axioms") in an "absolutely symbolic" way, then there must be a sphere or structure (however imaginary!) in which it is fulfilled — and which explains why you can't. I do not wish to be misunderstood: Lambert was not anticipating the *proof* of the completeness theorem. Rather he was simply demanding what any good follower of Leibniz would demand in his situation, namely, a *sufficient reason* for claims of non-derivability. What

obscures the situation in Lambert's case is that a few pages later, at the end of his treatise, he purports to refute the third hypothesis after all, by means of what we now call the axiom of Bolyai: that a circle passes through any three non-collinear points. We call it that simply because Bolyai was the first to be *conscious* of using it. The reason, presumably, that Lambert was not satisfied enough with this work to publish it was just that, in view of his "imaginary sphere," he suspected he may not yet have become, as Hilbert put it, "fully conscious of what he was doing." But the main point about (c) is that the possibility of formulating it at all in Lambert's language shows that the need to distinguish between "syntax" and "semantics," in order to prove an independence result, goes back to him. But I am afraid that it also lay buried with him until the twentieth century. Which is why, if Kant *did* ever read Lambert, he almost certainly would *not* have found in him the "possibility" of non-euclidean geometry: he would rather, or so I believe, have thought that Lambert had *refuted* it! But again, I leave it to the reader to judge.

[7] In connection with this it should be pointed out that the "truth" he admitted we all saw in the parallel postulate, and hence presumably also in his new axiom ($A$), was really a matter of "all the *consequences* that for all kinds of purposes are deduced from it, that these consequences together may be regarded in many ways as a complete *induction*" (*ibid.*, p. 160). Later Lambert says that "It appears to me that such a principle is very unexpected in geometry" (*ibid.*, p. 177) and compares it with the case of "physical hypotheses" (*ibid.*). In a moment we shall see him referring to some consequences of ($L$) as "inconveniences," and in fact, in his paper (1765) on *practical* geometry, he defines geometry as "a science of convenience" (*Bequemlichkeit*). You measure such and such lines or angles of a figure, and then you don't *have* to measure the rest. The main inconvenience he saw in ($L$) was for astronomy where Lambert thought that, since we could in practice measure only two angles of astronomical triangles, we could never determine stellar parallax — and hence "complete" the Copernican revolution — without the euclidean angle-sum. But this was only because he had not yet done what Lobachevsky did: develop a hyperbolic theory of parallels. But Lambert's main point, that the parallel postulate and its equivalent ($A$), however much we *believe* them, are only *hypotheses* whose truth, like those of physics, is only to be found inductively in all their convenient and verifiable consequences, has been repeatedly overlooked, first by Peters (1961), and more recently by Albrecht (1979) and Gray and Tilling (1978).

[8] Apart from this, the reflection ($r$) brings us face to face, as do few other passages in Kant, with the question of what he could possibly have meant by calling space and time "forms of intuition," as well as the closely related one of the dependence of their metaphysical expositions on geometry and chronometry respectively. Concerning space, we ask: in calling it "the form of outer intuition" are we saying anything more than simply that every object we apprehend in space has a form or shape? And in calling this form a "pure" intuition are we not simply claiming its independence from the *size* of such an object? If so, then the "form of intuition" was, in view of Kant's definition of a pure intuition as "the form of an empirical intuition," simply his term for our faculty of apprehending the forms or shapes of empirical objects. This seems to be implied by Mainzer's reading of ($r$); but this also makes the metaphysical exposition of space as an intuition depend on geometry, for as Humphrey (1973) has pointed out, ($r$) was Kant's final attempt to reflect clearly on the exposition of space as "an infinite *given*

magnitude" from which he inferred that it must be an intuition. If Mainzer is right, this property of space was inseparably bound in Kant's mind with its euclidean structure, and hence the parallel postulate. (But we shall see below evidence in other of Kant's reflections that he never became clearly conscious of the independence of the euclidean structure of space from its infinity and infinite divisibility.) But how could Humphrey cite (*r*) as part of his case for the independence of the metaphysical exposition of space from any specific geometry? Only, it seems to me, by sidestepping the question of the meaning of its second sentence. In any case, one can't help feeling that Kant was *trying* to say something more in calling space a form of intuition, though I don't believe that anyone has ever overcome the difficulties pointed out by Vaihinger (1892) in coherently doing so. The most serious effort in this direction known to me is by Allison (1983), who follows Humphrey's lead in claiming the independence of the ideality of space from geometry. But on p. 95, by way of explaining how the infinity of space could be "given," he quotes, as "clearly articulating" the basic idea, Schultz's claim that: "If I should draw a line from one point to another, I must already have a space in which I can draw it. And if I am to be able to continue drawing it as long as I wish, without end, then this space must already be given to me as an unlimited one, that is, as an infinite one." But this makes the infinity of space depend as much on the extension postulate as conversely: neither Schultz nor Allison explains *how the infinity of space could be given independently of apprehending the truth of the extension postulate of geometry.* In Kant's case, it was only the absence of the order axioms that sustained this illusion. In fact, this postulate is presupposed also in Euclid's own formulation of his parallel postulate, whose equivalence to the scale-invariance of form brings us again to the second claim of (*r*). Indeed, Allison says, in connection with Kant's claim that every "determinate space" must be included in a unique unbounded space, that "moreover, we can see from this that the claim that 'space must be represented as an infinite *given* magnitude' must be taken as a claim about the 'form,' or essential structure, of every determinate representation of space, not as a claim about a unique representation of this infinite space itself" (*ibid.*, p. 95). Fine: but what else about the "form" or "essential structure" of such a determinate subspace could Kant have possibly had in mind other than its independence from the size of that subspace? Until someone clearly explains this, there is no coherent sense that I can see to be made of Kant's idea of space as the form of intuition beyond the obvious one, namely, that space is euclidean and that our intuition of spatial things must consequently be independent of scale. But I don't want to leave the wrong impression: Allison has made the most thorough and masterful attempt that I know of to make something more of Kant's idea. In particular, he has shown that Kant's distinction between the "form of intuition" and "formal intuition" cannot by itself lead to a coherent interpretation of space as "an infinite given magnitude."

[9] The difficulty of doing this is perhaps reflected by how little we have progressed beyond Leibniz's indiscernibility axiom (*i*): when actually using it we saw that he took the "determinants" it speaks of to be angles, and hence that it was essentially Leibniz's own equivalent for the parallel postulate. In fact, (*i*) has recently found a more precise formulation in the *formprinzip* of Lorenzen (1984): figures resulting from the same construction are geometrically indistinguishable, where figures are "constructed" by a term $T(\theta, \phi)$ beginning with the points $\theta$ and $\phi$. Like Leibniz's principle, Lorenzen's

also contains a veiled reference to angles, as is clear from its evident equivalence to $(F)$: the angles of a figure constructed by $T(\theta, \phi)$ are independent of the points $\theta$ and $\phi$ – which, of course, simply denies the existence of the correspondence between angles and segments discovered by Lambert in his proof of $(L)$. Since the form of a figure depends on its angles, we see that Lorenzen has appreciated the significance that Kant missed in Leibniz's indiscernibility principles for geometry. Indeed, Leibniz's $(i)$ and Lorenzen's $(F)$ are their respective ways of saying what Kant probably *wanted* to say, namely, that geometry is the science of pure form, and not as he did say, the science of space. But Lorenzen follows Dingler in the further claim that the very possibility of a "reproducible technology of measurement" *presupposes* $(F)$, while leaving its explanation and defense to his followers. Thus Janich and Tetens (1985) say that the "arbitrary reproducibility of measurement" depends on the reproducibility of the measuring apparatus without the help of a prototype, and explain that this independence requires that in calibrating, say, a ruler, no subdivision of it is "distinguished" as an exemplary basic unit. They try to make this vague claim precise as follows: supposing the ruler $G1$ to yield measurements $\varphi_1(p_1)$ and $\varphi_2(p_2)$ of two objects $p_1$ and $p_2$ respectively, while $G2$ yields $\varphi_2(p_2)$ and $\varphi_2(p_2)$ when applied to them, they require that $\dfrac{\varphi_1(p_1)}{\varphi_1(p_2)} = \dfrac{\varphi_2(p_1)}{\varphi_2(p_2)}$, which they call the condition of "scale-invariance." But a simple example shows that this fails to distinguish euclidean from hyperbolic geometry: suppose $p_1$ and $p_2$ are the perpendicular edges of a table top and we get $\varphi_1(p_1) = \varphi_1(p_2)$. Then obviously hyperbolic as well as euclidean geometry will say that we should also get $\varphi_2(p_1) = \varphi_2(p_2)$. Indeed, in hyperbolic as in euclidean geometry we choose arbitrary units of length and can change them without affecting the ratio of directly measured segments. The difference between them emerges only when we ask for the length of further segments depending on them, e.g. the one connecting the end points of $p_1$ and $p_2$. Euclidean geometry says that (i) it must be incommeasurable with $p_1$ and equal to $\sqrt{2}$, while hyperbolic geometry says that (ii) it must be greater that $\sqrt{2}$ and *can* be commeasurable, depending on the size of $p_1$. In fact, (i) and (ii) are equivalents of the euclidean and hyperbolic parallel postulates respectively, showing how extraordinarily theoretical and hair splitting they are from any "practical" point of view, and bringing us back to Lambert's idea that such postulates are only abstract hypotheses to be justified by their convenient and verifiable consequences. This is more compatible with Lorenzen's position than it seems since (i) could neither be verified by any direct measurement nor possibly have any practical relevance for a carpenter. The fact that its denial implies that he could not possibly realize a perfect square will hardly interest him if he's already convinced that he couldn't realize a perfect equilateral triangle or circle either. As for the ruler maker, we must wonder how on earth his 'norms of calibration' could presuppose a measurement that no possible calibration could ever register. One could meaningfully remind him to make his rulers as *rigid* as possible, but it is surely fantastic to imagine that in calibrating them he could in any sense follow a norm like $(F)$. He can choose his units and subdivisions 'arbitrarily,' but the suggestion that he could follow a norm precluding their geometrical definability in Lambert's sense seems to me utterly incoherent. One might just as well tell the sextant maker to make sure his sextants always yield $2R$ for the angle sum of a triangle.

## REFERENCES


Adickes, E.: 1924. *Kant al Naturforscher*. Vol. 1, Berlin: de Gruyter.

Albrecht, W.: 1979. *Geometrie des Anschaunsraumes*. Bayreuth: Ellwanger.

Allison, H.: 1973. *The Kant-Eberhard Controversy*. Baltimore: Johns Hopkins University Press.

Allison, H.: 1983. *Kant's Transcendental Idealism*. New Haven: Yale University Press.

Barrow, I.: 1734. *Mathematical Lectures read in the Public Schools at the University of Cambridge*. London: S. Austen.

Buroker, J. V.: 1981. *Space and Incongruence: The Origin of Kant's Idealism*. Dordrecht: D. Reidel.

Coffa, A.: 1981. 'Russell and Kant,' *Synthése* **46**, 247—263.

Einstein, A.: 1961. 'Relativity and the Problem of Space.' In *Relativity, the Special and General Theory*. New York: Crown, App. 5.

Enskat, R.: 1978. *Kants Theorie des geometrischen Gegenstandes*. Berlin: de Gruyter.

Euler, L.: 1950. 'Reflexions sur l'espace et le temps,' *Mémoires de l'académie des sciences de Berlin, 1748* **4**, 324—333.

Friedman, M.: 1985. 'Kant's Theory of Geometry,' *Philosophical Review* **94**, 455—506.

Gauss, K.: 1900. *Werke*, Vol. 8, Göttingen: in Commision bei B. G. Teubner in Leipzig.

Goe, G.: 1962. 'Kästner, Forerunner of Gauss, Pasch, and Hilbert,' *Proceedings of the 10th International Congress of the History of Science* 2. Ithaca.

Gray, J. and L. Tilling: 1978. 'Joachim Heinrich Lambert, Mathematician and Scientist, 1728—1777,' *Historia Mathematica* **5**, 13—41.

Hilbert, D.: 1922. 'Neubegründung der Mathematik. Erste Mitteilung.' *Abh. math. Sem. Hamb. Univ.* **1**, 155—177.

Hintikka, J.: 1981. 'Russell, Kant, and Coffa,' *Synthése* **46**, 264—269.

Hintikka, J.: 1982. 'Kant's Theory of Mathematics Revisited,' in J. N. Mohanty and R. W. Shehan (eds.), *Essays on Kant's Critique of Pure Reason*, Norman: University of Oklahorna Press, 1982, pp. 201—215.

Hume, D.: 1978. *A Treatise of Human Nature*. 2nd edn. Oxford: Oxford University Press.

Humphrey, T.: 1973. 'The Historical and Conceptual Relations between Kant's Metaphysics of Space and Philosophy of Geometry,' *Journal of the History of Philosophy* **11**, 483—512.

Janich, P. and H. Tetens: 1985. 'Protophysik. Eine Einfuhrung,' *Philosophia Naturalis* **22**, 3—21.

Kant, I.: 1902—56. *Kants gesammelte Schriften*. Herausgegeben von der Preussischen Akademie der Wissenschaften zu Berlin. 23 vols. Berlin: de Gruyter. Supplemented by *Vorlesungen*, Vols. 24—29. (Cited as 'Akademie' by volume number in the text.)

Kant, I.: 1982. *Critique of Judgement*. Trans. J. C. Meredith. Oxford: Oxford University Press.

Kant, I.: 1953. *Prolegomena to any Future Metaphysics*. Trans. P. G. Lucas. Manchester: Manchester University Press.

Kant, I.: 1965. *Critique of Pure Reason*. Trans. N. K. Smith. New York: St. Martin's.

Kant, I.: 1967. *Philosophical Correspondence, 1759—99*. Trans. and ed. A. Zweig. Chicago: University of Chicago Press.

Kant, I.: 1968. *Selected Pre-Critical Writing*. Trans. and intro. by G. B. Kerford and D. E. Walford. Manchester: Manchester University Press.

Kant, I.: 1970. *Metaphysical Foundations of Natural Science*. Trans. J. Wellington. New York: Bobbs-Merrill.

Kant, I.: 1973. 'On a Discovery According to Which Any New Critique of Pure Reason has Been Made Superfluous by and Earlier One.' In Allison (1973).

Kästner, A. G.: 1790a. 'Was heisst in Euclids Geometrie möglich?' *Philosophisches Magazin* **2**, 391—402.

Kästner, A. G.: 1790b. 'Über den mathematischen Begriff des Raums,' *Philosophisches Magazin* **2**, 403—419.

Kästner, A. G.: 1790c. 'Über Geometrie Axiome,' *Philosophisches Magazin* **2**, 420—430.

Lambert, J.: 1765. 'Anmerkungen und Zusatze zur practishen Geometrie,' *Beiträge zum Gebrauche de Mathematik un desen Anwendung* **1**. Berlin.

Lambert, J.: 1771. *Anlage zur Architektonik order Theorie des Einfachen und Ersten in der philosophischen und mathematischen Erkentnis*. Riga: J. F. Hortknoch.

Lambert, J.: 1794. *Neues Organon*. Leipzig: Johann Wendler.

Lambert, J.: 1895. 'Theorie der Parallellinien.' In F. Engel and P. Stakel (eds.), *Theorie der Parallellinien von Euclid bis auf Gauss*. Leipzig: B. G. Teubner.

Laplace, P.: 1880. *System of the World*. Trans. H. H. Harte. Dublin: Longmans.

Leibniz, G. W.: 1903. *Opuscles et Fragments inedits de Leibniz*. Ed. L. Couturat. Paris: Olms.

Leibniz, G. W.: 1956. *The Leibniz-Clarke Correspondence*. Ed. with intro. and notes by H. G. Alexander. Manchester: Manchester University Press.

Leibniz, G. W.: 1973. 'On the Principle of Indiscernibles.' In G. H. R. Parkinson (ed.), *Philosophical Writings*. London: Dent.

Leibniz, G. W.: 1976. *Philosophical Papers and Letters*. Trans. L. E. Loemker. Dordrecht: D. Reidel.

Leibniz, G. W.: 1981. *New Essays on Human Understanding*. Ed. P. Remnant and J. Bennett. Cambridge: Cambridge University Press.

Lobachevsky, N.: 1898. *Zwei geometrischen Abhandlungen*. Ed. F. Engel. Leipzig: B. G. Teubner.

Lorenzen, P.: 1969. *Normative Logic and Ethics*. Mannheim: Bibliographisches Institut.

Lorenzen, P.: 1984. *Elementargeometrie. Das Fundament der analytischen Geometrie*. Mannheim: Bibliographisches Institut.

Mainzer, K.: 1978. 'Der Raum im Anschluss an Kant', *Perspektiven der Philosophie*, Neues Jahrbuch 4.

Mainzer, K.: 1980. *Geschicte der Geometrie*. Mannheim: Bibliographisches Institut.

Mellin, A. W.: 1801. *Worterbuch der Kritischen Philosophie* 4. Jena und Leipzig: Friedrich Frommann.

Newton, I.: 1976. 'Geometry.' In D. T. Whiteside (ed.), *Mathematical Papers*, Vol. 7. Cambridge: Cambridge University Press, pp. 248—561.

Newton, I.: 1962. *Principia. The Motion of Bodies*, Vol. 1. Trans. A. Motte, rev. by F. Cajori. Berkeley and Los Angeles: University of California Press.

Parsons, C.: 1980. 'Mathematical Intuition,' *Proceedings of the Aristotelian Society*, New Series, 80, 145—169.

Peters, W. S.: 1961. *Lamberts Konzeption einer Geometrie auf eine imaginarien Kugel.* Bonn: Rheinische Freidrich Wilhems Universität.

Peirce, C. S.: 1958. *Collected Papers of Charles Sanders Peirce*, Vol. 8. Ed. by A. W. Burks. Cambridge, Mass.: Harvard University Press.

Rehberg, A. W.: 1792. 'Über die Natur der Geometrische Evidenz,' *Philosophische Magazin* **4**, 447—460.

Schubring, G.: 1982. 'Ansatze zur Begrundung theoretischen Terme in die Mathematik — Die Theorie des Unendlichen bei Johann Schultz,' *Historia Mathematica* **9**, 441—484.

Schultz, J.: 1792. *Prüfung der Kantischen Critik der reinen Vernunft.* Königsberg: Zweiter Teil.

Schütz, C. G.: 1785. [Review] In *Allgemeine Literatur-Zeitung* **162**, 41—45.

Schwab, J.: 1791. 'Über die geometrischen Beweise, aus Gelengheit einer Stelle in der A. L. Z.,' *Philosophische Magazin* **3**, 397—407.

Stent, G. S.: 1978. *Paradoxes of Progress.* San Francisco: Freeman Press.

Vaihinger, H.: 1881, 1892. *Commentar zu Kants Kritik der reinen Vernunft.* 2 vols. Stuttgart: Union Deutsche Verlagsgesellschaft.

Wolff, C.: 1736. *Philosophia Prima Sive Ontologia.* Frankfurt: Officina Libraria Rengeriana.

Wolff, C.: 1965: *Mathematische Lexicon*, Hrsg. und bearb. von J. E. Hofmann. Hildesheim: Olms.

ABNER SHIMONY

# COMMENT ON WEBB

This is a paper of great merit, quite apart from its relevance to the aims of the present book. Webb meticulously, fairly, and imaginatively analyzes Kant's arguments concerning geometrical knowledge. He brings to bear on this project a large body of information concerning the Kantian texts, the geometrical writings of Kant's predecessors and contemporaries, and the great harvest of results on the foundations of geometry in the nineteenth and twentieth centuries. Webb deploys this information not only for critically assessing Kant's theses, but also for exploring possible defenses of them which lie beyond the technical resources which were available to Kant.

In addition, Webb's paper makes several important contributions to the study of naturalistic epistemology.

First, among all the philosophers who stand in opposition to naturalistic epistemology (as the term is understood in the Introduction to this book) Kant is the one who looms the largest historically. There was strong philosophical motivation for his program of establishing a body of synthetic *a priori* knowledge and (even from a skeptic's viewpoint) considerable plausibility in his execution of the program. If, however, a careful, well-informed, and sympathetic examination of his treatment of geometry — which is one of the most promising domains for the synthetic *a priori* — concludes with a negative assessment, then indirect support is given to an epistemological program which freely utilizes a posteriori considerations, as all varieties of naturalistic epistemology do.

Second, Webb extends his critique of Kant's treatment of geometry to that of the protophysicists (Dingler, Lorenzen, and followers), whose 'practical idealism' was influenced by Kant though it retreated from some of his most ambitious claims. Since this school presents some of the most serious and technically competent opposition to naturalistic epistemology (witness Sagal's paper in this volume), Webb's critique is directly relevant to certain current issues.

Third, even if there is no synthetic *a priori* knowledge, and the central theses of the Kantian epistemology are rejected, he still would have to be recognized as a philosophical giant, and not just because of



*Abner Shimony and Debra Nails (eds.), Naturalistic Epistemology*, 71—72.

his immense historical importance. Although Kant undoubtedly would have found repugnant a 'psychologized' version of his transcendental philosophy, such a version has much intrinsic interest and in many ways anticipates the recent developments of cognitive psychology. Much of Webb's discussion of the productive imagination, the forms of intuition, and of the relation between intuition and concepts (the schematism) can be applied directly to the construction of a systematic 'psychologized' Kantianism, which would be a species of naturalistic epistemology.

Fourth, in any systematic naturalistic epistemology an important concern is the meshing of knowledge from different sources or of different modalities. A very important special problem of this sort is the meshing of visual information about space with information which is distilled from experience with mobile rigid (or quasi-rigid) bodies. Webb has some very interesting remarks about this problem in the course of his discussion of Kant's treatment of congruence.

Finally, even though I believe that evolutionary considerations are crucial for a fully coherent naturalistic epistemology, I also think that it is illuminating to see how much can be worked out with no explicit mention of biological evolution. It is important not to conflate naturalistic and evolutionary epistemology. In this respect, Webb's analysis, which never touches upon biology, throws some light upon the structure of a naturalistic epistemology.

*Boston University*

PETER SKAGESTAD

# PEIRCE'S CONCEPTION OF TRUTH: A FRAMEWORK FOR NATURALISTIC EPISTEMOLOGY?*

## 1. THE DILEMMA

This essay will address a dilemma in the foundations of naturalistic epistemology. As the title indicates, the focus will be on one of the great nineteenth-century founders of this interdisciplinary enterprise, to wit, Charles Peirce. I shall attempt to illustrate the dilemma by noting a certain ambiguity in Peirce's thought; thereafter I shall outline, and discuss the viability of, what I take to be Peirce's resolution of the dilemma.

Whatever naturalistic epistemology is, I assume that the following two extremes are excluded. On the one hand, an epistemology that is merely consistent with the results of natural science does not on that account qualify as naturalistic. On the other hand, an empirical investigation of knowledge processes which is purely descriptive in the sense of non-normative, does not quite amount to epistemology. From a naturalistic point of view, cognitive psychology, comparative linguistics, history of science, and sociology of knowledge are epistemologically relevant disciplines. Epistemology, however, has to provide a focus of its own, in addition to drawing on the resources of these and other special disciplines. And this focus must be provided by the customary meaning of the word 'epistemology'; otherwise there would be little point in retaining this not very respectable term, whose meaning is inextricably bound up with its somewhat checkered career. I shall take it, therefore, that naturalistic epistemology is the empirical inquiry into those natural processes whereby we are led to knowledge, *rather than* to illusion, error, or ignorance. Theoretically there could be a skeptical naturalistic epistemology interested in those natural processes which prevent us from attaining knowledge. What there cannot be is a purely agnostic naturalistic epistemology. An inquiry that studies knowledge processes in total indifference to which results count as knowledge and which do not, would not be epistemology — although such an inquiry might well be epistemologically *relevant*.[1]

In the sense specified, the present essay is an essay about, rather



*Abner Shimony and Debra Nails (eds.), Naturalistic Epistemology*, 73—90.

than in, naturalistic epistemology. Our central problem is this: in what precise sense can (say) the history of science be held to be epistemologically relevant? A preliminary answer, naively formulated, would be that this history provides resources for empirically investigating the processes whereby science produces truth. This formulation is naive because it assumes that the question *whether* science in fact produces truth can itself be answered empirically; if not, there can be no empirical inquiry into how science produces truth, rather than illusion, error, or ignorance. But is there any empirical question here? The enterprise of naturalistic epistemology, as defined above, requires an affirmative answer; yet when we look at two prominent rival paradigms in contemporary epistemology, we are in each case led to a negative conclusion.

In Karl Popper's realist approach, 'truth' means simply correspondence of statements with facts. Since all statements are fallible and may in principle turn out to be false, we can never know that any one of our statements actually corresponds with the facts (Popper, 1962, p. 226). We can, in Popper's view, attach meaning to the notion of verisimilitude — of getting nearer to the truth — but this notion, like truth, has no epistemic content, and there is no way of knowing whether we have in fact got nearer to the truth:

> Our idea of approximation to truth, or of verisimilitude, has the same objective character and the same ideal or regulative character as has the idea of objective or absolute *truth*. It is *not an epistemological or an epistemic idea* — no more than truth or content. (Popper, 1962, p. 234)

In this view, we can meaningfully ask whether science approximates towards the truth, and we can believe that it does, as Popper clearly believes. But since neither the notion of truth nor that of verisimilitude has any epistemic content, this is one belief that cannot conceivably be empirically confirmed or refuted.

In Willard Quine's pragmatism, we find a perspective in some respects diametrically opposite to Popper's — although they both habitually claim Tarski as their intellectual forbear. In Quine's view, to say that a particular sentence is true is simply to assert that sentence, something which we always do within the framework of existing theory:

> Where it makes sense to apply 'true' is to a sentence couched in the terms of a given theory and seen from within the theory, complete with its posited reality. . . . To say

that the sentence 'Brutus killed Caesar' is true, or that 'The atomic weight of sodium is 23' is true, is in effect simply to say that Brutus killed Caesar, or that the atomic weight of sodium is 23. (Quine, 1960, p. 24)

From this perspective, it makes no sense to ask whether there is a truth towards which science is progressing. Science can be improved, and no matter what improvements are made, scientists will regard the new, improved sentences as true; that is, they will be prepared to assert them. Nor does it make much sense to ask whether science has produced any truth to date. Of course it has, since it has produced the scientific beliefs which we hold today, and which are true simply in the sense that these are the beliefs we hold and assert today (Quine, 1969, p. 25). There is no empirical question here, over and above the myriad empirical questions *in* the sciences.

In Popper's outlook, then, there can indeed be epistemology, but it is not clear in what sense this epistemology can be naturalistic. Quine's outlook is thoroughgoingly naturalistic, but does not appear to leave any room for epistemology — a point which Richard Rorty (1979, esp. pp. 221—230) has developed in some detail. An interesting attempt at staking out a middle ground between Popper and Quine has been made by the philosopher-psychologist Donald Campbell. Having discussed Campbell's programme at length elsewhere, I shall here confine myself to a couple of remarks.[2] The dilemma noted above has been formulated with great clarity by Campbell (1974, p. 139): "How can we claim a fit between the theories of science and the real world, when we know that real world only through the theories of science?" Having posed the dilemma, however, Campbell resolutely refuses to let it bother him further: "In seeing the fit, and the puzzle, I do so only on presumptions which go beyond my capacity to verify or compellingly to demonstrate to another person" (1974, p. 140). This gambit is part of the perfectly respectable and frequently useful strategy of *divide et impera*; by putting the foundational problem to one side Campbell has been able to concentrate on the methodological and substantive problems of naturalistic epistemology. The strategy incurs a cost; to wit the cost that the fundamental dilemma of naturalistic epistemology has been left on the back burner. At the same time, the general drift of Campbell's programme appears to me sound; a resolution of the dilemma must combine Popperian objectivism with Quinean pragmatism. What this means is that it must assign pragmatic meaning to the concept of an objective and absolute truth.[3] Put that way, it sounds like an impossible task, and in what follows I dare not be dogmatic about the success of Peircean pragmatism in reconciling these desiderata. What I hope to have made clear so far is that this task is neither more nor less difficult than that of naturalizing epistemology without eliminating it *as* epistemology.

## 2. PEIRCE'S ANTI-FOUNDATIONALISM

Despite the enormous volume of his writings, Peirce's claim to fame as an epistemologist rests very largely on two series of journal articles, the first published in the *Journal of Speculative Philosophy* (hereinafter *JSP*) in 1868, the second in *Popular Science Monthly* (hereinafter *PSM*) in 1877—78. The first series provides what may well be the most profound and original critique of "foundationalist" epistemology to date; the second series formulates, all but in name, the philosophy later known as pragmatism.

In *JSP*, Peirce draws on the psychology of Wilhelm Wundt to argue in detail that there are no such things as intuitions; knowledge is not grounded in anything immediately given, whether to the intellect or to the senses.[4] His strategy is to examine, one by one, the various psychological phenomena which seemingly testify to the existence of intuitions, and to show in each case that the phenomenon in question is more simply explained as the result of an inferential process (e.g. CP, 5.236—237). Among Peirce's general conclusions, the following two are of particular interest. First, Cartesian doubt, so far from being an indispensable prolegomenon to philosophy, is in fact a perfectly futile exercise. We all have beliefs which we are unable to doubt, simply because it does not occur to us that they *can* be doubted. These beliefs carry no special epistemic warrant, but we cannot get rid of them by pretending to doubt everything. The person who goes through the motions of facing the skeptical dilemma will not in fact consider the dilemma resolved until he has recovered all the beliefs he pretended to give up at the outset — i.e. until he has found some highly doubtful beliefs with which to shore up beliefs that he never doubted in the first place. This is

> as useless a preliminary as going to the North Pole would be in order to get to Constantinople by coming down regularly upon a meridian. . . . Let us not pretend to doubt in philosophy what we do not doubt in our hearts. (CP, 5.265)

Here, in a nutshell, is the outlook that Peirce was later to baptize 'critical common-sensism'. There is only one vantage point from which to philosophize, and that is the point at which we actually find ourselves when we begin the study of philosophy (CP, 5.265). Epistemology, in particular, has no access to a fixed point of departure from which it can underwrite or undermine all our knowledge claims; it can be pursued only from within the body of our current beliefs, such as they are.

Second, the Cartesian criterion of truth, *viz.* subjective certitude, is vacuous. Peirce formulates this criterion as follows: "Whatever I am clearly convinced of, is true," and he goes on: "If I were really convinced, I should have done with reasoning and should require no test of certainty" (CP, 5.265). The Cartesian criterion, then, becomes available only at a time when there is no longer any need for it. Furthermore, the criterion runs counter to the spirit of modern science, where the subjective certainty of one individual counts for nothing unless he can persuade others to agree with him. A novel theory is held on probation until it has secured the consensus of all inquirers; *until* this consensus has been reached, the author's certainty counts for nothing; *after* it has been reached, this certainty becomes irrelevant because there is nobody left who bothers to raise the question of certainty. Here Peirce hints, rather than states, an alternative criterion of truth, to wit, consensus within the community of inquirers:

> We individually cannot reasonably hope to attain the ultimate philosophy which we pursue; we can only seek it, therefore, for the *community* of philosophers. Hence, if disciplined and candid minds carefully examine a theory and refuse to accept it, this ought to create doubts in the mind of the author of the theory himself (CP, 5.265)

Let us pause to take stock of what Peirce *appears* to be driving at. So far, he sounds deceptively like Quine — and, by the way, like Neurath.[5] Epistemology can be pursued only as part of natural science. There is no Archimedean point from which we can contemplate both reality and our knowledge of it to see how far the two correspond. Since reality is not available to us except through our fallible knowledge of it, correspondence to reality cannot serve as a criterion of truth. It is tempting to conclude that truth cannot *mean* correspondence to reality; all it can mean, apparently, is whatever is agreed upon by the community of inquirers. And if we take 'truth' in this sense, science will produce truth to exactly the extent that it produces agreement. Epistemology then becomes simply a part of the sociology of science.

## 3. PEIRCE'S PRAGMATISM

The conclusions just adumbrated are not the ones Peirce actually drew. Before we complicate matters, however, we shall note that this original impression is initially confirmed by a look at the articles published in *PSM* a decade later. In the first article, 'The Fixation of Belief' (FOB), we learn first of all that "the sole object of inquiry is the settlement of opinion" (CP, 5.375). We may fancy that the end of inquiry is to attain *true* opinion, but this is mere self-deception. To have a particular belief *is* to believe that that belief is true; hence, in practical terms, there is no difference between searching for true belief and searching for belief, i.e. settled opinion. The desire for a true belief is satisfied the moment I have attained belief: "The most that can be maintained is, that we seek for a belief that we shall *think* is true. But we think each one of our beliefs to be true, and, indeed, it is mere tautology to say so" (CP, 5.375). Here we note more than a whiff of the assertive redundancy theory of truth. Why, in Peirce's view, should we bother with inquiry at all? Why not simply cling to whatever beliefs we have come by, by hook or by crook? Peirce's forthright answer is that, if you can do that, there is no reason why you should not (CP, 5.377). The trouble with the pre-scientific methods of fixing belief — tenacity, authority, and the *a priori* method of believing whatever is agreeable to reason — is that each one of them is, as a matter of sociohistorical fact, ineffective in securing a long-term settlement of opinion. This can be achieved only by adopting the scientific method of surrendering one's beliefs to the judgement of experience (CP, 5.387).

In the second *PSM* article, 'How To Make Our Ideas Clear' (HTMOIC), we are presented with the pragmatic theory of meaning, which holds that the meaning of a term consists solely in the conception of the conceivable practical consequences of affirming or denying the term of an object (CP, 5.402). Thus, Protestants and Catholics may imagine themselves in disagreement over whether the elements of the sacrament are bread and wine or flesh and blood possessing all the sensible qualities of bread and wine. Since, however, they agree on all the observable facts, neither side is asserting anything denied by the other, and the dispute between them on this score is a meaningless one (CP, 5.401). Similarly, by calling a diamond hard we mean only that it will resist scratching. So long as the diamond rests on a bed of cotton it makes no sense to ask whether it is, *at this moment*, hard or soft. We may say that it is hard all along, or that it is soft until brought under pressure, whereafter it grows harder with increasing pressure. Neither statement asserts any observable fact denied by the other, and so they do not really differ in meaning (CP, 5.403).

By saying, somewhat obscurely, that meaning consists in conceived practical consequences, Peirce should undoubtedly be taken to mean that it consists in the expected observable consequences of conceived actions — what Israel Scheffler (1974, p. 78) has termed 'action-contingent consequences'. As Peirce was to make clear later, in 1905, the meaning of a scientific term consists in the specification of an experiment together with the prediction of an experimental outcome (CP, 5.411). This theory has an interesting consequence which is frequently overlooked, although it was brought out a number of years ago by Hjalmar Wennerberg (1962, p. 147), namely that the meaning of a word will change with our methods or habits of inquiry. A theory which when first formulated is untestable and so devoid of pragmatic meaning may acquire such meaning through changes in experimental techniques or in background theory, or simply through an increased willingness to employ experimental methods. In HTMOIC Peirce illustrates the idea of conceptual change by recalling to the reader the four methods of inquiry discussed in FOB, claiming now that each method embodies a distinct conception of truth. To Peter Abèlard, who practiced the method of tenacity, the truth was "simply his particular stronghold" (CP, 5.406), i.e. truth to him *meant* only the opinion which he was prepared to defend against all comers. A century later, when the Church had discovered the method of authority, "truth meant little more than the Catholic faith". To a rationalist metaphysician practicing the *a priori* method, truth meant only the internal consistency of his beliefs. To practitioners of the scientific method, finally, truth means *not*, as one might have expected, the opinions now accepted by the scientific community, but "the opinion which is fated to be ultimately agreed to by all who investigate" (CP, 5.407).

This is *prima facie* puzzling. Why bring a nebulous entity like fate into the matter? What is the force of the elusive 'ultimately'? And why clutter things up by putting the definition in the future tense? Why not say that; to a scientist, the truth is simply the set of beliefs currently held by the scientific community? There is more going on here than has so far been brought to the surface. Let us go back to the *JSP* articles from 1868 and fill in some of the missing pieces.

## 4. PEIRCE ON TRUTH AND REALITY

In 1868 Peirce argued that there is no fixed epistemological vantage point outside the web of beliefs; congruently with this conclusion he also argued, in a Berkeleyan vein, that we can have no conception of something absolutely incognizable, i.e. of a reality in principle inaccessible to thought. However, while Peirce gave short shrift to the notion of a thing in itself, he did not dismiss the concept of an independent reality. That some things are real, as opposed to imaginary, is indeed something which nobody doubts, and which a purportedly commonsensical philosophy has got to accommodate somehow. What this means, however, depends on what actual work the word 'real' performs in our cognitive life — specifically on what experienced facts require the real/unreal distinction as part of their explanation. Peirce maintained that there is one and only one such fact, namely the fact that we are sometimes forced to correct ourselves. We explain the renunciation of our former opinion by saying that it was erroneous or, alternatively, that what it asserted to be so was not really so. The real, then, is the complement of that which is asserted to be so in an erroneous opinion, and our conception of reality will depend on what we mean by error. There are, in general, two signs from which we infer that a belief is erroneous. The first sign is that the belief does not last; I find myself compelled — by perceptual facts or whatever — to abandon it. (Such compulsion, of course, can only be psychological compulsion; there is no logical compulsion, unless the belief be self-contradictory.) The second sign is that I am the only one holding the belief, and remain the only one after prolonged discussion; if I can see what others cannot, we call it hallucination (CP, 5.402, n. 2). An erroneous opinion is transient and individual; a true opinion, by contrast, is a permanent and universal one, and the real is that which is asserted to exist in such an opinion. In Peirce's words:

> The real, then, is that which, sooner or later, information and reasoning would finally result in, and which is therefore independent of the vagaries of me and you. Thus, the very origin of the conception of reality shows that this conception essentially involves the notion of a COMMUNITY, without definite limits, and capable of a definite increase of knowledge. And so those two series of cognitions — the real and the unreal — consist of those which, at a time sufficiently future, the community will always continue to re-affirm; and of those which, under the same conditions, will ever after be denied. Now, a proposition whose falsity can never be discovered, and the error of which is absolutely incognizable, contains, upon our principle, absolutely no error. (CP, 5.311)

I take Peirce here to be saying that transiency and individuality are each a sufficient condition of error, and that permanence and universal acceptance are separately necessary and jointly sufficient conditions of truth. As Peirce himself makes clear, this conception of reality is an idealist one (CP, 5.310),[6] and the last sentence quoted certainly recalls Berkeley's idealism. There is, however, one signal difference between Berkeley and Peirce; to Peirce, *esse* is not simply *percipi*, but rather the possibility of *percipi*. Reality, in Peirce's view, does not mean whatever we, the community, now believe to exist, but neither does it mean something essentially outside the realm of cognition and therefore inaccessible to cognition. Reality is the object of true opinion; and, although we believe each one of our opinions to be true, a fallibilist does not believe that all his opinions are true, so truth does not, to him, *mean* whatever he believes.

A few words need to be said at this stage about the meaning of 'meaning'. When Peirce defines 'truth' as "the opinion which is fated to be ultimately agreed to by all who investigate" and 'the real' as the object of that opinion, there is no suggestion that these words have no other meaning. Peirce agrees that 'reality' means "that whose characters are independent of what anybody may think them to be" (CP, 5.405), and that 'truth' means "the correspondence of a representation with its object" (CP, 5.553). These are formal definitions, they help relate the ideas of truth and reality to other ideas, and so they convey some understanding of these words. But we understand words *better* the more we know about the practical implications of their affirmation or denial. If we know all there is to know (something we never do) about the conceivable practical consequences of affirming or denying some predicate of an object; we understand everything that is meant by that predicate. Thus the pragmatic definition of 'reality' is intended as a further elucidation of what is meant by the formal definition; it does not conflict with the formal definition, an objection which Peirce considered:

> [The] answer to this is that, on the one hand, reality is independent, not necessarily of thought in general, but only of what you or I or any finite number of men may think about it; and that, on the other hand, though the object of the final opinion depends on what that opinion is, yet what that opinion is does not depend on what you or I or any man thinks. (CP, 5.408)

Similarly, the definition of truth as the ultimate or final opinion of inquiry does not conflict with the definition of truth as correspondence with facts. It is an attempt to spell out more precisely what the expression 'correspondence to facts' means in relation to the practices of someone who has adopted the scientific method of belief-formation.

This completes our jigsaw puzzle — provisionally, at any rate. On the one hand, there is no vantage point outside the body of knowledge from which we can directly inspect a transcendent reality. Since this claim boils down to the truism that we cannot know anything without knowing it, it follows that we cannot even have a conception of such a transcendent reality, except in the degenerate sense in which we can also have a conception of a square circle. On the other hand, the concept of reality and the correlated concept of truth are now seen to play a role *within* the realm of cognition. The pragmatic acid test does not eliminate the notions of truth and reality, but rehabilitates them from the disrepute into which they have been brought by the tradition of representative realism. Epistemology is thus naturalized, but it remains epistemology.

The question whether science does or can produce truth can now be meaningfully raised and investigated. It turns out to be an empirical question, which can be attacked by the resources of the history and sociology of science. Pragmatically understood, the question is whether science has evolved procedures which have produced, or can be reasonably expected to produce, long-run universal consensus among all inquirers. Any answer to this question is subject to Hume's strictures on all inductive conclusions. We cannot, now or ever, point to any particular consensus which is sure to last forever; for all we know, any actual consensus may be disrupted tomorrow. However, as I have argued elsewhere, the project is not that of giving an inductive justification of induction (Skagestad, 1981b, Ch. 5). The project is that of applying certain inductive procedures (those of history) to examine the results of other inductive procedures (those of physics, chemistry, biology, etc.). This is not circular; the unity of science has not arrived, and historical method as we know it owes very little to the methods of natural science. Nor is the project hubristic towards science; nobody expects historians of physics to tell physicists what methods to use or what conclusions to reach, but we do expect them to tell the rest of us how to locate physics in the intellectual geography of our culture.

What can we conceivably learn from such a project? If the history of science evinces a trend towards ever broader and longer-lasting agreements, and if this trend is correlated with the application of an identifiable set of procedures, we may tentatively and inductively infer that the further application of these procedures will perpetuate the trend. Conversely, if the historical evidence were to convince us that there has been no such trend, we should have no grounds for expecting any trend to emerge.[7] If there is a trend, but no identifiable set of correlated procedures, we should have no reason to expect the trend to continue. Either way, we should have no reason to regard science as a truth-producing enterprise, and we might have no reason to regard it as an epistemologically central enterprise. Of course, the historical questions spelled out are today subject to debate. The participants to this debate seem to agree *that* the debate is epistemologically relevant. My argument so far has been that Peirce provides something that is largely missing from the present-day debate, namely an explanation of precisely *how* the debate may be seen to be epistemologically relevant.

## 5. THE ENLIGHTENMENT INVERTED?

Peirce's position is further elucidated by a brief look at its historical setting. The consensual definition of truth was not, in fact, Peirce's brainchild; it was something of a commonplace among nineteenth-century experimentalists. We find it for instance in the writings of Claude Bernard and Karl Pearson, people with whom Peirce had rather fundamental disagreements.[8] The earliest formulation which I have happened across is Sir John Herschel's formulation from 1831, from a book which was for decades the authoritative primer on scientific method: "The grand and indeed only character of truth is its capability of enduring the test of universal experience and coming unchanged out of every possible form of fair discussion" (Herschel, 1851, p. 10). This formulation is itself the distorted mirror image of John Milton's famous expression of what was to become the fundamental article of faith of the Enlightenment:

> And though all the windes of doctrin were let loose to play upon the earth, so Truth be in the field, we do injuriously by licencing and prohibiting to misdoubt her strength. Let her and Falshood grapple, who ever knew Truth put to the wors, in a free and open encounter. (Milton, 1974, p. 242)

What has happened between Milton and Herschel is that the theory in which 'truth' functioned as a primitive term — the theory of epistemological optimism — has come to serve as an operational definition of 'truth', thus rendering the theory vacuous. The truth will survive every free and open encounter — yes, of course it will, provided we *mean* by 'truth' that which comes unchanged out of every possible form of fair discussion. We can have a pragmatic conception of truth or we can have epistemological optimism, but can we have both? Peirce's postponement of the relevant consensus into the indefinite future suggests that we can. A true belief is one which we shall never have to abandon. Never is a long time; in no finite future will we be in a position to say with certainty that we have attained any true belief, or even that there is any such thing. By defining truth as a consensus which will last in the long run, Peirce is not equating the truth with any consensus that has been reached, or which we know will be reached. Whether there will ever be a lasting consensus is a question still left open, a question which we may sensibly seek to answer inductively.

I should add, for the sake of completeness, that this is a subject on which Peirce's thinking underwent a certain transformation. Briefly, in 1878 Peirce believed that for any sequence, induction would invariably discover whether the sequence converges towards a limit and, if so, what the limit is. If it has no limit, he also believed, the sequence does not as such exist, and there is no undiscoverable truth about it. This view appears indistinguishable from Herschel's. Later in his life, Peirce came to realize that a sequence might not converge towards a limit, and we might never discover the fact, even if induction be continued indefinitely. Thus, in 1908 Peirce could write to Victoria Lady Welby: "I cannot infallibly know that there is any Truth."[9] That confession is consistent with the interpretation offered herein.

## 6. THREE OBJECTIONS

The mention of limits in the last paragraph is bound to serve as a reminder to some readers of some classic criticisms of Peirce's definition of truth. Peirce occasionally defined truth as a limit; thus, in 1901:

> Truth is that concordance of an abstract statement with the ideal limit towards which endless investigation would tend to bring scientific belief, which concordance the abstract statement may possess by virtue of the confession of its inaccuracy and one-sidedness, and this confession is an essential ingredient of truth. (CP, 5.565)

The standard criticism of this definition is, of course, Quine's (1960, p. 23) observation that, since the concept of limit is defined for measurements but not for theories, this definition precludes meaningful discussion of the truth or falsity of theories. An earlier, less known, criticism is Russell's objection to including the confession of inaccuracy in the definition of truth. In Russell's words: "If Peirce is to be interpreted strictly, he must mean that a statement is 'true' because it says it is inaccurate. This would enthrone Epimenides as the only sage."[10]

First, Quine's objection is not one which Peirce would recognize as an objection; indeed, Peirce himself made substantially the same point in his 1903 Lowell Lectures. With reference to qualitative induction Peirce wrote:

> The second order of induction only infers that a theory is very much like the truth, because we are so far from ever being authorized to conclude that a theory is the *very truth* itself, that we can never so much as understand what that means. (CP, 7.118)

We understand fairly precisely what it means for a number to be the true value of a parameter (of this, more below); with respect to a statement that does not contain quantitative parameters, such as 'light is electro-magnetic vibrations', we have no equally precise notion of truth. (To avoid misunderstanding, we do, of course, have a precise *semantic* concept of truth equally applicable to all kinds of statements; what we lack is a precise pragmatic concept applicable to theories.) We can understand its truth pragmatically only by analogy with that of quantitative statements; that is, we can understand it as being very much like the truth. It is of interest to observe in this context that in recent years the concept of truthlikeness, or verisimilitude, has come under fire primarily because of its imprecision.[11] Apparently that is exactly why Peirce chose it to denote the relation of theories to truth. Since we do not know precisely what it means for a theory to *be* true, we can also have no precise concept of the relation of an experimentally confirmed theory to the truth.

It is a corollary of Peirce's pragmatism that the epistemology of imperfect knowledge must elude precise formalization, since the degree of possible precision is a function of the state of our knowledge. If one believes, with Peirce, Hertz, and others, that nature is essentially continuous, it follows that only differential equations have a fighting chance at being true of nature.[12] In that case, there is a straightforward sense in which we get nearer to the truth by substituting differential equations for subject-predicate statements like 'light is electro-magnetic vibrations'. Where Quine's objection becomes troublesome is in the case of theoretical revolutions in mature sciences when, among other things, the parameters of the differential equations get redefined. This is a problem for all who want to speak meaningfully of progress towards truth across scientific revolutions, independently of how one defines 'truth'.

Second, while Russell's objection has been less influential than Quine's, the answer to it will reveal some extremely interesting features of Peirce's concept of truth. It is a well-known fact that in any precise measurement of a physical magnitude, one will never come up with the same value every time. (This might actually be a contextual definition of what it means for a measurement to be 'precise'.) For instance, if the weekly calibrations of the metal chunk NB 10 at the National Bureau of Standards were reported in integers, NB 10 would consistently weigh in at 10 grams. But the calibrations are reported with 6 decimals, of which the last 3 fluctuate considerably (Freedman, Pisani, and Purves, 1980, p. 91). At this level of precision there is no one authoritative reading. So what is the true weight of NB 10? It is the arithmetical mean, $\mu$, of the individual readings, defined as $\Sigma xp(x)$. Since we do not know the value of $p(x)$, we do not know the value of $\mu$. We estimate it by the sample mean $\bar{X}$, defined as $(\Sigma X_i)/n$. Now, there are three comments to be made. First, this distinction between the true value and the best available estimate is not a meta-distinction belonging to epistemology; it is a crucial distinction *in* statistics. One of the first things the novice learns in statistics is that $\mu$ is a constant while $\bar{X}$ is a random variable, and that the two therefore have quite different algebraic properties. Second, on the standard definition of probability as the limit of relative frequencies, $\mu$ is the limit of $\bar{X}$ as $n$ approaches infinity. So, when the word 'true' occurs within scientific discourse, it is used precisely in the sense of the limit of an estimator. Third, one cannot — logically cannot — report the true value of $\mu$ by giving simply a number; a mean, reported only as a single number, will be universally interpreted as the value of $\bar{X}$. The value of $\bar{X}$, or more generally the estimated value of any parameter, can be reported exactly; the value of $\mu$, like the true value of any parameter, can be reported only as a number $\pm$ the probable error. Omit the probable error, and you will no longer be talking about the true value. This, too, is not a meta-statement about statistics; "$\mu = \bar{X} \pm$ probable error" is an equation in statistics (cf. Wonnacott and Wonnacott, 1977, p. 199). The value of $\mu$ may happen to be, say, 9.999603 grams, but in the absence of the probable error you do not succeed in giving this number as the value of $\mu$, but only as the value of $\bar{X}$.

So, in the literal application of the limit definition to truth in measurement, there is nothing mysterious about the 'confession of inaccuracy and one-sidedness'; it is only by making this confession — by reporting the probable error — that we succeed in specifying the true value, as distinct from the estimated one. This is not a philosophical doctrine; to know this is part of knowing statistics. The philosophical point consists in extending this doctrine, by analogy, to statements in general. And the point is this: for the purpose of simply making categorical assertions about reality, the notion of truth is indeed redundant (CP, 5.375). We need the notion of truth only to express our awareness that reality itself poses constraints on what we may or may not believe about it (CP, 5.384). And these constraints express themselves in the ever-present possibility that our assertions may be false due to the potentially distorting perspective from within which they are made. It is only to the extent that I express my awareness of this possibility that I succeed in communicating my intention to make a *true* statement, in the sense in which that intention places my belief-formation under external constraints. It is only by making this 'confession of inaccuracy' that I succeed in making the conceptual distinction between the truth and whatever I happen to believe.

My response to Russell, finally, implicitly answers a fairly commonly heard objection to pragmatism, namely that it confuses the question of what 'truth' means with that of how one goes about ascertaining truth. I take it that Peirce would agree that the correspondence definition — e.g. in Tarski's formulation — completely specifies the semantics of the word 'truth' (CP, 5.553—554). That is, it specifies the conditions under which truth can be truly predicated of a statement.[13] But one does not stretch the meaning of the word 'meaning' by taking it to encompass also the pragmatic dimension of a word; i.e. the function played by the word within the matrix of purposive activities. And the most obvious function played by the word 'true' within the scientific activity is the one described by statistics; i.e. that of a limit towards which the value of our best estimator converges as our sample approaches infinity. This was also the obvious function which the word played for Peirce, whose chief professional pursuits were photometry and gravity research. Today's philosophy teachers do not spend much time on the job taking measurements; we typically do spend a lot of time teaching introductory logic courses. If we teach logic truth-functionally (as do most introductory textbooks), we need the students to forget the pragmatic dimensions of the word 'true' and focus solely on the semantic dimension. As we professionally use the word, it functions essentially as an unknown, to be totally divorced in the student's mind from all epistemic considerations. To the extent that we use the truth-table method to introduce logic, we commit ourselves to telling the students over and over that 'truth' means nothing but correspondence to facts. The function which the word 'truth' has in truth-functional logic courses is to have no other meaning than this. Whether it is all right to talk this way in the classroom can be left for debate to another day.[14] But to insist outside the classroom that 'truth' has no other meaning than correspondence to facts is rationally on a par with insisting that the truth-table definition of the conditional gives the whole and only meaning of the English word 'if'.

## 7. CONCLUSION

If the history of science is to be epistemologically relevant, there must be a way of empirically investigating the question of whether science does, or can, lead to truth. When we apply Peirce's pragmatic maxim to this question, we are led to reformulate it as the question of whether there is, in the history of science, any discernible trend towards a permanent and universal consensus. The latter question is more precise than the former in that it indicates the lines of inquiry that would have to be undertaken for the purpose of answering it. But other, equally precise reformulations are no doubt possible. This one has been rendered precise in a particular direction for a particular purpose, namely that of answering it through empirical inquiry into natural knowledge processes. This purpose is adopted at a cost; what is lost is the vagueness of the original formulation, a vagueness which permits alternative interpretations excluded by the reformulation. There are advantages to vagueness, and we might want to preserve those; even if we want greater precision, we might want it for some other purpose than the one mentioned. Unless we adopt the above-mentioned purpose, and unless we are prepared to pay the cost, the pragmatic reformulation of our question can exert no logical compulsion on us. But in that case, we should have to find some other way out of the dilemma with which this paper opened.

*University of Lowell*

## NOTES

* The author is indebted to Abner Shimony for valuable comments on an earlier draft.

[1] This view appears to be shared by the most prominent present-day agnostics; cf. Barry Barnes & David Bloor (1982), p. 23.

[2] Peter Skagestad (1978) focuses on my reservations to Campbell's programme; my (1981a) provides a more balanced overview of our points of agreement and disagreement.

[3] Another way out of the dilemma might seem to be provided by the causal theory of reference. I take it that Hilary Putnam has shown that the causal theory cannot *define* 'reference', and that the concept of reference therefore is parasitical on the concept of truth. Cf. Putnam (1980).

[4] For a good discussion of this subject, cf. Bruce Altshuler (1980).

[5] Cf. Quine (1960), pp. 3—4, for Quine's use of Neurath's famous analogy.

[6] The complex relationship between idealism and realism in Peirce's thought is discussed in Skagestad (1981b), Ch. 2, c and d. Cf. also Altshuler (1980).

[7] A somewhat stronger falsifying condition is represented by Hilary Putnam's famous 'meta-induction'; cf. Putnam (1978), p. 25. For a comparison of the Peircean perspective with Abner Shimony's 'hypothesis of verisimilitude', cf. Skagestad (1981a).

[8] Claude Bernard (1927), p. 220; Karl Pearson (1892), p. 30. On Peirce's critical response to Pearson, cf. Skagestad (1983).

[9] Charles S. Hardwick (1977), p. 73. For details of Peirce's views on induction, cf. Abner Shimony (1970).

[10] Bertrand Russell (1939), p. 145. To Russell's further criticism that Peirce's definition involves much 'disputable sociology', p. 146, one can only reply that this can hardly be helped if what one wants to specify is the social function of the concept.

[11] Cf. e.g. Larry Laudan (1977), pp. 125—126, and footnote citations.

[12] Peirce's *synechism* found frequent expression, but especially in CP, 6.169—213. For Hertz's formulation of this view, cf. Heinrich Hertz (1956), pp. 36—37.

[13] The idea that a wide variety of definitions may satisfy Tarski's Criterion T is argued in Putnam (1978), pp. 15—16.

[14] Theoretically, the natural deduction approach is undoubtedly superior. Whether the truth-functional approach has compensating pedagogical advantages is an open question in my mind.

## REFERENCES


Altshuler, Bruce: 1980. 'Peirce's Theory of Truth and his Early Idealism,' *Transactions of the Charles S. Peirce Society* **16**:2, 118—140.

Barnes, Barry and David Bloor: 1982. 'Relativism, Rationalism and the Sociology of Knowledge.' In M. Hollis and S. Lukes (eds.), *Rationality and Relativism*. Oxford: Basil Blackwell, pp. 21—47.

Bernard, Claude: 1927. *Introduction to the Study of Experimental Medicine*, tr. H. C. Greene. New York: Dover.

Campbell, Donald T.: 1974. 'Unjustified Variation and Selective Retention in Scientific

Discovery.' In F. Ayala and T. Dobzhansky (eds.), *Studies in the Philosophy of Biology*. London: Macmillan, pp. 139—161.

Freedman, David, Robert Pisani, and Roger Purves: 1978. *Statistics*. New York: W. W. Norton.

Hardwick, Charles S. (ed.): 1977. *Semiotic and Significs: The Correspondence between Charles S. Peirce and Victoria Lady Welby*. Bloomington, Indiana: Indiana University Press.

Herschel, John: 1851. *Preliminary Discourse on the Study of Natural Philosophy*, Second edition. London: Longman, Brown, Green & Longmans.

Hertz, Heinrich: 1956. *The Principles of Mechanics*, tr. D. E. Jones and J. T. Walley. New York: Dover.

Laudan, Larry: 1977. *Progress and its Problems*. London: Routledge & Kegan Paul.

Milton, John: 1974. 'Areopagitica', in C. A. Patrides (ed.), *John Milton: Selected Prose*. Harmondsworth: Penguin Books, pp. 196—248.

Pearson, Karl: 1892. *The Grammar of Science*. London: Walter Scott.

Peirce, Charles S.: (CP), *Collected Papers of Charles Sanders Peirce*. Vols. 1—6 ed. by C. Hartshorne and P. Weiss, vols. 7—8 ed. by A. Burks, Cambridge, Massachusetts: The Belknap Press of Harvard University Press, 1935, 1958.

Popper, Karl R: 1962. *Conjectures and Refutations*. New York: Basic Books.

Putnam, Hilary: 1978. *Meaning and the Moral Sciences*. London: Routledge & Kegan Paul.

Putnam, Hilary: 1980. 'Models and Reality', *Journal of Symbolic Logic* **45**, 464—482.

Quine, Willard V. O.: 1960. *Word and Object*. Cambridge, Massachusetts: MIT Press.

Quine, Willard V. O.: 1969. *Ontological Relativity and Other Essays*. New York: Columbia University Press.

Rorty, Richard: 1979. *Philosophy and the Mirror of Nature*. Princeton: Princeton University Press.

Russell, Bertrand: 1939. 'Dewey's New *Logic*'. In P. A. Schilpp (ed.), *The Philosophy of John Dewey*, The Library of Living Philosophers, vol. 1. Evanston, Illinois: Northwestern University, pp. 135—156.

Scheffler, Israel: 1974. *Four Pragmatists: A Critical Introduction to Peirce, James, Mead, and Dewey*. New York: Humanities Press.

Shimony, Abner: 1970. 'Scientific Inference'. In R. G. Colodny (ed.), *The Nature and Function of Scientific Theories*. Pittsburgh: University of Pittsburgh Press, pp. 79—172.

Skagestad, Peter: 1978. 'Taking Evolution Seriously: Critical Comments on D. T. Campbell's Evolutionary Epistemology', *The Monist* **61**: 4, 611—621.

Skagestad, Peter: 1981a. 'Hypothetical Realism'. In M. Brewer and B. Collins (eds.), *Scientific Inquiry and the Social Sciences: A Volume in Honor of Donald T. Campbell*. San Francisco: Jossey-Bass, pp. 77—97.

Skagestad, Peter: 1981b. *The Road of Inquiry: Charles Peirce's Pragmatic Realism*. New York: Columbia University Press.

Skagestad, Peter: 1983. 'Peirce and Pearson: Pragmatism vs. Instrumentalism', in R. S. Cohen and M. W. Wartofsky (eds.), *Language, Logic, and Method*, Boston Studies in the Philosophy of Science, Vol. 31. Dordrecht: Reidel, pp. 263—282.

Wennerberg, Hjalmar: 1962. *The Pragmatism of C. S. Peirce*. Lund: C. W. K. Gleerup.

Wonnacott, Thomas & Wonnacott, Ronald: 1977. *Introductory Statistics for Business and Economics*, Second edition. New York: John Wiley.

MILIČ ČAPEK

# THE PHILOSOPHICAL SIGNIFICANCE OF PIAGET'S RESEARCHES ON THE GENESIS OF THE CONCEPT OF TIME

## I. PIAGET'S GENETIC THEORY OF KNOWLEDGE

In the introduction to his book *The Child's Conception of Time* (hereafter, *CCT*), Piaget (1969, p. VII) wrote that his researches on the development of the concept of time were inspired by Albert Einstein who fifteen years before the publication of that book (i.e. in 1931) presided over the international course of lectures on psychology and philosophy at Davos in Switzerland. Einstein suggested to Piaget a number of questions such as the following ones: "Is our intuitive grasp of time primitive or derived? Is it identical with our intuitive grasp of velocity? What, if any, bearing do these questions have on the genesis and development of the child's conception of time?" Piaget then continued:

> Every year since then we have made a point of looking into these questions, at first with little hopes of success because, as we quickly discovered, the time relationships constructed by young children are so largely based on what they hear from adults and not on their own experiences. But when, after trying to apply the idea of 'groupings' to the development of the child's conception of number and quantity, we went on to apply it to the concept of motion, velocity and time, we discovered that the problems of duration and temporal succession had become greatly simplified. The results are presented in this volume.

But before we shall describe and analyze Piaget's ingenious and careful experiments, let us try to place him in a wider framework of the history of epistemological ideas. He clearly belongs to the tradition of empiricism, even though not to the classical pre-evolutionary empiricism of David Hume and John Stuart Mill. Piaget's epistemology belongs to the family of *biologically* oriented theories of knowledge as they began to appear with the coming of the Darwinian era. In truth, Herbert Spencer's *Principles of Psychology*, which for the first time applied an evolutionary and genetic approach to the human mental structure, was published in 1855, i.e. four years before *The Origin of Species*; and his



*Abner Shimony and Debra Nails (eds.), Naturalistic Epistemology*, 91—111.

article 'The Development Hypothesis', which outlined the idea of evolution in a far more general way than Darwin, appeared as early as 1852. Other outstanding protagonists of the biological and evolutionary approach to the theory of knowledge were Ernst Mach, Hermann Helmholtz and — what is less widely known — Friedrich Nietzsche in Germany, and in this century, in France Henri Poincaré and Henri Bergson;[1] even more recently Hans Reichenbach and Karl Popper. There are some important differences among these thinkers which I shall mention later; but they all agree in one important point that, to use Bergson's (1944, p. xxiii) words, "theory of knowledge and theory of life are inseparable", and that what we call 'reason' in man is not an immutable and ready-made structure, but a result of a long process of evolutionary adaptation. In this sense Jean Piaget belongs clearly to this group.

For the same reason Piaget's genetic approach is opposed to — and opposed by — not only by those who regard the human cognitive functions as immutable, as the traditional rationalists and idealists do, but also by those who, while they accept the evolutionary view of human mind, still regard it as epistemologically irrelevant. To the second group belong Russell, Wittgenstein and the neopositivists. According to Wittgenstein (*Tractatus Logico-Philosophicus* 4.1122), "Darwin's theory has no more to do with philosophy than any other hypothesis in natural science" and Russell said (according to an English follower quoted by Piaget (1960, p. 19)), in order to prove the uselessness of research on thought in children, that "the logician is interested in true ideas, while the psychologist finds pleasure in describing false ones." Such an attitude is characteristic of mathematicians and logicians in general, although there are significant exceptions such as Professor Quine (1969, pp. 126—127), who concedes that the applicability of our logic to reality can be explained only as a result of evolutionary adaptation.

It is difficult to understand how any serious epistemological theory can disregard such an important and wide area of facts as that of evolutionary biology. I believe that there are two different, although complementary, explanations for such neglect: one psychological and one historical. The minds prone to and trained in abstract thinking feel increasingly uncomfortable with concrete facts which do not yield easily to formal and deductive treatment; and the more complex and concrete experience is, the greater their discomfort becomes. Obviously, differential psychology, i.e. the psychology of individual differences, does play a certain role in philosophical preferences.

But probably even more important a reason for the neglect of the genetic approach is a historical one. At the end of the last century and the beginning of the twentieth century there was a powerful reaction against classical positivism and evolutionary naturalism in general. While the fact of evolution in biology, psychology and other areas has not been denied, it was declared to be philosophically and epistemologically irrelevant. Every view and every theory should be taken on its merit; what is important is its truth-value, that is, whether it is true or false — its psychological genesis is unimportant. The very term 'psychologism' became a term of reproach and its opposite view — the so-called logicism — definitely began to prevail on the continent as well as in the English speaking countries. 'Psychologism' was accused of commiting the 'genetic fallacy' in confusing *questio facti* with *questio juris* by diverting the attention of epistemologists from the only important question of the validity of theories to the irrelevancies of their psychological origin. Logicism can be traced to Bernard Bolzano and his concept of 'truth in itself' (*Wahrheit an sich*), and its development culminated in neoKantianism, in particular in the school of Marburg which also decisively influenced the phenomenology of Husserl and Meinong. But logicism also definitely influenced the neopositivism of the Vienna school which, in contrast to the classical positivism of Spencer, Mach and others, showed considerable indifference to the genetic and evolutionary approach to epistemology; hence its name — 'logical positivism'. Its attitude is summed up in the quotation from Wittgenstein given above.

I already mentioned one objection which logicism, at least in its extreme form, cannot satisfactorily answer: no responsible theory of knowledge can safely ignore any wide area of human experience, in particular such significant facts as those of evolutionary biology and psychology. To use William James's word, empiricism is either *radical*, that is *complete*, or it is not empiricism at all; a one-sided or truncated empiricism is not worthy of its name.[2] Second, logicism cannot easily answer the objection raised by Piaget (1960, pp. 20—21) himself: if the *a priori* rational categories exist or, as Russell said, subsist timelessly, independently of psychological processes and mental development, why are they not intuited by us at once, at the time of our birth? Why does it take a certain time for them to manifest themselves in our consciousness? Plato had his answer — his theory of gradual anamnesis of the unconscious, pre-existing knowledge; this Russell could not accept and, as Piaget correctly noted, it was not accidental that he eventually gave up his Platonism. But the example of Plato is especially instructive since it shows clearly that even an extreme anti-genetic theory cannot altogether avoid the question of *the origin* of knowledge — even if the answer it gives is clearly mythological. A few logicists today follow Plato; in most instances they prefer to ignore the question entirely. The third and the last objection to antigeneticism of logicists is equally important: although it is true that for methodological reasons the context of discovery should be separated from the context of justification, it remains nevertheless true that the knowledge of the origin and of the development of many concepts and theories sheds often a decisive light on their adequacy. In the history of science and, more generally, in the history of ideas, the examples abound — we would face *embarras de choix* to select them. How many fallacies could be avoided by both scientists and philosophers with a greater knowledge of the history of ideas and, in particular, with a greater insight into the psychological sources of certain fallacies which crop up again and again! We should remember Ortega y Gasset's words: "By ignoring the past we are bound to live it again."

A full understanding of Piaget is impossible without determining more accurately his position within the wider trend of evolutionary and biologically oriented epistemologies. As stated above, there are significant differences among their individual representatives, but they fall into two main groups: one which chronologically more or less coincided with the first impact of Darwin's writings in the second half of the nineteenth century — the impact which lasted until the beginning of this century. The second group consisted of those whose ideas coincided with, or even were directly influenced by, the incipient impact of new discoveries in physics. Both groups agreed in one essential point: that the structure of our reason is not immutable, but is itself a result of a long evolutionary process by which our mind gradually adjusted itself to its environment. This was nothing but a mere application of the general theory of evolution to the cognitive functions of man i.e. perception, imagination and reason; for these functions, as the parts of a psycho-physical organism, must be subject to the same evolutionary process of which the organism itself is a result. The apparently *a priori* and immutable character of our reason is due to our limited perspective in time; its apparent immutability during a few millennia of recorded history is compatible with its development through the past long geological periods.

But there is one essential difference separating these two groups: the first one, characterized by the names of Spencer, Mach, Avenarius, Helmholtz, and even Poincaré, regarded the process of cognitive adaptation as *closed and complete*, at least in its essential features; in other words, according to them, the environment to which our intelligence is now adapted, coincides with the total cosmic environment. This was the prevailing view of not only evolutionary epistemologists, but also of the whole scientific and philosophical community at that time; as Marcellin Berthelot (1885, p. 151) exclaimed a century ago, "the world was without mysteries". More specifically, the basic concepts of classical science such as the Newtonian time, Euclidean space, the Laplacean type of causality and the homogeneous matter were regarded as accurate replicas of the objective features of reality. On this point there was a remarkable, though only apparently paradoxical, agreement between the classical evolutionists and apriorists; although they held that what is *a priori* for the individual is not *a priori* for the species, they accepted the adequacy of classical physics as fully as Kant and the neoKantians. The idea of the *absence* of evolution in Kant led to the same conclusion as the idea of *already completed evolution* in Spencer and his contemporaries.

The second group, whose ideas were expressed most clearly by such diverse twentieth century thinkers as Bergson and Reichenbach and occasionally hinted at by Piaget, differs from the first in that they insist that the process of evolutionary adaptation of our intelligence is still open and incomplete not only in its insignificant details, but also in its very basic features. For our mental structure mirrors only a certain sector of reality to which it is adjusted. It is a very wide segment — the zone of the middle dimensions and of low velocities — which until the recent explosive technological development was the only zone biologically important for man; it would have been thoroughly uneconomical for the original *homo sapiens* to be adapted to the region which did not affect him nor was affected by him. The triumphs of classical science, both theoretical and technological, were due to this fact. But for the very same reason our classical intellectual habits lose their applicability outside the limits of the region to which they were originally adjusted. The so-called crisis of modern physics is basically the crisis of our imagination, especially of our visual imagination, which was so decisively shaped by its prolonged contact with the world of middle dimensions.

Piaget's researches about the development of the child's intelligence are important contributions to the evolutionary epistemology. They established beyond doubt that such allegedly *a priori* concepts as 'object', 'number', 'space', 'time', 'causality' are the results of the development in the early months and years of the child's life. These concepts are not *a priori* either in the Kantian or in the Spencerian sense. Spencer based his view that "what is *a priori* for the individual is *a posteriori* for the species"[3] on the doubtful assumption of the heredity of acquired characteristics which he accepted together with Darwin, whose fanciful hypothesis of pangenesis was based on the same assumption.[4] Although the dispute on this problem is not entirely closed — let us recall such interesting objections as those raised by Karl von Frisch (1963, pp. 77—78), the discoverer of the language of the bees — the heredity of acquired characteristics is neither postulated nor implied by Piaget's genetic theory. For Piaget's researches clearly established that the concepts referred to above *are* originally absent, as they are gradually formed in a very early period under the continual pressure of the macroscopic environment. The concept of time is one among them.

I underline the term 'concept'; it is obvious that the vague and undifferentiated awareness of temporality exists prior to the formation of the concept of time and as such it belongs probably to the primordial awareness in general. If there is any opening in Piaget's epistemology for any kind of *a priori*, it should be looked for in the pre-conceptual, pre-metrical and pre-dimensional awareness of time. I shall return later to this question, at least briefly, although it is obvious that we shall tread on a very uncertain ground. The concept of time itself is formed much later than the concept of permanent object, which is acquired before the end of the second year; we can say rather facetiously that the man at the age of two is implicitly ready for the metaphysics of substance. On the other hand, as you will see, the child does not grasp fully the concept of objective time before the age of seven or eight; it does not become implicitly Newtonian before that age.

## II. PIAGET'S EXPERIMENTS

I am going to concentrate on Piaget's book *Le developpement de la notion de temps chez l'enfant*, translated into English by Margaret Cook with the preface of Gardner Murphy under the title *The Child's Conception of Time*; in particular, I discuss the experiments described in Parts II and III. Inevitably, because of space limitations, I deal only with some of them; but since they are all very similar, each of them being fairly representative of a particular stage, their choice will be irrelevant; there will be no danger of unrepresentative sampling. After describing the experiments, I shall discuss their significance, which, in my view, is much more far-reaching than Piaget realized.

The first type of experiments which I consider are those dealing with the succession and simultaneity of perceptible events. The experimental arrangement is fairly simple: two small figures or mechanical snails of different color moving with either the same or different velocities, either continuously or by fits and starts, each start being accompanied by a rap on the table. According to Piaget (*CCT*, p. 90), the latter method is generally preferable "as there can be no failure to perceive the synchronism of two runs or the order of succession or simultaneity of the final stopping points." In the first experiment two figures start at the same moment, move on parallel tracks with the same velocity, and stop at the same time. No difficulties and no errors or confusions occur when the child is asked what happened; all questions are answered correctly concerning the simultaneity of the beginnings and the stoppings of both motions as well as the equality of the durations. In this case the spatial and temporal order coincide: simultaneity is a coincidence in time *as well as in space*, and the duration of motion coincides with the length of both trajectories which are equal. The difficulties arise for the children of the age 4—7 years when the spatial and temporal order do not coincide, that is when two figures do not move with equal velocities and do not stop at the same time. Here is the experiment:

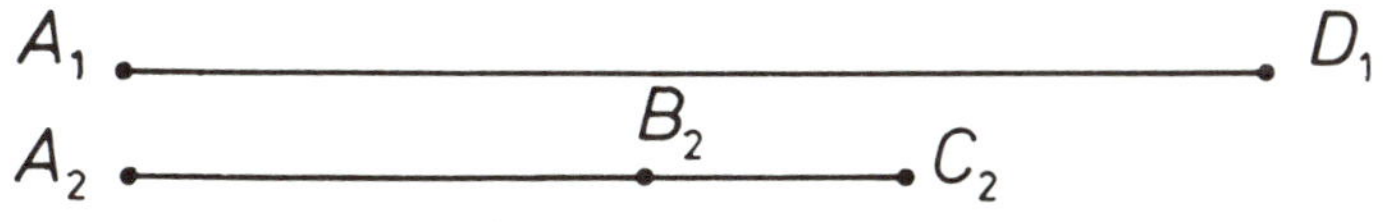


Fig. 1.

There are again two figures, yellow (*I*) and blue (*II*) moving on parallel tracks, but with different velocities: (*I*) moves faster so it reaches the point $D_1$ when *II* (the blue figure) is only at the point $B_2$;

but while the yellow figure (*I*) stops at $D_1$, the blue figure continues its motion until it reaches the point $C_2$. From the answers the child gives, it is clear that it has no difficulty in grasping that when (*I*) stops at *D*, (*II*) keeps moving, but nevertheless fails to conclude that (*I*) stopped before (*II*). Sometimes the child even goes so far as to assert that the duration $A_1D_1$ is greater than that of $A_2C_2$ simply because $D_1$ is farther away from the starting point than $C_2$. Piaget (*CCT*, p. 91) then concludes:

> The question of whether (*I*) was still moving when (*II*) stopped and *vice versa*, thus involves only one, clearly defined aspect of time: one might call it perceptive time by contrast to the intellectual time which is constructed at the moment of perception and which alone concerns us in this book. Perceptive time, in effect, is exclusively involved in what can be directly distinguished as being successive or simultaneous, but involves no comprehension of these concepts . . . .

I prefer to rephrase Piaget's words in the following way: the child constructs a *wrong intellectual time* and in this way it gives a distorting intellectual interpretation of its correct perceptual time. We shall see what is the source of this fallacy and also that the contrast between the perceptive time and its intellectual misconstruction is not confined to children only.

Piaget anticipates one important objection to his experiments: that the confusion is basically nothing but a linguistic one. The word 'before' is ambiguous; it may mean 'In front' as well as 'earlier'. This ambiguity can be avoided by explaining to the child that (*I*) stops at lunch-time and then ask if (*II*) stopped *before* or *after* lunch. An analysis of one concrete example will show the character of the answers given by the children far more clearly than any general statements:

> Hes (4,5). The yellow figure (*I*) stops in $D_1$, while the blue figure (*II*) keeps moving on from $B_2$ to $C_2$: Did they stop at the same time? *No.* Which one stopped first? *The blue one* (*II*). Which moved longer? *The yellow one* (*I*). When (*I*) stopped it was lunch time, so did (*II*) stop before or after lunch? *Before lunch.* But which stopped first? *The yellow one* (*I*). *No, it was the blue one, the yellow one* (*I*) *went longer.* Let's do it again. (The race is re-run.) *The yellow one* (*I*) *stopped first, the blue one was still moving, so the yellow one went on longer.* But did one stop before the other? *The blue one* (*II*). (*CCT*, p. 93)

This is an interesting example since it shows that the child was on the verge of grasping correctly the true order of succession, but after the race was re-run, it returned to the original wrong conclusion. In other instances such hesitations are usually absent. Allow me one more example:

Dom (6,6): Did they stop at the same time? *No.* (*II*) *stopped before, sooner, first* etc. Which went on longer? *That one* (*I*) *because it went further.* When (*I*) stopped was (*II*) still moving? *Yes.* And when (*II*) stopped was (*I*) still moving? *No.* So which one went on longer? *The yellow one.* Why? *Because it was further away.* If (*I*) stopped at lunch, did this one (*II*) stop at lunch as well or before or after? *Before lunch.* (*CCT*, p. 93)

It is important to stress that the children's errors are not due to the errors of perception: *all* subjects agreed that when figure (*I*) had stopped, the figure (*II*) was moving and yet they all obstinately asserted that the latter (*II*) stopped 'sooner', 'first' or 'before'. Why? It is impossible to disagree with Piaget's conclusion: the child does not differentiate between the spatial and temporal order since the differences in velocity are not taken into account; its attention is centered on the resulting static trajectories and thus the spatial 'before' means 'sooner' or 'earlier', and duration is identified with the distance covered. I would rephrase Piaget's conclusion in the following way: *the visual perception of the terminal situation with relative positions of now stationary bodies makes the child forget the perception of motions which were correctly perceived immediately before.* In other words, the immediate memory of what preceded is inhibited, or suppressed by the visual perception of the terminal static situation. Hence the confusion of duration with a spatial trajectory and of the temporal order with the spatial order. It can be said equally correctly that the mnemic span of the child's consciousness is still very limited; hence the failure to reconstruct correctly what was correctly perceived in its immediate past. (We shall see that the mechanism of this fallacy is not confined to children only and its significance is far greater than Piaget apparently realized; at least he does not stress it enough.)

The correctness of the previous analysis is confirmed by Piaget's own experiments on the subsequent stages in the child's development of the notion of time. There are, according to his experiments, three stages in this development; the first one which I have just discussed, the intermediate one, when the child begins to dissociate the temporal order from the spatial relations, and the final one when the original confusions are completely eliminated. In the second stage 45% of the children grasp correctly the true order of events, but still they evaluated durations exclusively by the distances covered; another 45% correctly evaluate the durations because they begin to comprehend the inverse relation between the speed and the distance covered, but at the same time they tend to confuse the temporal order and the spatial order; about 10% showed hardly any progress at all. Piaget characterizes the second stage as "beginning of differentiation between temporal and spatial order and articulated intuitions of time"; but these articulated intuitions still remain uncoordinated; beginning of their coordination takes place in the substage IIB, but only in the last stage is their coordination achieved fully and without any hesitation. One example will show it clearly:

Dani ($6\frac{1}{2}$): Did they stop at the same time? *No*, (*I*) *stopped first*. Why? *Because when it stopped the other one was still going*. Which went on longer? (*II*) Why? *Because it kept going after the other one had stopped*. (*CCT*, p. 103)

Note the reason which the child gives for his answer: "it kept going after the other one had stopped." Obviously, the child *remembers* what was going on before the motion of both figures was over; his attention *ceases* to be fixed on the terminal situation when both figures are both at rest and spatially apart. This fixation on the terminal static phase is called by Piaget 'egocentric centration' and the process by which the child succeeds in overcoming it is called by him 'decentering', I would simply call it 'an increase in the mnemic span'; the child's immediate memory *retains* what was going on in the whole temporal interval and it *resists* being suppressed by the obtrusive force of the present sensory perception of the final static situation.

Piaget's experiments dealing with the genesis of the concept of simultaneity led to the same results. The experimental setup was only slightly modified; the same two figures racing on parallel tracks were used, starting simultaneously and stopping at the same moment. There was no problem when both figures moved with the same speed; the simultaneity of departures and arrivals was directly perceived and recognized as well as the equality of synchronous durations. Simultaneity in this case was a coincidence both in space and in time and the equality of durations was literally visualized by the equality of both trajectories. But there were great difficulties when both figures, while starting and stopping at the same time, moved with different speeds. Then they obviously stopped in different points and their tracks were unequal. The simultaneity of the departures was recognized since the initial positions coincided both in space and time; but the simultaneity of arrivals was *resolutely* and *repeatedly denied* as well as the equality of synchronous durations. The children 4—6 years old claimed that the blue figure (*II*) which moved more slowly than the figure (*I*) stopped sooner than (*I*) and that (*I*) went longer "*because it went further away*". This was the first stage characterized by the denial of simultaneity at different points in space and by the alleged proportionality of duration to distance. In other words, a complete identification of the spatial and temporal relations as in previous experiments. Again it must be stressed that the children's answers were due neither to the lack of perception nor to verbal confusions; the *sound signals* marked the moment when both bodies stopped — yet, simultaneity, although directly perceived, was denied because the coincidence in time was, so to speak, 'negated' by a lack of coincidence in space. It would be otiose to repeat what was said about the failure to differentiate space from time in the previous experiments (*CCT*, pp. 106—110).

The subsequent two stages are also entirely analogous to those found in the previous experiments (*CCT*, pp. 110—117). In the second intermediary stage there is an incipient differentiation of space from time, but incomplete and inconsistent: they recognize simultaneity, but deny the equality of synchronous durations or *vice versa*. The third group of children, which at first appears the most backward since they deny both the simultaneity of arrivals and equality of durations, shows nevertheless a new and correct insight (wrongly applied, of course): *inverse ratio between time and velocity* (*or the space traversed*), which is not always recognized by the first two groups. In their words, the figure (*II*) which, according to them stopped first, went on for a longer time because it was "dawdling" or moved "more slowly". An obvious contradiction from the standpoint of an adult person, resulting from two primitive uncoordinated centrations: *a*) 'before' in time = 'before' in space *b*) 'slower' = 'left behind'. (The latter is often true, but not always; even the fast running Achilles is for a very short interval *behind* the tortoise before overtaking it.) In this particular case a longer lasting motion would imply that the slower moving figure stopped *later* — but this is what the same subject denies! Only in the third and the last stage there is a direct coordination of simultaneity and synchronism (equality of durations), i.e. a direct insight that they both imply each other; the constructive operation becomes reversible. The answer which comes again and again with only insignificant differences in wording is: "Two of them were moving together because they started and stopped at the same moment." The attention is *decentered* from distances and speeds and thus the despatialized notion of time underlying the motions of different speeds — what Piaget (*CCT*, pp. 117—119) calls different *co-displacements* — is finally firmly grasped. The child is then ready to accept the macroscopic time of classical physics which is independent of any particular bodily motion since it contains them all; it is the time of Newton which "of itself, and from its own nature, flows equably without relation to anything external, and by another name is called duration."

## III. THE PHILOSOPHICAL SIGNIFICANCE OF PIAGET'S RESEARCH

The philosophical and epistemological consequences of Piaget's research fall into three distinct groups. The first is immediately obvious from what has just been said: the concept of the universal, metrically uniform and all-embracing time is *not* an *a priori* form, but a schema which is gradually formed in the human mind in the early years of life by the pressure of macroscopic environment. The primitive notion of time has a purely local character; we can say that the child originally has a *number of local times* each of which is associated with a particular motion. The child's confusions result from the inability to integrate them into a single, universal time, underlying them all. This is the meaning of Piaget's psychological definition of time as a scheme of co-ordinated *co-displacements*. More specifically, the concept of absolute simultaneity of spatially separated events is not only unknown to, but even difficult for the child to grasp. I am not sure whether Einstein was acquainted with Piaget's results, but he certainly should have been pleased by them. For the concept of simultaneity of distant events, of 'Everywhere Now', is precisely a link by which the classical space was connected with the classical time in what may be called 'Newtonian space-time'. This is such a deeply rooted concept in our adult mind that it was — and still remains — the main source of intellectual inhibitions even among those who overtly and verbally profess to accept relativity. It was precisely because of its obstinate and psychologically almost irresistible character that the concept of classical time was viewed as an *a priori* possession of mind.

But this certainly does not mean that the child has no sense of time whatever before its *conceptual* awareness is formed; this is the second important consequence of Piaget's research. He investigated this pre-logical and pre-conceptual awareness of time in *La construction du réel chez l'enfant* (Piaget 1971).[5] This should require another lengthy and detailed analysis which would go far beyond the dimensions of this article. For our purpose suffice it to say that even in the earliest period, to quote Piaget, "the nursling shows himself capable of two operations which concern the elaboration of temporal series. In the first place, he knows how to coordinate his movements in time and to perform certain acts before others in regular order." But, as John Locke already observed, "the succession of perceptions is not the perception of succession", and Piaget is fully aware of that:

> But that does not prove that the sequence as such is perceived by the subject, that is, that it gives rise to a consciousness of sequence. Or if this consciousness exists, nothing proves that it is related to the sequence of external events (gestures as physical displacements or movements of things) and is not solely related to the development of internal states, objectified and conceived as filling the universe of perception. (*CRC*, p. 365)

On the other hand:

> The preceding considerations do not in any way demonstrate that awareness of time is absent in the first two stages of intellectual evolution. All that we say is there are not yet concepts of time applying to external phenomena nor is there a temporal field encompassing the development of events in themselves independently of personal action . . . Hence the primitive time is not time perceived from without, but duration experienced in the course of the action itself. (*CRC*, p. 368)

This duration is mingled with impressions of expectation and effort, with the development of the act, experienced internally; it fills the child's whole universe, since there is no distinction between the 'internal' and 'external'; it does not comprise either 'before' or 'after', which are always relative to external events or a measure of intervals, which also depends on the formation of relations between the actions and the guidemarks of the external world. "It is therefore time itself in its immediacy and imprecision: simply the feeling of a development and of sequential directions immanent in the states of consciousness . . .", "qualitative extension of the past into the present" (*CRC*, pp. 368–369).

All this sounds like the Bergsonian '*durée réelle*', and the similarity is, indeed, striking. If Bertrand Russell were still alive, he would promptly remark that "the babies are the best Bergsonians" — the honor which in 1912 (p. 323) he conceded only to bees and ants. But philosophical controversies are not settled by jokes; nothing would be easier than to ridicule some strange pronouncements of Russell. To say that the babies better understand the nature of time is about as correct — or as incorrect — as the claim that the infants have a better insight into the nature of the microphysical world because they do not have any notion of permanent object. But as Piaget pointed out in a very interesting article 'The Child and Modern Physics' (Piaget, 1957), there is a rather startling affinity between the absence of permanent objects in the perceptual field of infants and the non-permanent, flash-like, vanishing entities, improperly called 'particles', of modern microphysics. What is important is the fact that the nature of human mind is far more flexible and pliable than the apriorists and the traditional rationalists generally believed and that it should not be confused with its present rigid form, with that 'logic of solid bodies' which was imposed on it by the pressure of macroscopic environment. This original plasticity can be recovered by a conscious effort, by what Bergson called "*l'élargissement des cadres de l'intelligence*", when it becomes obvious that the old categories of understanding do not apply any longer. There is no need to be afraid that a physicist or a philosopher will ever descend to the pre-linguistic or pre-conceptual level of infants; for there is a fundamental difference between the *preconceptual* mental level of infants and the *deconceptualized* or rather *re-conceptualized* imageless thought of a philosopher or a physicist, even though the latter can find some interesting cues on the preconceptual level. This leads us to the second group of consequences of Piaget's investigations.

## IV. FALLACIES OF SPATIALIZATION

We have seen that the failure to differentiate the temporal from the spatial order is very common in children under the age of eight. But while children eventually get rid of this fallacy, some philosophers never do and they are even proud of it! It is true that their fallacy looks far more sophisticated, since it is concealed by a technical and abstruse language; but it is not difficult to see that from Zeno on to Bertrand Russell yesterday and Adolf Grünbaum today, it is basically the same confusion. Its source is the persistent tendency to symbolize time by a geometrical line. As long as we keep in mind that it is a mere symbol, there is nothing wrong in it; but as soon as we — quite often! — forget it, i.e. when we confer some or all the properties of a static geometrical symbol to its dynamic *designatum* or content, we are virtually returning to the infantile fallacy. In the first place, any geometrical line is essentially *static*, that is, it consists of coexisting, juxtaposed, simultaneous parts, and unless we keep in mind that it is a mere symbol, it tends to obscure the very nature of every temporal process, i.e. its *successive* character; thus succession is converted into juxtaposition, the difference between the past, present and future is abolished as it becomes, according to Einstein, who followed in this respect Spinoza and Laplace, a mere human illusion. (I must admit that, fortunately, Einstein vacillated on this point, in particular in the last years of his life;[6] but there is no question that his deep commitment to classical determinism which, as the names of Spinoza and Laplace indicate, has been almost invariably associated with the timeless view of reality, made his thought move in the same direction.)

Second, since every geometrical line is continuous in a mathematical sense, it is infinitely divisible, and the only indivisibility belongs to the dimensionless points whether these points are regarded as the constituting parts of a line or mere ideal limits of continued division which are never attained. Now by the same confusion of the static geometrical symbol with its dynamic designatum, the notion of the durationless mathematical instant is obtained and an infinite divisibility of a geometrical line is conferred on every temporal process. Hence the endless conceptual difficulties and paradoxes from the time of Zeno which, in my view, never have been successfully resolved. For no matter how much we increase the number of points-instants, we never obtain anything except the stationary durationless instants, and the intervals separating them forever elude us. It is true that in increasing the number of intermediate instants without limit, we hope eventually to succeed in filling every intermediate interval and thus to eliminate any kind of 'flux' or dynamic transition as an idle myth. But would we *really* succeed? Only if we make the number of instants actually infinite — and the concept of actual infinity is as difficult as that of durationless instant (cf. Čapek 1971b, 1974). But can we make it infinite — can what is by its own nature *incompletable* ever be completed? Couturat, a fellow infinitist of Russell, insisted that an infinite aggregate can be enumerated: "Give me an infinite time and I take upon myself to enumerate an infinite aggregate." He apparently did not realize that to say: "I can complete this operation after an infinitely long time" is equivalent to: "I shall never complete this operation" (Couturat, 1896, p. 462; cf. Čapek, 1971a, pp. 386—388). Furthermore in this case we are dealing with finite stretches of time; we would have to claim that a finite stretch of duration is equivalent to an infinitely large set of durationless instants. There is no time here to discuss the controversy between infinitists and finitists; let us remember only that Russell himself vacillated on this point, as I tried to show, and that he himself conceded that the notion of extensionless instant is inapplicable in psychology since the very notion of instantaneous experience, i.e. experience having 'zero-duration' is, in his own words (Russell, 1915, pp. 216—217), absurd (cf. Čapek, 1971a, pp. 335—345). In other words, even he conceded that there are certain types of temporal processes to which the notion of instant remains inapplicable. In the light of the recent physical discoveries about the quantum-like character of the microphysical processes there is increasing evidence that 'instants' are mere conceptual fictions, inapplicable even to physical processes. If it came as a surprise, it was only because the prestige of infinitesimal calculus was so great and its applications in classical physics so successful that its unlimited extrapolation seemed to be assured. It was another instance of an overconfident extrapolation of the mental habits applicable to a very large, although still limited zone of reality beyond the limits of their applicability. I believe that only the genetic theory of knowledge can explain this fact.

Time has been systematically represented by a straight line only after the astronomical revolution in the sixteenth and seventeenth centuries; not infrequently, especially in early Greek philosophy, it was assimilated to a circle or rather, to a circular motion. But this will be better analyzed when we turn our attention to the third group of fallacies which identify time with some kind of *motion*. These fallacies are generally less crude since the incomplete character of time is still preserved, yet, they are misleading enough.

## V. THE FALLACIES OF "KINEMATIZATION"

The fallacy of spatialization was usually denounced by the process oriented philosophers, in particular by Bergson; but one that I just mentioned — I would call it the *fallacy of kinematization* — is criticized much less frequently. In truth those who criticize it often jump to the conclusion denying time completely or — which is the same — assimilating it to a static fourth dimension.[7] Piaget's investigations showed that progress in understanding the meaning of time consists in the process of 'decentering', i.e. in diverting the attention from any particular motion and focussing it on what underlies all various motions going on together (or, as he says *co-displacements*), whether they are uniform or accelerated, rectilinear or curvilinear. As I said, the child is then virtually ready to accept the Newtonian absolute time, independent of any particular motion. Yet, the kinematic metaphor persists, even in the thought of Newton who believed — and with him very nearly all classical physicists — that the most adequate representation of time is *inertial* motion. But which inertial motion? It is clear that such a metaphor cannot be taken literally; if it is, its absurdity becomes obvious. For instance: with what particular speed does time flow? Clearly a meaningless question since there is no frame of reference with respect to which such speed could be determined. Or take the question of 'the direction' of time. Does time flow forward from the past to the future or *vice versa*? Does the present move toward the future or does it stand still, "like a rainbow over the waterfall", to use Schopenhauer's metaphor, and do the future events stream through it to become the past? Which of these descriptions is correct? The only possible answer is *neither of them*. Both are metaphorical attempts to translate into spatial and kinematic terms the elusive nature of time. As soon as we try to illustrate the nature of time by comparing it to motion, the principle of kinematic relativity of motion will sooner or later appear in our diagrams; hence two apparently contradictory answers concerning the alleged 'direction' of time: we can either regard the present moment stationary, and future moments 'moving' to it; or the future moments stationary, and the present 'advancing' to them. (Cf. Čapek, 1969, pp. 345—349, 396—397.)

Yet, the term 'direction of time' is taken quite seriously and sometimes almost literally by those who uphold the relational theory of time, for instance by Hans Reichenbach (1956). The title of his last book is quite characteristic. If the 'direction' of time is defined by the direction of a certain physical motion or a physical change such as an increase of entropy, then some puzzling questions may be raised. Would time flow backwards if the direction of all physical motions would be reversed or the entropy of the universe begins to decrease? The implications are not only puzzling, but they cannot even be stated in a self-consistent language. Reichenbach was vaguely aware of this difficulty when he considered the possibility of temporary reversals of the entropy curve which would imply, according to him, temporary reversions of the direction of time. But a more attentive analysis would disclose the impossibility of such a view. For the exceptional intervals of entropy decrease would still be *intervals of time*, i.e. they would *follow* the intervals of the previous normal entropy increase and they would *precede* the *subsequent* intervals in which the entropy curve again would have its normal upward slope. This Reichenbach conceded when he admitted that the allegedly counterdirected segments of time are still contained in *super-time*. This clearly would imply their succession and thus the asymmetrical 'before-after' relation would be re-instated on the super-time level (Reichenbach, 1956, pp. 128—129; cf. Čapek, 1969, p. 349, and 1959). Reichenbach tries to circumvent this consequence by claiming that super-time would have only ordinal properties, but no direction. But such an artifical distinction between temporal betweenness and temporal direction is based again on the deceptive analogy of the 'course of time' with a geometrical line. How far more convincing is the view of the late Professor Bridgman (1955, p. 65):

> But in no case is there any question of time flowing backward, and in fact the concept of backward flow of time is absolutely meaningless . . . . If it were found that the entropy of the universe were decreasing, would one say that time was flowing backward, or would one say that it was a law of nature that entropy decreases with time?

Similar difficulties are inherent in the circular *theory* of time, also known as the theory of eternal recurrence, *die ewige Wiederkunft des Gleichen*. If world history consists of the succession of completely identical cycles, we face the following alternative. If these cycles are *completely* and *in all details* identical — and this is what Nietzsche, following in this respect some ancient Greek philosophers, insisted upon — then they are only *numerically* different; then the only thing which differentiates them is their *succession*, their different dates, i.e. their different 'positions' in time — and thus we are back to the universal, irreversible time or 'super-time'. Or — and this would be consistent from the standpoint of the relational theory of time with which the cyclical theory is usually connected — there is only *one* cycle; but in that case it would be meaningless to speak of either repetition or return. In truth, there would not be even one circle since within that single 'cycle' there would be no repetition at all; therefore such time could be equally well represented by a finite straight segment! Obviously, it is extremely difficult to escape the idea of irreversible and in this sense absolute time. The theory of eternal recurrence itself is a result of incongruous overlappings of two incompatible theories — the absolutist and the relational theory. Cf. Čapek (1960, 1967, 1980).

This may sound suspiciously like a defense of Newtonian time, but it is not. Bridgman quoted above was certainly not an antirelativist. Newton was certainly wrong in upholding the reality of one universal *metrical* time. But does the diversity of the metrically discordant temporal series, proposed by the relativity theory, necessarily exclude *any* kind of unity of time? Such unity was proposed by Whitehead who distinguished between various time-series and what he called "creative advance of nature." In his words,

> The various time-series each measure some aspect of the creative advance, and the whole bundle of them express all the properties of this advance which are measurable. (Whitehead, 1920, p. 178)

But, as he says, the creative advance itself is, properly speaking, "not serial at all," i.e. non-metrical. In other words, what Whitehead defends is not the universal *metrical* time of Newton, but what may be termed the unity of time in a non-metrical, or, as he calls it, 'non-serial' sense.

We should not be deceived by Whitehead's poetic term — 'creative advance' — nor by his overconcise formulation which may not appear quite convincing. But how meaningful his concern about the non-metrical, i.e. *topological* unity of time is can be seen precisely in the light of Piaget's research. The present situation in physics seems to be curiously similar to that described above when Piaget found that children at first experience a great difficulty detaching their 'egocentric attention' from any particular motion; at that stage they recognize only the plurality of local times each of which is identified with a particular local motion. Only by 'de-centering' their attention from any particular motion do the children attain the unity of macroscopic time which underlies all 'co-displacements', being their co-ordination. Are we not facing a similar situation today — naturally on a higher, more complex and more sophisticated level? Are we not facing a similar task today of de-centering our attention from the plurality of metrically different relativistic times in order to grasp their *non-metrical* (*topological*) *unison* which, being their co-ordination, underlies them all? For different world-lines despite their metrical discordance (which is particularly striking in the so-called 'twin paradox') are certainly 'synchronized' or *contemporary* as they — *sit venia verbo* — move together 'in the same direction' to the future. But this would certainly require a more detailed conceptual elaboration as well as some new definitions, in particular a definition of 'contemporaneity of intervals' in contradistinction to the simultaneity of instants. This would impermissibly increase the dimensions of this article and thus must be a subject of separate study.

## NOTES

[1] Cf. more detailed treatment of these trends in Čapek (1971a, Part I, pp. 3—80); on Reichenbach, in Čapek (1957); on Mach and Poincaré, in Čapek (1963).

[2] It is certainly fair to include William James among the protagonists of the biological theory of knowledge. See in particular 'Necessary Truths and The Effects of Experience,' the final chapter of James (1890).

[3] Spencer (1885, §§ 53—54) and (1897, II, esp. §§ 426—431) his polemic against Mill's pre-evolutionary empiricism.

[4] Cf. C. Darwin, *The Variation of Animals and Plants under Domestication* (New York, Appleton and C., 1896), II, Ch. XXVII 'Provisional Hypothesis of Pangenesis,') pp. 349—399. About Darwin's growing tendency toward Lamarckism cf. Erik Nordenskiöld, *The History of Biology*, trans. by Leonard Bucknall Eyre (New York, Knopf, 1927), p. 472.

[5] See especially Chapter 4, 'The Temporal Field,' pp. 363—394.

[6] According to Rudolf Carnap's testimony, Einstein around 1952 felt that the character of 'now' and the difference between the past and the future are not adequately dealt with by physicists. Cf. Carnap (1963). I am indebted to Professor Abner Shimony for calling my attention to this particular passage.

[7] For instance Smart (1949), in contrast to his later writings, concludes with warnings against the spatialization of time and substantialization of events.

## REFERENCES

Bergson, Henri: 1944. *Creative Evolution*. Tr. Arthur Mitchell. New York: Random House.

Berthelot, M.: 1885. *Les origines de l'alchemie*. Paris: G. Steinheil.

Bridgman, P. W.: 1955. *Reflections of a Physicist*. New York: Philosophical Library.

Čapek, Milič: 1957. 'The Development of Reichenbach's Epistemology,' *Review of Metaphysics* **11**, 42—67.

———: 1959. [Review of Reichenbach, 1956], *Philosophy and Phenomenological Research* **19**:3.

———: 1960. 'The Theory of Eternal Recurrence in Modern Philosophy of Science with Special Reference to C. S. Peirce,' *The Journal of Philosophy* **57**, 289—296.

———: 1963. 'Ernst Mach's Biological Theory of Knowledge,' *Synthese* **18**. (Reprinted in R. S. Cohen and M. W. Wartofsky (eds.), *Proceedings of the Boston Colloquium for the Philosophy of Science, 1966—1968. Boston Studies in the Philosophy of Science*, Vol. 5. Dordrecht: D. Reidel, 1969, pp. 400—420.)

———: 1967. 'Eternal Return.' In Paul Edwards (ed.), *The Encyclopedia of Philosophy*, Vol. 3, pp. 61—63. New York: Collier-Macmillan.

———: 1969. *The Philosophical Impact of Contemporary Physics*. 2nd ed. Princeton: Van Nostrand.

———: 1971a. *Bergson and Modern Physics. Boston Studies in the Philosophy of Science*, Vol. 7. Dordrecht: D. Reidel.

———: 1971b. 'The Fiction of Instants,' *Studium generale* **24**, 31—43.

———: 1974. 'Two Types of Continuity.' In R. S. Cohen and M. W. Wartofsky (eds.), *Logical Epistemological Studies in Contemporary Physics. Boston Studies in the Philosophy of Science*, Vol. 13. Dordrecht: D. Reidel, pp. 361—374.

———: 1980. 'Cyclic Time.' In G. J. Whitrow (ed.), *The Natural Philosophy of Time*. 2nd ed. Oxford: Clarendon, pp. 39—41.

Carnap, Rudolf: 1963. 'Intellectual Biography.' In *The Philosophy of Rudolf Carnap. The Library of Living Philosophers*, Vol. 11, ed. by Paul Schilpp. Evanston, Ill.: Open Court, pp. 37—63.

Couturat, Louis: 1896. *De l'infini mathématique*. Paris: Alean.

Einstein, Albert: 1949. 'Autobiographical Notes.' In *Albert Einstein: Philosopher-Scientist. The Library of Living Philosophers*, Vol. 7, ed. by Paul Schilpp. Evanston, Ill.: Open Court.

Frisch, Karl von: 1963. *Man and the Living World*. Tr. Elsa B. Lowenstein. New York: Harcourt, Brace & World

James, William: 1890. *Principles of Psychology*. 2 vols.

Piaget, Jean: 1957. 'The Child and Modern Physics,' *Scientific American* **197**, 46—51.

———: 1960. *Psychology of Intelligence*. Tr. Malcolm Piercy and D. A. Berlyne. Paterson, N. J.: Littlefield, Adams and Co.

———: 1969. [*CCT*]. *The Child's Conception of Time*. Tr. A. J. Pomerans. Boston: Routledge and Kegan Paul.

———: 1971. [*CRC*]. *The Construction of Reality in the Child*. Tr. Margaret Cook with preface by Gardner Murphy. New York.

Quine, W. V. O.: 1969. *Ontological Relativity and Other Essays*. New York, Columbia University Press.

Reichenbach, Hans: 1956. *The Direction of Time*. University of California Press.

Russell, Bertrand: 1912. 'The Philosophy of Bergson,' *The Monist* **22**.

———: 1915. 'On the Experience of Time,' *The Monist* **25**.

Smart, J. J. C.: 1949. 'The River of Time,' *Mind* **58**, 483—494.

Spencer, H.: 1896. *The First Principles*. 4th ed. New York.

———: 1897. *The Principles of Psychology*. 3rd ed. New York.

Whitehead, A. N.: 1920. *The Concept of Nature*. Cambridge: Cambridge University Press.

ABNER SHIMONY

# COMMENT ON ČAPEK

If 'naturalistic epistemology' is broadly construed to mean the investigation of human cognition as a natural phenomenon, then Piaget's work should be recognized as the most massive contribution to the discipline made by any single person. One wishes to know, however, what the relevance of Piaget's work is to a more narrowly construed naturalistic epistemology — to a normative discipline which is devoted to judging epistemic claims, but which does so partly by means of scientific information about man's place in nature. Some of Čapek's remarks help to answer this question, even though he has not undertaken to do so systematically.

Čapek notes, in effect, that Piaget offers a *via media* between the traditional poles of nativism or apriorism on the one hand and empiricism on the other. The cognitive schemata postulated by Piaget are indeed genetic and hence *a priori*, but their manifestations depend upon the maturation of the child and upon its interplay with the environment. Development proceeds by assimilation and accommodation, the former having an affinity to apriorism and the latter to empiricism; the central theme of Piaget is that both ingredients are indispensable in cognitive development and also functioning. "What is important," Čapek points out, "is the fact that the nature of human mind is far more flexible and pliable than the apriorists and the traditional rationalists generally believed" (p. 104).

Čapek also credits Piaget with offering a valuable tool to phenomenologists, who are seeking to discern the given of experience which lies beneath the trappings of conventional interpretations and articulations. Locke, Hume, Husserl, and almost every other philosopher engaging in this enterprise have relied primarily upon introspection, conceptual analysis, and linguistic analysis — all of which are exercises in armchair psychology. Čapek's summary of Piaget's experiments on time suggests that the labile and uncontrolled data of introspection can be corrected by systematic observations of the behavior of children.

Čapek's use of phenomenology, corrected and guided by genetic epistemology, as an instrument for evaluating scientific theories is a



*Abner Shimony and Debra Nails (eds.), Naturalistic Epistemology*, 112–114.

quite intricate matter, which calls for some sorting out. I agree with him that in principle phenomenology can recover information which is lost in customary observation reports and nevertheless is of great significance for fundamental natural science. I also agree that the phenomenological account of temporal becoming is a case in point and that Weyl (1949) was wrong in his famous statement:

> The objective world simply *is*, it does not *happen*. Only to the gaze of my consciousness, crawling upward along the life-line of my body, does a section of the world come to life as a fleeting image in space which continuously changes in time. (p. 116)

In this passage, as elsewhere in Weyl's philosophical writing, there is a residue of Kant's thesis that the transcendental subject is not an entity in nature. If, however, one adopts a naturalistic epistemology inconsistent with this thesis, then the history of the subject's consciousness is a process in nature, and therefore subjective becoming is very difficult to understand unless there is a corresponding objective becoming. In other situations phenomenology may not be a reliable guide to fundamental natural principles. From the naturalistic point of view our phenomenological presentations are the resultants of the construction of the nervous system, of the interplay of brain and mind (which we hope can eventually be understood within the compass of the natural sciences), and of the influence of the environment. Consequently, temporal phenomena like the specious present may reflect the 'engineering' of the brain and the mind more than the fundamental structure of time. Bacon was right that there are Idols of the Tribe, resulting from the generic constitution of the human organism, which are hindrances to knowledge. Unless further arguments are presented on the basis of physical theory, we are not entitled to make inferences about the microstructure of time from the peculiarities of the phenomenological time sense. Čapek's appeal to quantum mechanics at this point (p. 106) is unconvincing, mainly because there are unequivocal physical means for measuring temporal durations many orders of magnitude shorter than the minimum intervals discriminable subjectively. My agreement with Čapek concerning becoming and my disagreement concerning the microstructure of time are both consistent with naturalism: in the latter case I can see, at least in outline, how to derive the relevant phenomenological features from current physical theory together with some reasonable psychophysical assumptions, whereas in the former case I do not see how to do this (and in particular I am not convinced by the argument of Grünbaum (1973, pp. 314—329).

Finally, I discern an inconsistency in Čapek's judgments concerning the philosophical significance of the stages of a child's cognitive development. When the nursling exhibits a sensitivity to *durée* which the older child to some extent loses, then earlier is more profound, along the lines of Wordsworth. On the other hand, when the maturing child achieves decentration and despatializes the notion of time, then later is more profound. The moral, I think, is that the search for correlations between a normal child's psychological development and the adequacy of its Weltanschauung is a will-o'-the-wisp. By adolescence a normal child has indeed achieved pragmatically efficacious powers of processing data and making inferences. The extent to which these powers suffice for a deep understanding of the nature of things depends upon many factors which lie beyond the developing child: upon the norms of the culture in which the child is raised, upon the educational institution in which it is placed, upon the history of scientific investigation in that society, and not least upon the structure of nature itself, especially upon the way in which deep-lying regularities are exhibited or covered up in phenomena on a human scale. We must keep in mind all of Bacon's Idols.

*Boston University*

## REFERENCES

Grünbaum, Adolf: 1973. *Philosophical Problems of Space and Time*. Dordrecht and Boston: D. Reidel.

Weyl, H.: 1949. *Philosophy of Mathematics and Natural Science*. Princeton: Princeton University Press.

MILIČ ČAPEK

# REPLY TO SHIMONY

Professor Shimony's fair comment on my paper provides me with an opportunity to clarify some points which because of their concise presentation may not appear entirely clear. I am glad to note that there are some large areas of agreement between us while the differences of view — some of them at least — are apparently due to misunderstanding rather than to the lack of agreement. I am going to comment on his comments in the same order in which he formulated them.

Shimony first correctly reproduces my characterization of Piaget's epistemology as an attempt at mediating between apriorism (nativism) and empiricism and quotes me — apparently approvingly — that Piaget's investigations proved that the structure of human mind is far more flexible than the traditional rationalists assumed. I have only one terminological objection to his restatement of my views when he speaks of 'phenomenologists' belonging to the tradition of Locke, Hume and Husserl. Now the term 'phenomenology' in its present usage has a rather special, technical meaning of Husserl's theory of knowledge; Husserl himself would very probably resent to be placed into the same group as Locke and Hume — precisely because of his anti-empirical and anti-genetic bias. What Shimony has in mind by using the term 'phenomenology' is 'introspective psychology' from the time of Descartes (cf. his Second Meditation) on up to the present. On the other hand, it is important to stress that it is not fair to speak of such psychology as a mere 'armchair science'; there was an enormous observational material even in the writings of those psychologists who did not perform any experimental work, such as (a random example) in Th. Ribot's opening monographies about the diseases of the will, memory, personality etc.; furthermore, introspective psychology was often combined with a physiological approach. Even some apparently 'pure' introspectionists such as Gestalt psychologists and the unjustly forgotten Würzburg school made significant empirical contributions: the significance of the discovery of the imageless thought, made just at the same time when physics began to turn away from the classical visual models, has not yet been fully appreciated.



*Abner Shimony and Debra Nails (eds.), Naturalistic Epistemology*, 115–118.

I greatly value Shimony's agreement with me that an introspective approach offers an important (though not the only) clue to the nature of time in general, including the time in physics. In other words, he rejects Weyl's view, shared by some contemporary philosophers of science, about the static, becomingless character of physical 'time'. (In fairness to Weyl it is important to stress that on this point he was fortunately not consistent, as I tried to show; the title of his 1956 book *The Open World*, in which he rejected the Laplacean timeless vision of the world, is certainly significant.) Professor Shimony is right in pointing out that "subjective becoming is very difficult to understand unless there is a corresponding objective becoming". This is also what I have always been stressing since I began to write about this problem (cf. Čapek 1965; 1975; 1985, esp. pp. xxxvii–xlvi).

On the other hand, Shimony cautions that "we are not entitled to make inferences about the microstructure of time from the peculiarities of the phenomenological time sense." He adds that my appeal to quantum mechanics at this point is "unconvincing, mainly because there are unequivocal physical means for measuring temporal durations many orders of magnitude shorter than the minimum intervals of time discriminable subjectively". There is not disagreement between us on this point; in truth, I am surprised to hear Shimony imputing to me the doubts about the extreme minuteness of the physical, in particular microphysical, events. All that I have written shows how strongly and insistently I stressed the fact that "microcosmos is also largely microchronos". All that I said in the paper was that the applicability of the concept of instant — which was given up in psychology long ago — appears doubtful in physics, in particular microphysics, precisely because of doubts about the infinite divisibility of microphysical time. The microinterval of $10^{-24}$ seconds (i.e. the hypothetical value of chronon) is, mathematically speaking, as far away from the strictly instantaneous zero-interval as the so-called 'specious present' — $10^{-2}$ seconds.

But I believe that the locus of disagreement between me and Professor Shimony is somewhere else: it concerns the true significance of what is usually called rather misleadingly 'specious present' and what I prefer to call simply 'psychological present'. He obviously regards such present as being *literally* specious, illusory, spurious. It is to him nothing but an 'illusion of consciousness' — the term frequently used by the last century's physiologically oriented psychologists and contemporary behaviorists — which, so to speak, hides to us the much shorter 'presents' of the neural processes, or, as he calls it "engineering of the brain" (he, more cautiously, adds: "and mind".) Obviously, we are touching here on the thorny mind-body problem and I did not have space enough to treat it in my paper, and I have even less space for it in my rejoinder. Allow me to repeat what I wrote before: to speak of 'illusion of consciousness' is an *epistemological nonsense*. It is meaningful to speak of *sensory illusions* when our illusory perception differs from the objective situation; or when in dreaming I do experience mental states to which nothing objective, i.e. outside of my consciousness, corresponds; yet, the fact of my own — often vivid — experience remains. William James (1940, p. 151) expressed it lucidly: "A material fact may indeed be different from what we feel it to be, but what sense is there in saying that a feeling, which has no other nature than to be felt, is not as it is felt?" The case of the psychological present is similar: since it is experienced as a *single, indivisible* feeling, it is *qua* feeling single and indivisible despite the fact that the corresponding neural process is extremely complex, involving thousands and thousands of particles and elementary physical processes. I believe this is one of the main difficulties of the identity or double-aspect theory: to account for the radical differences between the physical and mental (Čapek, 1969). The metaphors such as 'two sides of the same surface' or 'two translations of the same original' are just nothing but metaphors. The same is true for other qualities such as the sensation of green, a pang of pain etc.; how could they be *at the same time* complex motions of molecules, atoms and electrons in the neural tissue?

In the concluding paragraph Shimony's doubts about the epistemological significance of Piaget's researches are couched in such general and rather vague terms that they can hardly become the basis for discussion. Not a single experiment is discussed and its interpretation criticized; apparently their significance is summarily dismissed as a 'will-o'-the-wisp'. No, the egocentric 'temporal field' of the nursling is *not* more profound 'along the lines of Wordsworth' (Wordsworth was a poet of nature!); its only significance is in the fact that it shows that the Newtonian homogeneous time is *not* an inborn category of mind. The despatialized and decentered time of the maturing child is in a sense more profound, since it is better adjusted to its macroscopic environment. In other words, the maturing child is ready to accept the Newtonian time. But we know today that the Newtonian physics does not represent a complete adjustment to the total cosmic environment; that, as far as the concept of time is concerned, we probably are facing a new task of 'de-centering' on a far more complex and sophisticated level. Only a very general sketch of such a task was given in my paper (because of the lack of space): to recover the topological, non-metrical unity of time, underlying the plurality of the metrically discordant relativistic times.

## REFERENCES

Čapek, Milič: 1965. 'The Myth of Frozen Passage.' In Robert S. Cohen and Marx W. Wartofsky (eds.), *In Honor of Philipp Frank, Boston Studies in the Philosophy of Science* Vol. 2. Dordrecht and Boston: D. Reidel, pp. 441—463.

Čapek, Milič: 1969. 'The Main Difficulties in the Identity Theory,' *Scientia* **104**, 338—403.

Čapek, Milič: 1975. 'Relativity and the Status of Becoming,' *Foundations of Physics* **5**:4, 607—616.

Čapek, Milič: 1985. 'Introduction.' In E. Meyerson, *The Relativistic Deduction*, trans. D. A. Sipfle. Dordrecht and Boston: D. Reidel.

James, William: 1940. *Some Problems of Philosophy*. London: Longmans, Green and Co.

THEODORA J. KALIKOW

# KONRAD LORENZ AS EVOLUTIONARY EPISTEMOLOGIST: THE PROBLEM OF INTENTIONALITY[1]

## INTRODUCTION

Konrad Lorenz, winner of the 1973 Nobel Prize in Physiology or Medicine for his contributions to the founding of ethology, has been preoccupied with epistemological questions throughout his career. His writings on the subject began in the 1930s and continue until the present.[2] He has discussed the role of instinctual elements in human thinking and behavior and examined the implications for a society which is evolving faster than its individual members are evolving biologically (see Lorenz, 1966, 1974). Lorenz has also done a comparative examination of some features of human and animal '*a prioris*' and has argued that the results vindicate a correspondence theory of truth and justify the belief that human beings can approach to a knowledge of reality (see Lorenz, 1941).

But I am not going to discuss these themes here. Rather, I want to concentrate on another topic in Lorenz's work that is also crucial to naturalistic or evolutionary epistemology. This is his attempt, recently summarized in *Behind the Mirror* (1973),[3] to demonstrate many of the 'mechanisms' in the human perceptual and cognitive apparatus which are also present in other animals. Lorenz's aim is to show the constitutive features out of which human consciousness must have evolved, and of which it is still composed. This attempt is important because in naturalistic or evolutionary epistemology we need to answer the question: "How is it that creatures endowed with the consciousness that we know ourselves to possess could have evolved from ancestors apparently not so endowed?" (Shimony, this volume, p. 311)

I am going to argue that although the attempt is important, Lorenz gives an incomplete answer. He does not grapple with the phenomenon of intentionality. Indeed, he claims that human subjective life is not a fit subject for comparative psychological investigation.

Lorenz's thesis in *Behind the Mirror* is that human consciousness arises as an 'emergent' property of a novel combination of animal



*Abner Shimony and Debra Nails (eds.), Naturalistic Epistemology*, 119–144.

mental functions which singly or uncombined do not attain the full range of achievements of human consciousness. This useful idea assumes that elements present in human consciousness are also present in at least some other animals. Although Lorenz believes that human consciousness is unique because of the multiplicity and novelty of its capacities, he also assumes that other animals have 'mental' capacities which can be investigated comparatively, both for their own sakes and because they can shed light on human mentation.

So, especially in *Behind the Mirror*, Lorenz discusses many "mental" and "knowledge-producing" *mechanisms*[4] which are present in animals and which also operate in human perception and mentation. He describes such things as innate releasing mechanisms (IRMs), insight learning, Gestalt phenomena, feedback mechanisms, exploratory behavior and so on. But the result is a curiously unsatisfactory picture of human consciousness. One feels that the essence of being a human knower has eluded us. I think this is the result of a too-narrow view of what human consciousness is and what it does, and a concomitantly too-narrow view of the proper subject of comparative psychology. Lorenz implies that almost the only part of mentation that needs to be — or can be — explained is the 'objective' part, the part that 'sees' the external world, the world of facts. But this picture of mentation leaves out almost completely one of the most characteristic features of human mental activity: its intentionality, the fact that we produce a world of events fraught with meaning and value, and act within that world. We pay attention, choose to concentrate on this rather than that, reject in disgust or accept as significant or beautiful, fear this, love that, and so on. We do not just register the facts of the world, we select and evaluate them. We bring more to bear on the world than our perceptual apparatus.

Of what elements this 'more' consists is a question that has concerned many thinkers. Robert Solomon (1976), for instance, argues that our incessant evaluative view of the world has its basis in the emotions. Later in this essay I will sketch his views. His ideas provide a possible bridge from Lorenz's almost exclusively 'third-person' view of our mental activities and capacities to another goal of evolutionary epistemology: to be able to explain our 'first-person,' self-reflective capacities as well. Not only do we have a view (actually many views) of the external world, but also we have evaluative attitudes as part of that view. In what does this aspect of first-person activity consist? How did it come about?

These are questions that evolutionary epistemology must address, because if we are going to take seriously the task of explaining human consciousness, we cannot assume *a priori* that intentionality is a property that is unimportant, inexplicable or newly emergent. Rather, we must ask, what are the features of human intentionality, and do they have any precursors, analogues, shared features, etc., among (some) animals? Just as we believe that there are features of human perception/mentation that are present also in animals, though no other 'mind' of the full human sort, so we might best believe (as good Peirceans, not blocking the road to further inquiry) that there are features or precursors of human intentionality present in some animals, though perhaps not the full human panoply of values, emotions or judgments. Focusing on the perception of, or even knowing about, the facts of the world emphasizes only the third-person aspect of mentation. While this may be a necessary first step in evolutionary epistemology, it cannot be the only step.

I will argue, however, that neither Lorenz's model of mind nor his model of scientific psychology is suited to this task. This can be seen as we examine how he tries to deal with one aspect of intentionality: feelings and emotions. These are related to the phenomenon of intentionality in the following general way. One aspect of intentionality is the emotional and valuational responses we have to events and people. It has been argued since Aristotle that these emotional responses have their own anatomy. One of the indispensable components is bodily feelings, ones in some way consciously available to the subject. There has been much controversy over the details, but at least that much seems to be clear.

The study of feelings and emotions ought to be a fruitful area for comparative-psychological study. Work on the emotions has shown, for example, that some are more complex than others: some are little more than feelings, others are dependent on the agent's possession of a large number of facts and interpretations of facts. A comparative psychologist who also knew the conceptual anatomy of emotion might make interesting contributions to our understanding of behavior and its relations to feelings and emotions, and might add to our understanding of the differences in cognitive capacities among species. This would be helpful for the task of evolutionary epistemology.

But Lorenz doesn't seem to acknowledge this as a possibly useful line of inquiry. While as a good observer he recognizes that at least some animals have feelings, Lorenz often writes as if subjective phenomena can play no role in the explanation of behavior. This seems to contradict our experience and possibly also the cases of some animals. So one problem is that Lorenz's system implies that subjective phenomena are unimportant.

Another related problem is that Lorenz seems to claim at times that although subjective phenomena, including feelings, are the product of physical processes, we can never understand in detail how they occur. He thus downgrades them as scientifically relevant phenomena — they are incapable of serving in the explanation of behavior, mere epiphenomena.

Finally, in a more peripheral area but still one which can lead to problems, Lorenz sometimes mixes up feelings and emotions and draws misleading conclusions about both people and animals, attributing either too much or too little cognitive capacity to them.

These problems arise because Lorenz has the wrong model of consciousness or mentation. He sees it as information-processing only, just collecting the facts. The problem is *not* that he conceives of consciousness as amenable to mechanistic or materialistic sorts of explanation — it is that he has an overly narrow view of what has to be explained.

In this essay I first outline Lorenz's account of the evolution of human mental capacities. I mention his working assumptions and describe his model of consciousness as an information-processing device which has evolved out of simpler elements. Second, I discuss a wider view of human consciousness, involving intentionality and subjective aspects of consciousness such as emotions or feelings. Lorenz's problems in dealing with the subjective aspects of animals and human beings are then mentioned, including his claim in *Behind the Mirror* that subjective experience is logically impossible to understand. In the third and final part of the essay, I discuss some of the conceptual confusions in Lorenz's descriptions of animal feelings and emotions. I then sketch a more nearly adequate analysis of the concept of emotion and suggest how that analysis might be used in shaping further ethological theory and experiment.

## 1. EVOLUTION OF HUMAN MENTAL CAPACITIES

In *Behind the Mirror* Lorenz gives an account of how human mental capacities have evolved from those of lower animals. This is his 'natural history of human knowledge.' Three assumptions underlie his account. First, the theory of evolution: species interact with their environments over time and may, through random mutations, adapt better to those environments. "Evolution tests here and there with mutations and retains what is good."[5] Causal processes — mutation and the differential rate of survival of the better-adapted — can produce change in a species.

Second, Lorenz characterizes evolution as a complex reciprocal feedback process between acquiring energy and acquiring information. He claims, in fact, that our judgment of 'higher' and 'lower' creatures, and the direction of evolution itself, depends on our evaluation of how much information species have at their disposal, whether that information "has been acquired by natural selection, learning or exploratory investigation, and whether it is preserved in the genome, in the individual's memory, or in the traditions of a culture" (Lorenz, 1973, p. 28).

Third, what turns evolution into evolutionary *epistemology* is Lorenz's assumption that perception and mentation are the results of physical aspects of organisms. Thus, how we know and what we know can be illuminated by comparative-biological and evolutionary methods. "Our cognitive apparatus is itself an objective reality which acquired its present form through contact with and adaptation to equally real things in the outer world" (Lorenz, 1973, p. 7).

Lorenz applied these assumptions to produce his evolutionary epistemology as follows. His detailed knowledge of many species led him very early to the conclusion that not only do forms of innate behavior change over species, but the possibilites of perception, on which much behavior is based, must also vary. Lorenz argued, beginning in the 1930s, that many behavior patterns are 'released' — caused to be performed — by an organism's perception of a specific cue — a 'releaser' — in the environment. For example, a tick does not have any 'idea' what it should bite, but will fall onto and bite anything that smells of butyric acid and has a temerature of about 37° Centigrade, even a hot rock on which a perspiring human being has recently sat. Or a species of bird will not have a unitary concept of its offspring, but will react to certain of its features with appropriate behaviors. But if the situation is changed, for instance if the chick gets outside the nest, the parent will attack it or drive it away. The parent does not recognize the chick as 'the same' in the new context. We can build up a hypothesis as to an organism's world picture or Umwelt[6] by investigating the features of the world to which it responds. Lorenz discovered that just as instinctive behavior patterns varied across species, so did releasers. That is, the perceptual capacities of organisms also respond to the same evolutionary pressures as behavior and organs do.

Later, Lorenz generalized his views to include learning. What the organism can learn is also based on features of the nervous system: *what* and *how* it learns will thus be subject to environmental pressures. Learning will vary from species to species, and the capacity for it will be explained by its contribution to a species' survival capability.

All the ingredients are now in place for an evolutionary epistemology which explains the 'mental' equipment of any species. This equipment produces 'perception' and 'understanding' of just so much of the world as the species needs to survive in it. Whether the species has a restricted Umwelt like the tick's or a more complex one, all its 'mental' functions are the result of a complex evolutionary interplay between the species and the world. Thus. every species possesses a set of physical, CNS structures which are part of its evolutionary heritage, created over the course of its developmental history, and which are the preconditions and boundaries of its experience. The absolute, transcendental *a priori* of Kant has been transformed into a relative, psychologized concept which has developed *a posteriori* in the history of a species.

> One familiar with the innate modes of reaction of subhuman organisms can readily hypothesize that the a priori is due to hereditary differentiations of the central nervous system which have become characteristic of the species, producing hereditary dispositions to think in certain forms.[7]

Lorenz was convinced that if Kant had lived after Darwin and had known about animals, he too would have abandoned transcendental idealism for Lorenz's type of neo-Kantianism, which claims, with C. F. von Weiszäcker, that "Kant does not mean more than one important step in a long, and not yet finished, process of analysis [of human understanding]."[8] For Lorenz, of course, meant to speak not only of 'subhuman organisms' but also of human beings.

> The biologist convinced of the fact of the great creative events of evolution asks of Kant these questions: Is not human reason with all its categories and forms of intuition something that has organically evolved in a continuous cause-effect relationship with the laws of the immediate nature, just as has the human brain? Would not the laws of

reason necessary for a priori thought be entirely different if they had undergone an entirely different historical mode of origin, and if consequently we had been equipped with an entirely different kind of central nervous system? (Lorenz, 1941, pp. 182—183)

An important consequence of the evolutionary, psychologized *a posteriori* version of the *a priori* is that it can be investigated using the methods of comparative biology, just like an organ or a structure. As Lorenz (1941, p. 186) put it,

the a priori which determines the forms of appearance of the real things of our world is, in short, an organ, or, more precisely, the functioning of an organ. We come closer to understanding the a priori if we confront it with the questions asked of everything organic: "What for?" "Where from?" "Why?" These questions are, first, How does it preserve the species? second, What is its genealogical origin? third, What natural causes make it possible?

How does the human *a priori* preserve the species? In 'Kant's A Priori' Lorenz began by answering generally: it conveys the information needed for survival. In *Behind the Mirror* he is more specific, claiming that the human *a priori* is a novel set of CNS structures and processes which makes possible the momentous advance of the individual (and cultural) acquisition, storing and transmission of knowledge. Also in *Behind the Mirror* Lorenz answers the last two questions, "What is its genealogical origin?" and "What natural causes make it possible?"

The answers to the two questions can be given together. Lorenz's view of evolution includes the development of more complex *systems* out of simpler components. The simpler components retain their functions in the new totality, but the new 'higher' more complex system will have larger capabilities. In addition, the combination of simple components into a new whole may produce novelties that are unpredictable based upon knowledge of the components alone. Evolutionary processes may thus produce a series of changes in *kind* of functioning, not mere increases in complexity of organization. Lorenz uses system theory to argue for the sudden emergence of unpredictable novelties. He calls them *fulgurationes*, 'creative flashes,' and claims that their existence is compatible with materialism.

Cybernetics and systems theory have shown that the sudden emergence of new system characteristics has nothing to do with miracles, thereby absolving phylogeneticists from the reproach of vitalism. There is nothing supernatural about a linear causal chain joining up to form a cycle, thus producing a system whose functional properties differ fundamentally from those of all preceding systems. If an event of this kind is phylogenetically unique it may be epoch-making in the literal sense of the word. Some geological epochs have derived their names from such events. (Lorenz, 1973, p. 30)

Lorenz argues that human knowledge and mentation capacities are the result of such a *fulguratio*, one that is built out of simpler components in the perceptual-mentational systems of other animals. Tropisms, instinctive behavior patterns, insight learning, Gestalt perception, individual memory, all have developed separately and in simpler combinations in other animals. Human consciousness puts them together into a new entity which cannot be fully explained on the basis of its simpler precursors. It represents a new stage in the development of organisms, a new and vastly enhanced mechanism for processing information about the external world.

## 2. INTENTIONALITY AND THE SUBJECTIVE

### a. *The Human Life World*

Lorenz has indeed given us a natural history of *some* aspects of human consciousness. He has treated it as an information processor whose major task is to manipulate and deliver the traditional 'objective facts' about the world. While no one would claim that this traditional preoccupation of psychology and evolutionary epistemology is not important, it does not represent the entire characteristic range of human experience. And evolutionary epistemology ought to have as a goal the explanation of the total range.

Marjorie Grene, in discussing the work of Adolf Portmann, a contemporary comparative psychologist influenced by Husserl, describes a more inclusive view of human experience. She terms it the 'life world' and differentiates it from the narrower scientific view.

Each of us lives . . . in a primary life world, out of which and within which the world of science, or of any other highly articulate discipline, develops. Portmann has repeatedly emphasized the importance of giving due heed to *both* these aspects of our lives. By the world of primary experience he means, however, something more inclusive than the concept of a 'life world,' as I introduced it earlier, may at first suggest. The world in which, from infancy, we come to live, and the human world shared by members of all cultures, does, of course, include the surface of experience, the colors, sounds, the rhythms of movement that confront us on all sides. But it also includes our feelings, our desires, our dreams, the creative aspirations of artists, the vision of saints and prophets, even the delusions of the insane. No single term can adequately characterize this whole range of primary experience; perhaps we can still speak here of the 'life world' if we remember that it is much more than the plain, open order of 'common sense' to which we are referring. Such a life world, then, with all its opacities and ambiguities, stands in contrast with the limited but lucid sphere governed by the operations of the intellect — and that means, in our culture, by the operations of science and technology. Human nature comprises both and can dispense with neither. (Grene, 1968, p. 50)

Robert Solomon, in *The Passions* (1976), argues that the passions or the emotions provide the structure of this wider life world. They do it by providing perspective, not a physical one but a personal one, based on importance-to-me — "not physical but psychological distance" (p. 68) vis-à-vis the facts. Our world as lived, what Solomon calls surreality, is not just the facts but what they mean to me: and this meaning is produced by "values, demands, expectations and fantasies" (p. 69).

My world contains something *over and above* the (objective) world. In my world, objects are not only heavy and made of iron or canvas; they are also beautiful or ugly. In my world, people are not merely shaped so; they are attractive or repulsive. In my world, an action is more than an event; it is also important or trivial, admirable or blameworthy. In *my* world, some things *matter*; in *the* world, everything is as bland as Yorkshire pudding, substantial perhaps (even true) but tasteless and without spice or color. (p. 67)

How can evolutionary epistemologists come to grips with any of this? Assuming that this sketch of the human life world is even approximately correct, what can we pick out as its features? Besides the perception of some facts, we find evaluation of them in terms of wants and beliefs. The evaluation may be very complex or relatively simple. Emotions, or at least feelings, seem to be associated quite often with life worlds, either as products of evaluations or as factors that engender them. In any case, if Solomon is on the right track, life worlds have a felt component to them — more than a simple feeling of pleasure or pain, more like an emotional tone, a complex of feelings, judgments, values and beliefs. Not only are the felt aspects of an experience just as real to the subject as are the physiological givens about the external world, they combine somehow to direct and inform further actions and decisions, influence where one directs one's attention, and so on. If this is true (the interested reader is referred to Solomon for a fuller discussion), then clearly we need a means of dealing with life worlds.

### b. *Lorenz on the Subjective: Problems*

Assuming that evolutionary epistemologists wanted to deal with the human life world, one way to begin would be to ask Lorenz's questions: what natural causes make it possible? How did it come about? What good is it for human survival? Do we share any of this way of being with other animals?

Suppose we agree with Solomon that at least in part the passions or emotions "provide the structures for my world," and suppose that we also accept (pending further discussion below) that emotions are complexes of feelings, judgments, values and beliefs. Then we have the beginning of a way to approach the comparative psychological work that will be needed for this task of evolutionary epistemology. Do animals have feelings? All animals, or just some? Could we say that any of them have beliefs? Knowledge? Values? Emotions? Do any of them make judgments?

But even on the (perhaps) simplest of these questions, Lorenz throws up a roadblock: "the question ... 'Do animals undergo subjective experience?', is one which I am often asked to answer. The answer is: 'If I were able to give an answer, I would have solved the problem of body-and-mind'."[9]

Why does Lorenz think that he has to solve the entire mind-body problem in order to answer this question? Because he has posited an unbridgeable, unanalyzable gulf between the subjective and the objective. He has almost put them into different ontological categories, which is just what a materialist usually hopes to avoid when dealing with the mind-body problem!

But Lorenz has a special stake in this point of view. There are four reasons why he is concerned to deny that subjective phenomena can be investigated. First, in psychology in general, feelings have had a bad press ever since early psychologists decided to 'be scientific' on the idealized model of the physical sciences and to study only objectively measurable behavior or neurophysiological functions. Second, students of ethology carry their own additional burden to 'be scientific' by avoiding 'industrious ant' stories in favor of recording what organisms actually do. Third, Lorenz himself is violently opposed to vitalism — I suspect that any appeal to subjective experience smacks of vitalism to him. Finally, Lorenz has always worked with a narrow view of scientifically respectable concepts. For him, the canon of acceptable concepts includes only publicly observable behavior, changes of physical, physiological or neurophysiological state that are in principle directly observable, or theoretical models that are analogs of mechanical processes. But these neither represent the limits of what is empirically investigable, nor do they set the limits for good theoretical terms.

Here is an example of the lengths to which Lorenz goes to reject subjective phenomena:

> If I am walking along with a tame greylag goose which suddenly stretches, extends its neck and softly utters a harsh warning-call, I may say 'now it is alarmed'. However, this subjective abbreviation only means that the goose has perceived a flight-eliciting stimulus and that — in accordance with the principles of stimulus summation — its threshold values for other flight-eliciting stimuli have been markedly lowered. . . . In saying that the goose is alarmed, I am expressing the freely-admitted *belief* that subjective processes are taking place within the bird. We all believe that animals experience things; after all, we have animal protection laws and we do not unnecessarily cause pain to animals. However, the scientific content of my observation is restricted to the statement that a goose which behaves in the manner described is much more likely to fly away than usual. (Lorenz, 1963, p. 323)

Note that Lorenz is implicitly limiting what a subjective experience could be — it is only a simple feeling of pleasure or pain. But this is not a sufficient analysis of 'being alarmed.' If we were to describe a human being as being alarmed, we would be implying that the subject had perceived some facts about the world and evaluated them as being harmful in some way. We would further imply that a bodily feeling (which we would need to specify) and a tendency to behave in certain ways (depending on the situation) has resulted from the subject's assessment. But Lorenz has made both the evaluation of the surroundings and the feeling of the goose irrelevant. He has taken 'being alarmed,' an unacceptably subjective description, and translated it into behaviors and mechanisms: threshold lowering, likelihood of flying away. An interpretation or evaluation of external stimuli must still be done: however it is not done BY the goose but by mechanisms that operate IN the goose. The functions of the subject have been taken over by 'independently operating' physiological mechanisms.

Lorenz is *not* making one of the standard moves in discussion about the mind-body problem. He is not claiming that anything we can say about animal subjective experiences can be translated without residue into statements about physiological phenomena. Instead, he argues that 'mental' and 'physical' or 'subjective' and 'objective' are concepts which are irrevocably separate, a-logicially connected, complementary.[10]

In *Behind the Mirror*, Lorenz claims that human consciousness emerged from simpler physical information-processing systems in a creative flash, a *fulguratio* — a jump from one system level to a higher one. But because of the nature of this *fulguratio*, we are confronted with an unanalyzable gulf between the subjective and the objective.

> The gulf between subjective experience and objective-physiological reality, however, is of a different kind [from that beween the organic and the inorganic], for it is not caused by a gap in our knowledge but by a basic inability *ever* to know, an inability determined *a priori* by the structure of our cognitive apparatus. (Lorenz, 1973, pp. 169—170)

With the cognitive apparatus produced by a creative flash, we are trying to analyze that very cognitive apparatus. But if Lorenz's report of system theory is correct,[11] we can't analyze new system characteristics wholly in terms of their simpler components. "It is idle to look in these original systems . . . for those system characteristics which come into existence only at a higher level of integration" (Lorenz, 1973, p. 43). And since we have no meta-cognitive apparatus to "look down from," we're stymied.

There are several things wrong here. Lorenz gives only a program for arguing that human consciousness was produced by a *fulguratio*. He uses the analogy of combining two sorts of electrical circuits, one with a resistor plus a condenser and one with a resistor plus an inductance coil to produce a novel and unpredictable result — diminishing oscillation (Lorenz, 1973, p. 29—31). This, he says, is like combining a large number of component parts like Gestalt perception, tropisms, reactions to stimuli, etc., which are found in other animals into the unpredictable result of the full human consciousness with all its novel capacities for learning, memory, etc. But what is missing here is an argument that these novel capacities are and must be the result of a *fulguratio*, and a retrospective account of how the system might work. One might have been unable to predict diminishing oscillation from the separate properties of the two circuits described above, but one can retrospectively explain the new result in terms of circuit theory with its usual concepts. Perhaps this is a bad analogy — but Lorenz gives few if any better ones.

In the same discussion in *Behind the Mirror*, Lorenz claims that the problem may be more serious than an evolutionarily temporary transcendental barrier.

> The 'hiatus' between soul and body, on the other hand, is indeed unbridgeable, albeit perhaps 'only for us', as Nicolai Hartmann put it — that is, with the cognitive apparatus at our disposal. Yet I do not believe that this is a limitation imposed just by the present state of our knowledge, or that even a utopian advance of this knowledge would bring us closer to a solution of the problem. The autonomy of personal experience and its laws cannot in principle be explained in terms of chemical and physical laws or of neuro-physiological structure, however complex. (Lorenz, 1973, p. 170)

These claims are all contraries and they may all be false. That is, we might not need a meta-system to anaylze human consciousness. Or we might be able to invent one. Or we might be able to explain the subjective in physical terms. Or we might be able to invent a wholly new terminology in which one set of concepts would be adequate to explain both 'physical' and 'psychological' phenomena. (See Bertalanffy, 1968, pp. 220—221.) In one variation on this view, Susanne K. Langer (1967) argues that the basis on which mind is constructed, i.e., feeling, should be "regarded as a phase of a physiological process instead of a product (perhaps a by-product) of it" (p. 23). The reader is referred to Langer's trilogy for a demonstration of this approach.

Clearly Lorenz does not agree. He holds something like a psycho-physical complementarity theory in which the mind and body stand in some sort of unique unanalyzable relationship, one where subjective processes are in some ineffable sense *caused* by CNS processes. He thinks that although not all brain processes have a correlate in consciousness, consious processes always have a physical correlate. And the processes that can rise to consciousness can be simple or complex. (These ideas are all right — it's the larger context that has serious problems.) However the split runs, for Lorenz (1981, p. 72) it is unbridgeable: it is 'for us an impenetrable wall". He even says, perhaps somewhat fancifully, that "we have two completely different organs for comprehending two incommensurable aspects of the same reality" (p. 72). There is no way to cross the divide.

This topic needs more discussion. Lorenz's 'double aspect theory' needs to be clarified: is it psychophysical parallelism or complementarity? Can Lorenz's undoubted materialism be reconciled with his ideas about the unattainability of the subjective? If subjective phenomena are denied full ontological status, is this genuine complementarity? Is

Lorenz using an idiosyncratic interpretation of system theory to bolster ideas about the unknowability of the subjective? We can't answer these questions here. But we can conclude that Lorenz's treatment of the subjective systematically devalues it as a topic for ethological investigation and as a source of explanatory terms. It implies that subjective experience is not a suitable topic for evolutionary epistemology. This is too bad. Not only do we live in an 'intentional' world and have a wide range of feelings and emotions ourselves, but we incessantly attribute subjective phenomena to each other and to animals. This is a pervasive human trait that evolutionary epistemology should try to explain, not dismiss as inappropriate.

## 3. EMOTIONS AND ETHOLOGY

### a. *Lorenz on the Subjective: Again*

In fact it is much too strong to claim that Lorenz has completely closed the door on an investigation of the subjective. He claims that our consciousness was produced from simpler components, some of which have a pronounce subjective correlate. "The gulf between the physical and the mental lies elsewhere, perhaps at right angles to the gulfs which separate the organic from the inorganic, animal from lower animal and human being from animal — it goes vertically through" (Lorenz, 1981, p. 72). So the possibility of a comparative investigation of the subjective aspect of human and animal mentation is not totally ruled out. But because of certain conceptual confusions which we will discuss below, as well as the anti-subjective prejudices which we have already mentioned, Lorenz has never done the kind of acute comparative study of the subjective that he has done for some aspects of behavior patterns and some aspects of perception.

Although he claims that there is a vast gap between people and animals, one that deserves to be called the result of a fulguratio, Lorenz also believes that there is an area of the subjective that is shared. But here the conceptual problems begin. The basic subjective experiences, he claims, are the same for people and animals.

> The message of displeasure ('that was wrong') and of pleasure ('do it like that again') are probably the most powerful generalized and abbreviated pieces of information contained in our *Ego*. The 'ability to experience pleasure and sorrow', as Wilhelm

Busch so aptly says, is doubtless the primaeval form of experience. And this is exactly what I feel can be attributed to higher animals.[12] (Lorenz, 1963, p. 334, in Lorenz, 1971)

Lorenz thus seems to imply that the beginning of the subjective is a *fulguratio*, and the gap between people and animals is the result of another one. But the first one. 'the subjective', seems to be an area where systems function similarly, at least insofar as we are referring to the capacity for pleasure and pain. If we make a distinction between the broadly 'subjective' and the later *fulguratio* which produced full human capacities, then Lorenz's claim, that no subjective experience can ever be understood because human consciousness cannot be understood, loses much of its force. The way is then open to do a comparative investigation of feelings, analogous to what has been done for perceptions. In fact, this is discussed in *Leben und Lernen* and elsewhere in Lorenz's writings. However, the investigation is marred by a confusion between feelings and emotions, and by Lorenz's constant attempt to restrict the discussion as much as possible to 'objective' behavior and analogies based on behavior.

I cannot fully comprehend what the dog experiences when it gets angry and barks, because I have wholly different physiological organs for getting angry[13] and none at all for barking. I cannot do that at all. But I can, naturally, because I know many animals and many analogies, say with great certainty that the animal experiences pleasure [Freud] and pain [Leid]. Indeed, if it has two general expressions for pleasure and pain, as the goose in fact does — I am not talking at all about a dog — then I can say: it has something like a subjective experience. (Lorenz, 1981, p. 71)

The phenomena of pleasure and pain may well be examples of physiological processes that somehow rise to consciousness as feelings, but which can be explained primarily in physical terms. However, Lorenz easily makes the transition from pleasure and pain to anger (Ärger), which is an *emotion*. (He may do this also for joy and sorrow, but due to the ambiguity of the German words and the lack of context, I can't be certain that he does.) In the following passage, Lorenz makes the related mistake of discussing a physiological locus for emotions.

Subjective experience, the capacity for joy [Freud] and sorrow [Leid], is not at all different in a dog than in us. This is to be expected, because we certainly know the places in the brain where our emotions [Emotionen] are located. This is all located in the hypothalamus and in the limbic system, which is just the same in higher animals as in us. There is anatomically hardly any difference. (Lorenz, 1981, pp. 61—62)

However, emotions are more than simple bodily feelings. As Solomon puts it, one prerequisite for emotions is the Self. That is, there are varying cognitive capacities needed for having the various emotions. Fear, for instance, involves the belief that this is harmful to me; anger, that my interests have been adversely affected; shame, that I have done something that is wrong or bad. In every case, some degree of Ego, reflexive awareness, is present. There must be a psyche for a psychological perspective. But this requirement is often overlooked.

> Many passions make it a point to hide their self-involvement and present themselves as mere 'feeling', as unconceptualized and inarticulate, as existing before and independently of any expression or description. But this apparent primitiveness (which often passes for what is 'natural' or 'instinctive') is a sham . . . . (Solomon, 1976, p. 81)

There is much more work to be done in making precise just what cognitive capacities are involved in the having of emotions and in specifying the capacities of various species. This task will require both comparative methods and conceptual analysis. For example, we may want to attribute only the 'simpler' emotions (e.g., fear) to some animal species, because on the basis of behavioral study we find no evidence that it has the complex self-reflective ability that more complex emotions (e.g. shame) require. But the questions need to be asked, the studies need to be done. Assimilating all emotions to feelings is a dead end.

### b. *Emotions — A Short Analysis*

A somewhat more detailed analysis of feelings and emotions will be given to indicate the task ahead. There are several distinctions to begin with. First, there is the class of 'self-produced' feelings such as seasickness, stomach- or tooth-ache, cold, hunger, etc. These are mainly produced by internal physiological processes which are not under the control of the subject and which do not depend on any beliefs about the environment, although beliefs about the environment may help human beings to bear them or alleviate them. This is the class of feelings broadly described as 'pleasure' or 'pain' experiences.

Second, there is a class of experiences that is a kind of intermediate class, one that Lorenz has been primarily responsible for bringing to our attention. Beginning with 'The Innate Forms of Possible Experience',[14] Lorenz has described some innate 'emotional reactions' (positive or negative feelings, sometimes more articulated even than that) which human beings have or seem to have in certain situations, usually directed to the attributes or the behavior of conspecifics. Examples of these reactions include positive nurturing feelings and specific behavior toward human infants, and positive or negative aesthetic and ethical 'evaluation feelings' towards people based on their physical or behavioral characteristics (see, e.g. Lorenz, 1950, pp. 154 ff. in Lorenz, 1971).

Lorenz argues for the innateness of these feelings because they are incorrigible by reason and because they are 'set off' erroneously in us by other animals and even by inanimate objects. We see some animals or dolls as 'cute' if they share the characteristics of the 'baby schema' [Kindenschema], e.g., chunky cheeks, high domed forehead, small ears, cuddly texture. We see the eagle as 'noble', the camel as 'haughty', the ant as 'altruistic', because we automatically apply schemas 'meant' for human beings to them. Beliefs about the environment may help human beings to modify the deliverances of these schemas (if you want to know how a camel feels, look at its ears, not its face), but the schema will always deliver the same message in spite of our other beliefs, knowledge, etc. The camel will still look 'haughty' even though I know it's friendly.

Lorenz has identified a genuine kind of feeling, which shares with the first-mentioned self-produced feelings the characteristic of being the product of physiological processes not under the control of the subject. But they differ in that they depend on perceptions of the environment, perceptions that have a strong (perhaps overwhelmingly) innate component — where this is more than saying that like all perceptions they depend on physiological mechanisms. These feelings may depend on some innate CNS schemas. But much more work needs to be done in specifying this possible innate perceptual component and the situations in which it can be usefully said to be operating. As Lorenz uses it now it is too vague and has indefinitely expandable boundaries so that almost anything that has a felt component is likely to be lumped into this class of innately produced feelings. This is a problem especially with the 'ethical' and 'asethetic' feelings, which Lorenz discussed in 'Innate Forms of Possible Experience' and elsewhere, and which I have criticized elsewhere (Kalikow, 1976).

There is a third class of entities, emotions, which have an indispensable feeling component but which also depend on intentional elements to a degree which may vary with the particular emotion but which is never zero. As we have seen, Lorenz neglects this class almost entirely, thus attributing too little in the way of reason to animals and perhaps too little to human beings as well. He seems to collapse emotions into the first two classes of feelings we have discussed, treating all emotions as if they were entirely self-produced. But this is incorrect. I want to take the time to show some of the features that emotions seem to have.

A caveat. As Amélie Rorty (1980, pp. 1–8) says, the class of emotions is not a natural one and the philosophical study of emotions inherits all the problems of philosophical psychology. We don't know enough in general about the relations between cognition and affect: "it becomes clear that a proper account of the emotions requires a revision of the whole map of psychological processes and activities, and of their complex interrelations" (p. 3). What follows is only an outline of some of the factors that must be taken into account in the attribution of emotions to a person or animal. There will be variations depending on the emotion, the animal to which it is attributed, the circumstances, etc. And there is much that we don't know. But onward.

First of all, emotions are "directed toward a particular thing, person, or animal (or group of them)" (Pitcher, 1965, p. 327), what George Pitcher calls agent-directed emotions. Common examples are anger, disappointment, happiness, jealousy, resentment, love, hate, fear, sadness, shame, unhappiness. When we use emotion words of this type, what is implied?

When we ascribe an agent-directed emotion to someone we are doing more than describing bodily feelings or behavior. We are interpreting actions. We are giving an implicit complex causal structure which explains why a person or animal is behaving in a way we need to account for. Some true statements about bodily changes and behaviors are probably necessary conditions for the truth of the description, '*X* is angry', but they are by no means sufficient. (This is why Lorenz can tell that the dog is sad, but not why.) Of course we must have some reason to think of ascribing an emotion (as opposed to something else) to a person or animal, and this is usually that they behave or even 'react' physiologically in some way. But these behaviors or reactions are not usually univocal clues to specific emotions (although in some animals they might be). What more than behavior is needed to be sure about emotion ascriptions?

We also need the specification of an external context of which the emoter is aware. In order for emotion ascriptions to be made, it has to be the case that the emoter can be said to have certain sorts of 'thoughts' (Thalberg, 1964) about the situation. First, the emoter must have certain sorts of knowledge (Gordon, 1969). *X* cannot be angry with me for insulting her unless she knows I did insult her. Fido cannot be sad that his master is gone unless he knows that his master is gone.

There must also be some evaluative attitude on the part of the emoter, perhaps one formed in the light of other things desired or valued (Gordon, 1974; Walter, 1972). If Fido is sad, he must value his master's presence.[15] Sometimes it is also necessary to take into account 'producers' of emotions (Donnellan, 1970), i.e., to explain what other conscious experiences made the emoter have certain beliefs and therefore certain emotions. I may be afraid of what you'll say next, although you obviously haven't said it yet. The producer of this emotion may be quite complex. Finally, the emoter must be affected bodily in some way: otherwise, whatever the cognitive state, the person or animal is not undergoing an emotion (Lyons, 1974).

Our apparently simple use of emotion words is really the use of a fairly complex description. The complexity of this analysis is justified partly because it explains the fact that we often recognize from behavior, bodily symptoms and contextual information when someone else is undergoing an emotion and what emotion it is (Gordon, 1974; Green, 1972). If we see that Fido is sad, we assume that there msut be a context for it, even if we cannot discover it right away. We can empirically test emotion ascriptions because they are causally related, in a complex way of course, to a person's or animal's experience of the world. The relationships may be simpler for animals than for people, and it will depend greatly on the animal species, but if we choose to describe an animal or a person as having a certain emotion, then we are compelled by the nature of emotion concepts to ascribe some intentional properties such as thoughts, beliefs, knowledge, values — as well as bodily feelings — to the emoter. Exactly which ones and how much of each will depend on the emotion and on the animal. Ascriptions of emotions function like theories, not like names.

We do not have to choose emotion descriptions. Alternatives might not imply that a person or animal had any mental activity at all. Compare 'Fido was sad' to 'Fido's tail and ears drooped, the corners of his mouth turned down, he crawled into a corner and did not eat.' Although we need not choose to ascribe agent-directed emotions to a subject, we may feel that something has been left out if we don't. The cognitive elements attributed to the emoter vanish, and so does the emotion description. Emotions are not temporally and causally prior to reason — it is the other way around.

This analysis shows that the emotions are not 'in' the limbic system, the brain or the hypothalamus (Lorenz, 1981), although the capacity for emotional responses may correctly be said to have one of its proximate causes in these systems. So if we say that a goose or a dog experiences joy or sorrow, not merely pleasure or pain, we are claiming that something about their mental capacities has played a causal role in the experiencing of an emotion, It is too simple to think of emotions as something that people or animals 'have' like they 'have' toothache. And emotions are more than reactions to stimuli. There is a 'reaction' quality to the having of many emotions, and just because there are cognitive elements involved this does not mean that emotions are under total rational control! But the problem with considering emotions as just like self-produced feelings responding to a given stimulus is the 'given stimulus'. Who or what gives the given? In the kind of emotions I have been discussing, the knowledge, beliefs, expectations and values of the emoter make an indispensable contribution.

### c. *Emotions and Ethological Explanation: What Next?*

This analysis shows that just as Lorenz and others have done valuable work in showing that the perceptual Umwelten of creatures are different but all related to Reality, there is an analogous job to be done for the feelings and the emotions. Lorenz may have assumed that work on the information-processing capacities is the most basic, fundamental to any understanding of other cognitive properties, because it builds on the basic function of 'fact-seeing'. But this is only an assumption. The capacity to view the world as Objective may not be the entire basis for being human at all (though nobody will argue that it's not important). Susanne Langer, for example, claims that this capacity is probably a very late evolutionary achievement of human beings. The intentional view of the world, intertwined heavily as it is with emotions, may have been the original basis for human-being, and it may be an area where our commonality with other animals may be most fruitful to explore (Clark, 1982). This will be an even more complex task than the exploration of possibilities/modalities of perception and learning (though it will build on it), because it must focus, not on bodily feelings, but on the kinds of knowledge, judments or values a creature can be shown to have.

It is very important not to have too narrow an idea of what can be investigated. If attributions of emotions function as explanations of behavior, they can be tested. If we assume that *X* is undergoing an emotion, what do we have to attribute to *X*? Can we find ways to confirm the possession of these attributes? Will a simpler explanation do the job so that emotion ascriptions are not needed? This is the usual sort of ethological procedure that has been used so successfully to test perception and learning capacities, and which has so often shown that our uncritical attributions were wrong. This sort of investigation would undoubtedly teach us much about the nature of animal cognitive capacities. The concept of emotion too would be refined by stringent comparisons and attempts to produce adequate and elegant explanations of behavior.

I want to conclude with a suggestion for further work. It is a bit of a paradox because the phenomenon to be investigated is probably one of the reasons that ethologists like Lorenz have avoided the subjective in the first place.

One of the reasons ethology focused on behavior and tried to purify itself from talk of emotions, purposes and intentions was because of the 'Industrious Ant' stories; our incessant and apparently irresistible imputations of motives, values, purposes, feelings, emotions and so on to ants, bugs, bees, birds and so on. I recently heard a woman discussing a spider whose web had gotten wet — she said, "You could just see how disgusted he was at the water." What's going on here? Lorenz himself has taught us that the attribution of a particular emotion or attitude to animals may be the result of a misapplication of an innate schema — the haughty camel-proud eagle phenomenon. The interpretive perception 'meant' for another human has been automatically and mistakenly applied to an animal.

But with the industrious ant phenomenon, rather than having a narrow schema misapplied, we have a general one operating, I think — what Lorenz called, following Karl Bühler, *Du-Evidenz*. It was characterized by Lorenz as an *a priori* component of human thought that leads us inexorably to the belief that our fellow human beings have subjective experiences:

> my knowledge about the subjective experiences of my fellow men and my conviction that higher animals (such as a dog) also experience things are two quite closely related phenomena. They are *not* based on analogization, as has long been assumed by philosophers. One of the greatest acheivements of my respected teacher, the late Karl Bühler, was the unequivocal demonstration that the assumption of other human subjects with similar experiences is an inescapable and compulsive train of thought. *It is a genuine a priori necessity of thought and interpretation*, which is just as evident as any axiom. Bühler therefore spoke of '*Du-Evidenz*', the evidence of 'you'. (Lorenz, 1963, p. 324)

I think Lorenz is right about *Du-Evidenz* as an a priori component of human thought. Not only does it help to explain our attributions of emotions, etc. to our fellow human beings, its operation also explains our descriptions of animals, and also plants and even inanimate objects, as endowed with human purposes, feelings and emotions. It is not clear to me whether this is a mistaken spill-over, or an example of the operation of an *a priori* necessity of thought that we inherited from and that 'works on' other animals. It may also have been a good thing evolutionarily speaking that it 'worked on' animals as well as people (Clark, 1982). *Du-Evidenz* may also provide (if it's needed) a pre-linguistic basis on which a common language of emotions and subjectivity could have been built. It lets us escape 'my own case' and the uniformly weak arguments from analogy for the existence of other minds. And it gives us a biological basis for Wittgensteinian arguments against private language — see Langer for extensive discussions of the possible evolution of language from the commonality of shared emotions and their expression.

It is striking that Lorenz nowhere gives or even attempts an evolutionary account of *this* 'a priori necessity of thought' as he has done for so many other aspects of the *a priori*. While he says that we are 'driven' to assume that people and animals undergo experiences, he does not ask his usual questions: Where from? What for? Why? as he has done, for example, for the concepts of space and causality.

The only hint Lorenz gives as to how we could test the functioning of *Du-Evidenz* is to tell us to use physiological analogy. Here again his bias toward the physiological and the 'objective' is evident.

> The physiological-psychological parallelism or 'isomorphy' of processes which I objectively and subjectively observe in myself permits the conclusion that a fellow human being, whose physiological functions are analogous to my own, also experiences in a manner analogous to myself when subject to the same physiological events. When

> applied to animals, analogization is much less valid. The greater the difference in structure of sense-organs and nervous systems from that of my own, the greater will be the difference in functions. The nature of the accompanying experience is fundamentally inaccessible to me, and remains so even when an inescapable piece of evidence drives me to assume that my dog undergoes some kind of experience. The lower we descend in the organic series, the greater the invalidity of analogization, and at low levels of organization, even the evidence of the 'you' [*Du-Evidenz*] is muted. Even I can kill mussels without a trace of sympathetic feeling. (Lorenz, 1963, pp. 324–325)

But Lorenz is probably wrong about this. It is not physiological knowledge that produces the action of *Du-Evidenz*. I would guess that what evokes it is the appearance of individually controlled behavior which makes sense relative to an environment — what Marjorie Grene (1968, p. 27) calls 'centricity'. Otherwise it would be hard to explain why people who have no physiological knowledge to speak of have trouble killing their first frog in Biology 101. Live frogs jump around and catch flies — mussels just lie there. Degree of physiological organization and centricity don't always necessarily match, either. Even H. S. Jennings (1906), a great observer, certainly sometimes wrote as if he wanted to think that one-celled creatures knew what they were doing. Indeed, as long as an organism — however simple — behaves as if it 'knows' what it is doing in its environment, our first prescientific impulse is to credit it with an entire complex capacity of knowing, believing, feeling, intending, emoting, planning, etc. This impulse should not be ignored: it should be studied.

A great scientific advance in the study of animal and human behavior, due in large part to ethology, was to show that much 'intelligent' behavior can be described and explained by the use of concepts such as reactions to stimuli, tropisms, taxes, and so on. That is, ethology has often told us when our naive 'intentional' descriptions are inappropriate and unnecessary. This has been a fruitful and fundamental advance. It might possibly be an advance of comparable magnitude to discover when to let intention-descriptions back into scientific discourse about animals.

*University of Northern Colorado*

## NOTES

[1] Part of the research on which this article is based was done under National Science Foundation Grant no. SOC78-11404.

[2] Many of Lorenz's works with a strong epistemological emphasis are found in collections of his translated papers. The two major collections are Evans (1975) and Lorenz (1970, 1971). In addition to a selected bibliography provided in the References section of this paper, Lorenz has a forthcoming volume on 'the maladies of the human mind' (cf. Lorenz, 1973, pp. 5, 243).

[3] Although the copyright date is 1973, the book did not actually appear in English translation until about 1978.

[4] From his earliest days Lorenz has held a strong meta-theoretical commitment to materialistic mechanism as opposed to vitalism. See Kalikow (1975).

[5] Lorenz (1981, p. 20). All translations from this volume are by the author.

[6] The term seems to have been used first in this connection by another of Lorenz's predecessors, J. von Uexküll.

[7] Lorenz (1941, p. 183). All citations from this work are from the English translation in Evans (1975).

[8] Preface to the English Edition, Weizsäcker (1952, p. 7). Von Weizsäcker's main source of biological examples seems to have been Lorenz (1943a), which he cites in Weizsäcker (1949).

[9] Lorenz (1963, p. 323). All citations from this article are from the English translation in Lorenz (1971).

[10] Another theme runs through Lorenz's discussion of mind and body, subjective vs. objective phenomena: objective phenomena are Real and subjective phenomena are somehow only 'experience'. This can be seen in the citation from Lorenz (1973) which follows immediately in the main text.

[11] Lorenz appeals to Nicolai Hartmann's hierarchy theory for some of his ideas. But (at least in Hartmann, 1953) I find only very limited support for Lorenz. For instance, while Hartmann does claim that it is impossible to elucidate "the question of how physiological processes act upon psychic processes and vice versa" (p. 104), he also claims that one should not deny a causal connection between them, and further, we should "obliterate" the "ontic frontier" between them, as the practicing psychiatrist does (p. 105).

[12] A note on translation: As the subsequent discussion will make clear, it is a mistake to equate pleasure and pain with joy and sorrow respectively. The first pair are feelings (though their analysis is not unproblematic either) and the second pair are emotions. However, German (even German!) uses Freud and Leid for both feeling-emotion pairs, pleasure-joy (Freud) and pain-sorrow (Leid). So here is an example of the misleading nature of ordinary language! Does or doesn't Lorenz know that there needs to be a distinction? (Same question for his translator, Robert Martin.) Sometimes the context makes clear what is meant, other times not. In my translation I shall make the distinction as best I can — even if I happen to clear up Lorenz's ideas for him beyond what he did for himself, I don't think it will affect the argument.

[13] This seems to contradict Lorenz's assertion, cited below, that we share with animals the major physiological locus of the emotions, i.e., the hypothalamus and the limbic system.

[14] Lorenz (1943a). Elements of this work furnish important components of Lorenz's post-World War II work in human evolution, perception, epistemology, etc.
[15] I do not think this analysis of emotions begs the question of the necessary function of intentional elements if sometimes the values in question are 'innate'. The issue needs more investigation.

## REFERENCES


Bertalanffy, L. von: 1968. *General System Theory*. New York: Braziller.

Clark, Stephen R. L.: 1982. *The Nature of the Beast: Are Animals Moral*? Oxford: Oxford University Press.

Donnellan, Keith S.: 1970. 'Causes, Objects and Producers of the Emotions,' *Journal of Philosophy* **67**, 947–950.

Evans, Richard I.: 1975. *Konrad Lorenz: The Man and His Ideas*. New York: Harcourt, Brace, Jovanovich.

Gordon, Robert M.: 1969. 'Emotions and Knowledge,' *Journal of Philosophy* **66**, 408–413.

———: 1973. 'Judgmental Emotions,' *Analysis* **34**, 40–48.

———: 1974. 'The Aboutness of Emotions,' *American Philosophical Quarterly* **11**, 27–36.

Green, O. H.: 1972. 'Emotion and Belief,' *American Philosophical Quarterly, Studies in the Philosophy of Mind Monograph Series* **6**, 24–40.

Grene, Marjorie: 1968. *Approaches to a Philosophical Biology*. New York: Basic Books.

Hartmann, Nicolai: 1953. *New Ways of Ontology*. Tr. Reinhard Kuhn. Chicago: Henry Regnery Co.

Jennings, H. S.: 1906. *Behavior of the Lower Organisms*. New York: Columbia University Press.

Kalikow, Theodora J.: 1975. 'History of Konrad Lorenz's Ethological Theory, 1927–1939: The Role of Meta-Theory, Theory, Anomaly and New Discoveries in a Scientific "Evolution,"' *Studies in the History and Philosophy of Science* **6**, 331–341.

———: 1976. 'Konrad Lorenz's Ethological Theory, 1939–1943: "Explanations" of Human Thinking, Feeling and Behaviour,' *Philosophy of the Social Sciences* **6**, 15–34.

Langer, Susanne K.: 1967, 1972, 1982. *Mind: An Essay on Human Feeling*. 3 vols. Baltimore: The Johns Hopkins Press.

Lorenz, Konrad: 1941. 'Kants Lehre vom Apriorischen im Lichte gegenwärtiger Biologie,' *Blätter für Deutsche Philosophie* **15**, 94–125. (English translation in Evans, 1975).

———: 1942. 'Induktive und teleologische Psychologie,' *Die Naturwissenschaften* **30**, 133–143. (English translation in Lorenz, 1970).

———: 1943a. 'Die angeborenen Formen möglicher Erfahrung,' *Zeitschrift für Tierpsychologie* **5**, 235–409. (Unpublished English translation by the author, 'Innate Forms of Possible Experience.')

———: 1943b. 'Psychologie und Stammesgeschichte.' In G. Heberer (ed.), *Die Evolution der Organismen*. Jena: G. Fischer, pp. 105—127. (English translation of revised 1954 version in Lorenz, 1971).

———: 1950. 'Ganzheit und Teil in der tierischen und menschlichen Gemeinschaft,' *Studium Generale* **3**, 455—499. (English translation in Lorenz, 1971.)

———: 1959. 'Gestaltwahrnehmung als Quelle wissenschaftlicher Erkenntnis,' *Zeitschrift für experimentelle und angewandte Psychologie* **6**, 118—165. (English translation in Lorenz, 1971.)

———: 1963. 'Haben Tiere eine subjectives Erleben?' *Jahrbuch der Technischen Hochschule* (*München*). (English translation in Lorenz, 1971.)

———: 1965. *Evolution and Modification of Behavior*. Chicago: University of Chicago Press.

———: 1966. *On Aggression*. New York: Harcourt, Brace and World.

———: 1970, 1971. *Studies in Animal and Human Behaviour*. Tr. Robert Martin. 2 vols. Cambridge, Mass: Harvard University Press.

———: 1973. *Behind the Mirror*. New York: Harcourt, Brace, Jovanovich.

———: 1974. *Civilized Man's Eight Deadly Sins*. New York: Harcourt, Brace, Jovanovich.

———: 1981. *Leben und Lernen*. With Franz Kreuzer. München: Piper.

Lyons, William: 1974. 'Physiological Changes and the Emotions,' *Canadian Journal of Philosophy* **3**, 603—617.

Pitcher, George: 1965. 'Emotion,' *Mind* **74**, 326—346.

Rorty, Amélie: 1980. *Explaining Emotions*. (Ed.) Berkeley: University of California Press.

Solomon, Robert: 1976. *The Passions: The Myth and Nature of Human Emotion*. Garden City, N.J.: Anchor Press/Doubleday.

Thalberg, lrving: 1964. 'Emotion and Thought,' *American Philosophical Quarterly* **1**, 44—55.

Walter, Edward: 1972. 'The Logic of Emotions,' *Southern Journal of Philosophy* **10**, 71—78.

Weizsäcker, C. F. von: 1949. *The History of Nature*. Chicago: University of Chicago Press.

———: 1952. *The World View of Physics*. Chicago: University of Chicago Press.

WILLIAM A. ROTTSCHAEFER

# WILFRID SELLARS ON THE NATURE OF THOUGHT

Wilfrid Sellars is, I believe, exceptional among contemporary philosophers for the seriousness of his concern to recognize both the power of the scientific enterprise and the centrality of our ordinary understanding of ourselves as persons. Sellars holds both that "science is the measure of what is and what is not" and that "if man had a radically different conception of himself he would be a radically different kind of man," for "man is what he is because he thinks of himself in terms of this image [the manifest image, our ordinary way of understanding ourselves]" (Sellars, 1963a, pp. 6 and 15). Sellars' goal is the formulation of a synoptic vision of human beings in the world, a scientific vision that does justice to our ordinary view of ourselves as sensory, conceptual and intentional agents.

In this paper I shall focus on Sellars' account of human conceptual capacities, his theory of the nature of thought. I shall do so by considering one of several types of objections which have been raised against his views. First, though, I shall review briefly Sellars' account of the nature of thought conceived ordinarily and as we are beginning to look at it scientifically.

We can summarize Sellars' basic theses on our ordinary conception of thought as follows: I *Verbal Behaviorism:*[1] (1) Overt linguistic episodes are genuinely conceptual in an underivative sense that allows semantic categories to be applied to them in a primary way. The underived character and primacy in question is methodological or in the order of knowledge and not in the order of being; (2) Thoughts as inner conceptual episodes, thoughts-2, are understood on analogy with overt linguistic episodes thoughts-1; II *Theoretical Mentalism*: (3) Thoughts-2 are like theoretical entities; (4) Thoughts-2 are hypothesized on the basis of observable events, in particular, silent intelligent human behavior; (5) Thoughts-2 are postulated to explain certain observable events, in particular, silent intelligent behavior, and their explanatory power is extended to account also for the occurrence of overt intelligent behavior, thoughts-1; (6) Thoughts-2 are modeled on the characteristics of observable events and processes, specifically,



*Abner Shimony and Debra Nails (eds.), Naturalistic Epistemology*, 145–161.

thoughts-1; and (7) We can learn to refer to these inner conceptual episodes, though theoretical, in immediate reports just as we do in the case of the imperceptible entities of scientific theories.[2]

However, even if we grant that this is an accurate account of our ordinary view of our conceptual capacities and how we come to know about them, it is not in Sellars' view an unqualifiedly accurate account. For as a scientific realist he is committed to the view that it is through our scientific investigations that we will come to a more accurate understanding of our conceptual capacities. Thus our ordinary knowledge framework (*OKF*) conception of thought, though playing a necessary methodological role in the acquisition of scientific knowledge, is subject to addition, correction and even replacement by our emerging scientific conceptions of thought.[3] Two general constraints influence the kinds of theory which will replace our ordinary view of thought (Sellars, 1963d, pp. 183—186; 1963a, pp. 18—25; 32—34; 1967a; 1980). First, though we must recognize that the ideal of the unity of science is legitimate and that in fact the completed scientific framework represents such an ideal unification, this ideal does not destroy the independence of the various scientific disciplines. As a consequence, we must allow for the autonomy and irreducibility of theoretical psychological explanations of thought. Thus the entities and processes that such theories refer to will be a part of the completed scientific framework. Second, at the same time, the ideal of the unity of science will also require an integration of psychological theory with theories in other scientific disciplines, in particular, neurophysiology. There are also, in Sellars' view, two things that we can already say in general about thought from a scientific point of view. Using psychological theory we can view thoughts as fundamentally a representational system (Sellars, 1981b). And from the perspective of neurophysiology we can conceive of thought as neurophysiological processes (Sellars, 1963a, d; 1967e, 1980). Thus, to the extent that our current theories are accurate, thought conceived as a representational system constituted by a set of neurophysiological structures and processes has ontological priority. On the other hand, thought as conceived in *OKF*, that is, as overt linguistic episodes and as inner conceptual episodes and dispositions toward such episodes, has methodological primacy.

While Sellars' scientific realist commitments require that our *OKF* conceptions of thought be replaced by more adequate scientific conceptions, his commitment to a synoptic vision of persons in the world demands a continuity between our ordinary and scientific views (Sellars, 1963a). This continuity is provided in part by Sellars' functionalist account of thought (Sellars, 1968a, Chs. III–V; 1979; 1980). According to this account, the key semantical properties of thought, meaningfulness and truth, are given functional interpretations. Thus statements of meaning are regarded as classificatory in nature. In the first instance they classify linguistic expressions according to the role they play. Thus Sellars explicates

(1) '*S*' in *L* means *p*

as

(2) '*S*' in *L* is a ·*p*·

where '·*p*·' is a sortal predicate that applies to expressions in any language that play in their respective languages the same role played in the speaker's language by the sign design that occurs between the dot-quotes. Consequently, meaning statements are functional classificatory statements which assert that the utterance, '*S*' in *L*, belongs to the class of utterances in any language that plays the role '*p*' does in the speaker's language. And a truth claim is in the first instance an assertion that a statement is semantically correct according to the rules of the linguistic framework of the user. Consequently, the intentionality of thought, conceived of as the property of being about an object and asserting something about that object, is understood in the first instance as rule-governed linguistic activity. But the role players in thought, the subjects of the rulers which govern thought and which have the functions prescribed by these rules, may be made of an indefinite number of types of materials. Thus the vehicles of thought can be among other things, sounds, marks, gestures, artifacts, inner episodes, psychological and neurophysiological processes.

In Sellars' view the rules that govern thought concern three principal functions. These are, using somewhat different language than Sellars, (1) information gathering, (2) information coordinating, and (3) information using. In *OKF* these functions are performed in the first instance by overt linguistic episodes; (1′) takings-out-loud, (2′) inferrings-out-loud, and (3′) intendings (or volitions)-out-loud. In the second instance in *OKF*, the functions are instantiated by the corresponding inner conceptual episodes: (1″) perceptions, (2″) inferences, and (3″) intentions (or volitions). Conceived scientifically, these functions are accomplished by representational and neurophysiological processes in ways that we are only beginning to understand. Thus the conceptions of thought in *OKF* and in the emerging scientific theories are tied together internally and with each other by a functional characterization.

Given, then Sellars' complex set of commitments to the ontological primacy of our scientific conception of thought, the methodological primacy of our *OKF* conception of thought, and a synoptic vision adequate to both conceptions, an assessment of Sellars' position must take each of these theoretical aims into account. I shall focus on Sellars' account of our *OKF* conception of thought, in particular, the primacy it assigns to thoughts-1 in *OKF*. In Sellars' view the criteria of adequacy for any account of *OKF* in general, and of thought in particular, derive from the particular roles *OKF* plays in our cognitive enterprise as a whole. Our conception of thought in *OKF* provides us with (1) an understanding of ourselves as conceptual agents, (2) which itself provides the means for moving us to a more adequate scientific understanding of our conceptual capacities, and (3) which is used to understand and provide initial solutions for interframework philosophical problems.[4] Any adequate account of the *OKF* conception of thought must show how our *OKF* performs these functions. Though I cannot in this paper explore how well Sellars' Verbal Behaviorism and Theoretical Mentalism meet all these requirements, I want to outline briefly how in Sellars' view they satisfy part of the first requirement, namely, providing an understanding of ourselves as conceptual agents. This will give us a context for understanding some of the reasons why Sellars believes his account is preferable to alternative views and for situating the objection to Sellars' claims that I want to consider.

The *OKF* view of thought helps us understand ourselves as conceptual agents by describing and explaining our abilities (1) to determine what is and what is not the case, (2) to share understandings of the world and ourselves, (3) to use thought to guide our behavior, (4) to recognize each other as thinking agents who use thought to guide actions, and (5) to have privileged access to our own thoughts. This view of thought suggests and to some extent embodies implicit solutions for a number of classical philosophical problems which are interframework in character. These are problems of (1) objectivity, (2) intersubjectivity, (3) human agency, (4) other minds, and (5) privileged access.[5] Although these conceptual abilities overlap, the last two can be distinguished from the first three insofar as they concern the use of conceptual powers to recognize and understand oneself and others as beings who have the first three sorts of conceptual abilities. Let's focus on Sellars' account of our knowledge of others as conceptual agents.

Sellars believes that Verbal Behaviorism and Theoretical Mentalism best capture three central features of our *OKF* conception of our knowledge of others as conceptual agents: (1) the immediate accessibility of the thought and intentions of other persons, (2) their hidden, but accessible character, and (3) the role of overt thinking as a manifestation of inner conceptual episodes. Sellars contends that the ability attributed to persons in *OKF* to know each other as conceptual agents is best approached when thought is conceived of first as overt linguistic episodes directly accessible to us perceptually. For, as opposed to alternatives, such an account allows one to understand and readily explain how it is at times that we seem to know directly what a person is thinking. We can do so, in Sellars' view, because we know first and primarily what a person is saying or has said and do not have to ascertain, by some means or other, the meaning behind the words. The meaning is in the words — to continue the spatial metaphor. Moreover, this account allows us to explain how we can at times understand and explain directly the intelligent and not so intelligent behavior of others. For the linguistic behavior which precedes, accompanies, or follows such behavior and to which language users have direct access can be used to account for the behavior in question. Sellars' account can explain why it is that in some cases in the *OKF* view a person's intentions are immediately accessible. But since much intelligent, and not so intelligent, human behavior is not preceded, accompanied or followed by such out-loud explanations, it is reasonable to postulate that such silent behaviors be accounted for by inner conceptual episodes which are to be understood on analogy with the thinkings-out-loud that have been used to explain rational and irrational behavior.[6] And the theory of inner conceptual episodes can be extended to explain even thinkings-out-loud and the behaviors explained by the latter. The inner conceptual episodes can be considered to cause overt behavior, both linguistic and non-linguistic, and changes in dispositions to such overt behavior.

A number of methodological objections have been raised to Sellars' view of thought as conceived in *OKF*, thus to his Verbal Behaviorism and Theoretical Mentalism. I shall examine a criticism that attempts to show that Sellars' account implicitly presupposes the methodological primacy of thoughts-2. This type of criticism grants the profound analogies between thought and language but contends that the basic model for understanding the categories of intentionality and agency is thought as inner conceptual episode, not overt linguistic episode. The intentionality of the latter is derivative and based on the former. We can call this position the primacy of the mental thesis.[7]

Perhaps the most straightforward formulation of this objection is that Sellars' account is circular.[8] In its simplest form (Objection I) the objection goes something like this. Sellars claims that the meaningfulness of thoughts-2 is to be understood in terms of the meaningfulness of language, thoughts-1. But if we ask how we know that someone is speaking meaningfully and not merely parroting words, we find that speech is meaningful only if it is used to convey thoughts-2. Language is meaningful because it is an expression of thought. Otherwise it is merely a string of sounds, marks or motions. The Sellarsian reply to this objection is clear, given our earlier discussion. Meaningful speech can be distinguished from parroting by the patterned behavior that the former displays and the latter does not. These patterns reveal connections among (1) the linguistic response of the speaker and the speaker's environment, (2) the linguistic utterances themselves of the speaker, and (3) the linguistic utterances of the speaker and his/her subsequent behavior. Of course, in some cases we may not be sure of our assessment. Does Jonesella really understand what she is saying? Or do our next door chimps really intend to steal our bananas, even though that's what they're signing? But despite these doubts in certain contexts, we can usually distinguish language using from parroting by examining such patterned relations. We do not have to assume that language users are attaching the meaning of their thoughts to the words they are using or language hearers are attaching thoughts to the words they are hearing.

But this response leads us to a more penetrating criticism of Sellars' position (Objection II). It goes as follows. Genuine use of language reveals not just any sort of linguistic uniformities. If we are to take Sellars' notion of both meaning and truth seriously, then both involve uniformities which are exemplifications of rules. For the conceptions of meaning as rule-governed use and of turth as semantic assertibility imply the notion of rules. Thus the uniformities are not as Sellars himself argues *mere uniformities*. They are the results of rule-governed activity. But, so the objection continues, the ability to act on the basis of rules involves conceptual activity, in the sense of thoughts-2. For example, to act on the rule, 'One ought to do *A* in circumstance *C*', (an ought to do rule, *OTDR*) involves at least the following: (1) a grasp of what it is to do *A*, (2) a grasp of circumstance *C* and when it obtains, (3) a notion of agency, and (4) a notion of the force of a rule. Clearly then the attempt to characterize the meaning of thoughts-1 in terms of rule-governed linguistic uniformities is circular, for such uniformities depend on the conceptual activity that belongs to thoughts as inner mental episodes, thoughts-2. The Sellarsian reply to this objection is deceptively simple. The objection assumes that both the conceptual activity described above in (1) through (4) and the practical reasoning implied therein can only be accomplished via thoughts-2. But this is precisely the point at issue. According to Sellars, such conceptual activity and practical reasoning are first found in thoughts-1 (Sellars, 1973). This response, if correct, would indeed relieve the Sellarsian account of circularity. But one can expect that the champion of the primacy of the mental will not be satisfied. '*Gratis asseritur, gratis negatur*', is his likely response. The issue is how do we establish primacy in the order of knowledge. We will have to return to this issue shortly. Before we do so, let's look at another vicious circle in the clutches of which Sellars' account may also be caught. An examination of this new charge of circularity will help us to clarify both the methodological issues involved in the notion of primacy in the order of knowledge and the bases of Sellars' claim that meaning and truth are in the first instance characteristic of thoughts-1.

So to renew his attack the champion of the primacy of the mental argues as follows (Objection III): Let's grant Sellars his claim that conceptual activity can be discovered in thoughts-1, linguistic activity. For the speaker could say, 'I ought to do *A* in circumstance *C*' and then do *A* in *C*. But, in order to account for the conceptual character of the linguistic activity, Sellars cannot now invoke an *OTDR* under pain of vicious circularity. Indeed, Sellars realizes this and introduces a further distinction that he thinks will avoid the vicious embrace. As is well known, Sellars has introduced a tripartite distinction. Besides (1) *mere uniformities* (such as fire at *t*-1 and heat at *t*-2) and (2) *rule-governed uniformities* (behavioral uniformities resulting from *OTDR*s, for instance, 'One ought to do *A* in *C*'), there is another important type of uniformity, *pattern-governed uniformities*. These uniformities are governed by what he calls 'ought-to-be rules', *OTBR*s. Their form is: 'It ought to be the case that *A* is brought about in *C*'. For example, it ought to be brought about that Jonesella says no when the neighboring chimps try to take her bananas. Patterns of behavior exemplifying *OTBR*s are not mere uniformities, since the rules are in Sellars' view involved in the production of the uniformities. Yet the pattern is not the result of *OTDR*s because the behaviors are not performed with knowledge of the rules nor the intention to follow the rules. What concretely does Sellars have in mind? He has two favorite examples. First the case of language learners — or for that matter learners of any game. The responses of language learners follow the rules of language but not because they have appropriated those rules and apply them, but rather because their responses have been appropriately reinforced by trainers who do know the linguistic rules and how to apply them. Secondly, Sellars uses the example of certain innate activities, specifically the dance of bees, which informs about location of and distance from honey. The bees, Sellars argues plausibly, do not act in terms of *OTDR*s nor are their behaviors only an exemplification of mere uniformities (Sellars, 1963e, 1969).

So let us grant, as will the champion of the primacy of the mental, that there are pattern-governed behaviors in addition to mere behavioral uniformities and rule-governed behaviors and that these pattern-governed behaviors are exemplifications of and the results of *OTBR*s. Then it might seem that Sellars can avoid vicious circularity in his explication of conceptual activity by arguing that since all rules of language are not *OTDR*s — some are *OTBR*s — not all psychological concepts imply *OTDR*s. Some imply *OTBR*s. Thus conceptual activity is fundamentally and non-circularly explicated in terms of pattern-governed behaviors and *OTBR*s. To put the distinction in an alternative Sellarsian formulation, not all conceptual doings are *actions*, doings which are by their nature intentional. Some are *acts*, doings which are actualities as opposed to potentialities, in the Aristotelean sense of those terms. But they are not the result of intentional activity. Actions are guided by *OTDR*s and acts by *OTBR*s. This reply seems to release the circularity vise, but the relief is only apparent for there is another Sellarsian doctrine that seems to seal the grip of fate for Sellars' own program. It is the thesis that the concepts of act and *OTBR*s imply those of action and *OTDR*s (Sellars, 1969, pp. 508 and 511; 1973). We can grasp the former only because we can grasp the latter. Thus the introduction of acts and *OTBR*s has only staved off the closing circle by adding an extra notch to it. For if the concept of *OTBR*s implies that of *OTDR*s, then all psychological concepts, mediately or immediately, imply *OTDR*s. We have come full circle.

What is one to make of this objection to the Sellarsian attempt to explicate the notion of conceptual activity in linguistic terms? My view is that it misrepresents the Sellarsian project.[9] As Sellars puts it, the introduction of *OTBR*s serves to stop a *vicious infinite regress*, not to break a *vicious circle*. The difference may seem to be relatively unimportant, a debating point, but it is not. Let us see why. Objection III, as formulated, implies that Sellars project is one of conceptual reconstruction or explication.[10] On this reading Sellars' goal is to find the logically or analytically most basic conceptual components of the notion of conceptual activity or thought. Thus *OTBR*s seem to be the proper candidate for the role of the conceptually basic. They cannot, however, serve that function because, in Sellars' own view, they conceptually presuppose *OTDR*s. I believe, however, that such a reading of Sellars' purposes for the introduction of *OTBR*s is mistaken. *OTDR*s and actions are, in Sellars' view, conceptually basic in the order of knowledge. Our most basic starting points for understanding conceptual activity, either inner conceptual episodes or thinkings-out-loud, are actions and *OTDR*s. To put it in terms of Sellars' synoptic project, the *OKF* conception of thinking is basically linguistic actions governed by *OTDR*s. Why then introduce acts and *OTBR*s? The introduction is not for purposes of further conceptual analysis. Rather, it is the beginning of an explanatory move; it is a step toward that which is more basic, not in the order of knowledge but in the order of being. For the problem, once we come to the concept of conceptual activity as rule-governed linguistic activity, is not how to understand it in more basic terms. There are none more basic in the order of knowledge. The problem is to explain how there can be such rule-governed activity without a vicious infinite regress. The objection to such rule-governed activity is the familiar one against mental action and rule-governed behavior. How can it ever get started? How can it come to be? For in order to do *A* in *C*, I must follow rule *R*. But, then, in order to follow rule *R*, I must follow rule *R*' which states in which circumstances *R* is to be followed. And that in turn requires following rule *R*", *ad infinitum*. To stop the infinite regress, it must be the case that not all doings are actions. Some are acts. Acts do not require the following of rules. Thus, for example, language learners act in accord with rules that their trainers, not they, follow. So Sellars is arguing that in the order of knowledge the concepts of linguistic action and of *OTDR*s constitute our primary understanding of conceptual activity, linguistic and mental, and thus of such psychological concepts as intention and agency. But in the order of being acts and *OTBR*s are prior. Without acts there could be no actions. And the source of these acts is not and cannot be the agent's intentional activity. Rather these acts are the agent's because of the influence upon the agent of her natural and social environment and her genetic history.

Thus it is a mistake to view the introduction of *OTBR*s as an attempt to find what is most basic in the order of conceptual analysis. *OTBR*s have a priority in the order of being since without them there can be no *OTDR*s. But *OTDR*s have priority in the order of knowledge since *OTBR*s imply *OTDR*s in the sense that the former cannot be understood except by understanding *OTDR*s. The reason for this is, I believe, that *OTDR*s serve as a model for understanding *OTBR*s. To clarify this last claim consider the concept of natural selection. It is clear that natural selection implies no intentional activity on the part of nature. Yet the model for understanding natural selection is artificial selection, which does imply such intentional activity. Natural selection is prior in the order of being and without it there would be no agents capable of intentional activity including artificial selection. But artificial selection is prior in the order of knowledge and is used to give intelligibility to the notion of natural selection.

In sum my response to Objection III makes two points. First, though Sellars' explication of conceptual activity may appear to be just that, conceptual analysis, it is, in fact, something significantly different. For though Sellars is indeed interested in uncovering the basic concepts involved in the nature of thinking, especially the semantical notions of meaning and truth, he envisions this project not as the uncovering of the *a priori* or logically most basic understandings of thought and its semantic characteristics. Rather he is attempting to clarify the fundamental notion of thought and its semantical characteristics in *OKF*. As we have seen, he argues that thought and the semantical notions of meaning and truth are, in the first instance, discovered in thinkings-out-loud. They are subsequently applied to inner mental episodes. This brings us to our second point and to the question raised by the responses to both objections II and III. How does Sellars establish that thinking out loud is indeed the primary notion in *OKF* of thought?

A consideration of a parallel question about the primary conceptions of physical things in *OKF* will perhaps bring out the import of the question and suggest an answer to it. How might a Sellarsian establish that in *OKF* the primary conception of macro-physical things is that of colored extended objects in time and space with their properties and relations? Is this a matter of conceptual analysis, empirical observation or scientific theory? Conceptual analysis without extensive empirical observation will prove unsatisfactory, since as such it will reflect the cognitive limitations of its user, individual, cultural and historical. However, with empirical observation as a way to overcome these limitations, conceptual analysis can play an important clarificatory role. Nevertheless, it seems that conceptual analysis and empirical observation together will not suffice. What is needed in addition to these is scientific theory, specifically perceptual and congitive psychological and biological theories. This claim may appear quite paradoxical. For what it seems to be saying is that the most basic concepts of that framework which is defined in terms of our perceptual capacities are not themselves revealed by these capacities. Should we not be able to observe what concepts are most fundamental in our characterization of macro-physical objects as understood in *OKF*, the perceptual framework? An affirmative reply seems plausible, and the proposal that the discovery of these concepts is in part a theoretical affair appears implausible so long as we view perception itself as a process completely understandable on the level of *OKF*. Now clearly it is not so understandable. As with the macro-physical objects themselves so too with perception, our knowledge of the basic perceptual categories that we use to characterize macro-physical objects finds its beginning in observation, observational correlations, and conceptual analysis, but must be corrected by, enhanced, and enlarged by theoretical knowledge. The use of theoretical terms in perceptual reports helps us to understand this point. For the role of categorial and conceptual elements in perception is such that, though we are capable of a more analytic use of perception, very often in perception we make use of synthetic and derivative categories, indeed, theoretical categories (Shimony, 1978). Thus it is not clear that a careful and exhaustive collection and analysis of all perceptual reports about macro-physical objects would in itself reveal the basic categories in terms of which we conceive macro-physical objects in *OKF*. Perceptual and cognitive theories are also needed.[12] A similar argument can be used, I believe, in the case of the basic concepts which constitute thought in *OKF*. Consequently, establishing the claim that thought in *OKF* is in the first instance thinking-out-loud, where this is understood as rule-governed linguistic activity, is in part the task of scientific theory. Of course, conceptual analysis and empirical observation will have their place.

Thus I contend that the objections based on the primacy of the mental fail to establish that primacy. Specifically, they do not demonstrate that Sellars must appeal to the mental in order to overcome internal problems in his account. Indeed, it has not been established that Sellars' account is viciously circular. Does the *OKF* notion of conceptual activity fundamentally require anything more than rule-governed linguistic activity? Sellars replies in the negative. The champion of the primacy of the mental replies in the affirmative and contends that what is required is a notion of thought as inner mental episode. Our discussion has not resolved this issue.[13] All it has done is to point out that the Sellarsian mode of resolution will involve conceptual analysis, empirical observation and theory. Consequently, more than the resources of *OKF* will be required for its resolution. It should be apparent, therefore, that the mode of resolution itself will be controversial.

Before I conclude I want, in the light of what we have discussed, to reflect briefly on the characterization of Sellars' commitments to both the scientific enterprise and our *OKF* understandings with which I started this essay, since they give us an insight, I believe, into Sellars' conception of the goal of his enterprise, the synoptic vision of persons in the world.

How are we to understand Sellars' dark saying (1963a, pp. 6 and 15) that "if man had a radically different conception of himself he would be a radically different kind of man." Taken with the claims that "science is the measure of all things, of what is and what is not" and that "man is what he is because he thinks of himself in terms of this image [the manifest image]," it seems that the very process of discovering scientifically who we are leads to the remaking of ourselves. Or, more relativistically, we might conclude that since we view ourselves differently in the scientific and manifest images, we are different. Though, indeed, there is some truth to the claim that our successful and not so successful cognitive attempts to understand ourselves involve and lead to practices which in turn do change ourselves, we cannot accede to an idealist or relativist reading of Sellars. On the other hand,

Sellars is not maintaining that our ordinary conceptions of ourselves are definitive. His scientific realist commitments prevent such an interpretation. Our examination of Sellars' account of our *OKF* conception of ourselves as conceptual agents suggests another reading. If there is a sense in which our *OKF* conception of ourselves is radically false and, therefore, to be replaced by a more adequate scientific conception, there is another sense in which our conception of ourselves plays a necessary methodological role in our efforts at self-understanding.[14] We are, in other words, precisely the kind of being that comes by means of *OKF* concepts to understand ourselves as conceived scientifically. Thus our *OKF* conception of ourselves as conceptual and intentional agents, as persons, whose capacities and intentions are first manifested in the use of language, characterizes us in some essential way by describing the kind of cognitive agent we are. We necessarily make use of our *OKF* conception of ourselves and things in order to discover who we really are. And if, indeed, we ceased to make use of these conceptions, we would become other than who we are. But though such an evolutionary — or genetically engineered — change is not out of the question, it would nevertheless be precisely that, a change in who we are. Thus our identity does not change in discovering who we are, for we must use the *OKF* conception of persons to discover and understand who we are as measured by the scientific picture of the world (Rosenberg, 1975). But, in addition, our *OKF* conception of ourselves is itself only adequately known through scientific investigation.[15]

*Lewis & Clark College*

NOTES

[1] 'Verbal Behaviorism' is the term Sellars has adopted to describe his position in some of his most recent essays. See, for instance, Sellars (1980; 1979, Ch. 4). In his earlier essays he frequently used the expression, 'methodological behaviorism' to describe his position. Cf. for example, Sellars (1963d, pp. 183—186). However, he has also described his position as logical behaviorist in contrast with logical physicalism. Cf. Sellars (1969, p. 525). Sellars rejects *metaphysical* (*logical*) *behaviorism*, the claim that mental terms can be defined by means of observational terms. Cf., for instance, Sellars (1963d, 1980). He also rejects Skinnerian *radical behaviorism*, the view that though mental entities may exist, since they are not causes, but merely effects, of behavior, they need not play a role in scientific psychological explanations of behavior. Thus he does

not accept the radical behaviorist view that theories postulating mental entities and processes are unnecessary for an adequate scientific psychology. Cf., for example, Sellars' third Carus Lecture, 'Is Consciousness Physical' (1981a, pp. 66—90). Sellars' *verbal behaviorism* is best characterized, I believe, as a liberal *methodological behaviorism* in contrast with *narrow methodological behaviorism*. The latter position holds that though mental terms are not definable by means of observational terms, an operational characterization of such terms suffices for an adequate psychological account of the mental; and if theories about the mental are employed, they are not to be taken realistically. On the other hand, the former position claims that though mentalistic terms cannot be defined by means of observational terms, and though operational definitions of mental terms are necessary, an adequate account of the mental must employ theories about the mental and well-supported theories about the mental should be interpreted realistically. It is not immediately clear that Sellars' verbal behaviorism so characterized is anything more than a restatement of a commitment to good empirical methology and scientific realism applied to psychology. Though I cannot argue it here, I believe that it is. Cf. Rottschaefer (1983).

[2] I have drawn these theses together from a number of sources in the Sellarsian *corpus*, the most important of which are the following: Sellars (1963e, d, b; 1967a, b, e; 1968a, esp. Chs. 3—6; 1969; 1973; 1974a, b, c; 1968b, a correspondence with Chisholm; 1979, esp. Chs. 4—5; 1980; 1981a; 1981b) and Rosenthal and Sellars (1972). The *locus classicus* for Sellars' position is, of course, 1963d.

[3] Sellars discusses the relationships between ordinary knowledge and scientific knowledge in a number of places. Cf. Sellars (1963c; 1967c, d; 1963a, pp. 116—150; 1977).

[4] These criteria are, I believe, derivable from the requirements of a synoptic vision.

[5] These problems and their solution are at that heart of the issues discussed at length by Sellars (1968a).

[6] The classical Sellarsian presentation of this move is the myth of the accomplishments of the genius Jones among our Rylean ancestors. Cf. Sellars (1963d; 1968a, Ch. 6).

[7] The *locus classicus* of an encounter of Sellars with a proponent of the primacy of the mental thesis is the Sellars-Chisholm correspondence (1968b). David Rosenthal has examined this discussion with great care and insight and has concluded in favor of Sellars. (Cf. Note 2; Rosenthal and Sellars, 1972; Rosenthal, 1976). Julian Young (1973), on the other hand, has concluded in favor of the neo-classical position and the primacy of inner mental episode. I believe Young's conclusions follow only if one equates, as Young does, the views of Quine and Sellars at strategic points. Cf. Jay Rosenberg's (1974, Ch. 4) Sellarsian critique of Quine. Bruce Aune (1967, Ch. 8) has given a detailed examination of how the categories of intentionality are applicable to linguistic utterances from a Sellarsian point of view. Perhaps no one has paid more careful and critical attention to Sellars' views on the nature of thought than Ausonio Marras. In a series of papers Marras (1973a, 1973b) has challenged the adequacy of Sellars' verbal behaviorism and theoretical mentalism both in terms of its adequacy in the light of what we know ordinarily and scientifically. Sellars (1973) replied to the latter and Marras (1973c) presented a rejoinder. We shall be discussing his views in more detail below. Most recently John Bishop (1980) has assessed the Marras-Sellars interchanges, and though he finds fault with some of Marras' criticisms, nevertheless, sides with the neo-classical position in some important details. A second type of objection (Marras, 1977; 1978) attempts to force Sellars into giving up the methodological

primacy he attributes to overt linguistic episodes in understanding our conceptual capacities under pain of finding himself caught in a logical behaviorist account. I explore this objection in Rottschaefer (1983). A third type of objection denies that thoughts can be theoretical entities (Bishop, 1980; Sussman, 1975).

[8] Here I begin to reconstruct and interpret Marras' critique and the Marras-Sellars interchange. (Cf. 1963d, pp. 188—189; 1968a, p. 72).

[9] Indeed, Sellars (1973) intimates such a response. He contends that his purpose in introducing *OTBR*s is to avoid an infinite regress, not a vicious circle. But, then, instead of rejecting Marras' formulation of the objection, what in the text I have called Objection III, he raises the question of whether he really accepts the line of argument, the basis for what I have referred to as Objection II in the text. Sellars then proceeds to respond to Objection II along the lines I have indicated. But what about Objection III? Sellars leaves open the question whether he accepts as his own the line of argument it represents. I think he does not.

[10] The following quotation illustrates, I think, Marras' confusion concerning Sellars' methodology: "However, Sellars' claim is that although thought is causally and therefore existentially prior to speech, speech is conceptually (that is, in the order of rational reconstruction) prior to thought. . . . To state Sellars' position somewhat aphoristically, we could say that 'in the order of being' it is correct to say that speech is *thought-out-loud*, whereas 'in the order of conceiving' it is correct to say that thought is *silent speech*'" (Emphasis is Marras') (Marras), 1973a, p. 154). Cf. 1963d, pp. 188—189; 1968a, p. 72. The aphoristic formulation is incorrect; Sellars' view, as we have seen, is exactly the opposite of those attributed to him by Marras. And the conceptual priority attributed to speech by Sellars is not, I shall argue, achieved by rational reconstruction. Detailed textual analysis would be required to pin down my interpretation of Marras. Since I haven't the space for that here, I shall not identify the champion of the primacy of the mental with Marras.

[11] Sellars has noted the important connections between explanations in terms of natural selection and in terms of reinforcement. Thus *actions* achieve their ends by the intentional following of rules. *Acts*, on the other hand, attain their ends in the way reinforced behaviors attain their ends (without such intentional following of rules). See Sellars (1963, pp. 324—327). B. F. Skinner, among others, has also pointed out the connection between explanations in terms of natural selection and reinforcement. See e.g., Skinner (1971, pp. 14—17).

[12] I have argued for this position in several places. Cf., e.g., Rottschaefer (1976; 1978).

[13] In any case, I do not believe that Sellars' own naturalistic epistemology — or any naturalistic epistemology conceived generally as the thesis that a necessary condition for understanding human cognition is to grasp the place of human beings in nature and that the sciences must be used for solving epistemological problems (Cf. Shimony, in this volume) — rests on the theses of verbal behaviorism. It may be as Marras has argued, though I am not convinced of it, that the concept of the mental is a theoretical one. Cf. Marras (1977, 1978) and Rottschaefer (1983). But such a conclusion would not be incompatible with the theses of a naturalistic epistemology broadly construed as above. On the other hand, if per chance the classical introspectionist or the apriorist accounts turn out to be right, an unlikely occurrence in my estimate, the correctness of one of these accounts, since they do presuppose positions incompatible with a naturalistic epistemology, would put the latter in jeopardy.

[14] I have argued against what I take to be Sellars' position that *OKF* is in principle eliminable methodologically in Rottschafer (1978).

[15] For an excellent discussion of the circularity implicit in this kind of naturalistic approach and of its benign character, cf. Shimony (this volume).

## REFERENCES

Aune, Bruce: 1967. *Knowledge, Mind and Nature: An Introduction to Theory of Knowledge and Philosophy of Mind*. New York: Random House.

Bishop, John: 1980. 'The Analogy Theory of Thinking,' *Australasian Journal of Philosophy* **58**, 222—238.

Marras, Ausonio: 1973a. 'Sellars on Thought and Language,' *Nous* **7**, 152—163.

———: 1973b. 'On Sellars' Linguistic Theory of Conceptual Activity,' *Canadian Journal of Philosophy* **2**, 471—483.

———: 1973c. 'Conceptual Activity, Rules, and Linguistic Actions: A Rejoinder to Wilfrid Sellars,' *Canadian Journal of Philosophy* **2**, 495—501.

———: 1977. 'The Behaviorist Foundation of Sellars' Semantics,' *Dialogue* **16**, 665—675.

———: 1978. 'Rules, Meaning and Behavior: Reflections on Sellars' Philosophy of Language.' In Pitt, 1978, pp. 163—189.

Pitt, Joseph C. (ed.): 1978. *The Philosophy of Wilfrid Sellars: Queries and Extensions*. Dordrecht: D. Reidel.

Rosenberg, Jay: 1974. *Linguistic Representations*. Dordrecht: D. Reidel.

———: 1975. 'The Elusiveness of Categories, the Archimedean Dilemna and the Nature of Man; A Study in Sellarsian Metaphysics.' In Hector-Neri Castaneda (ed.), *Action, Knowledge and Reality: Studies in Honor of Wilfrid Sellars*. Indianapolis: Bobbs-Merrill.

Rosenthal, David: 1976. 'Mentality and Neutrality,' *The Journal of Philosophy* **73**, 386—415.

Rosenthal, David, and Wilfrid Sellars: 1972. 'Rosenthal-Sellars Correspondence on Intentionality.' In Ausonio Marras (ed.), *Intentionality Mind and Language*. Urbana, Ill.: University of Illinois Press, pp. 461—503.

Rottschaefer, William A.: 1976. 'Observation: Theory-Laden, Theory-Neutral or Theory-Free?' *The Southern Journal of Philosophy* **14**, 499—509.

———: 1978. 'Ordinary Knowledge and Scientific Realism.' In Pitt, 1978, pp. 135—162.

———: 1983. 'Marras' Critique of Sellars' Account of Thought: A Response.' *Philosophy Research Archives* IX, 511—533.

Sellars, Wilfrid: 1963. *Science, Perception and Reality*. New York: Humanities Press.

———: 1963a. 'Philosophy and the Scientific Image of Man.' In Sellars, 1963, pp. 1—40.

———: 1963b. 'Being Known and Being Known.' In Sellars, 1963, pp. 41—59.

———: 1963c. 'Phenomenalism.' In Sellars, 1963, pp. 60—165.

———: 1963d. 'Empiricism and Philosophy of Mind.' In Sellars, 1963, pp. 127—196.

———: 1963e. 'Some Reflections on Language Games.' In Sellars, 1963, pp. 321—358.

———: 1967. *Philosophical Perspectives*. Springfield, Ill.: C. C. Thomas.

———: 1967a. 'Physical Realism.' In Sellars, 1967, pp. 185—208.
———: 1967b. 'Notes on Intentionality.' In Sellars, 1967, pp. 308—320.
———: 1967c. 'Language of Theories.' In Sellars, 1967, pp. 321—336.
———: 1967d. 'Scientific Realism and Irenic Instrumentalism.' In Sellars, 1967, pp. 337—369.
———: 1967e. 'The Identity Approach to the Mind-Body Problem.' In Sellars, 1967, pp. 370—388.
———: 1968a. *Science and Metaphysics: Variations on Kantian Themes*. New York: Humanities Press.
———: 1968b. 'Intentionality and the Mental.' In *Concepts, Theories and the Mind-Body Problem. Minnesota Studies in the Philosophy of Science*, vol. 2. Minneapolis: University of Minnesota Press.
———: 1969. 'Language as Thought and Communication,' *Philosophy and Phenomenological Research* **29**. (Reprinted in Sellars, 1974, pp. 93—117.)
———: 1973. 'Reply to Marras,' *Canadian Journal of Philosophy* **2**, 485—493. (Reprinted in Sellars, 1974, pp. 118—127.)
———: 1974. *Essays in Philosophy and Its History*. Dordrecht: D. Reidel.
———: 1974a. '. . . this I or he or it (the thing) which thinks . . .' In Sellars, 1974, pp. 62—90.
———: 1974b. 'Metaphysics and the Concept of a Person.' In Sellars, 1974, 214—241.
———: 1974c. 'Empiricism and Abstract Entities.' In Sellars, 1974, 245—286.
———: 1977. 'Is Scientific Realism Tenable?' In F. Suppe and P. D. Asquith (eds.), *Philosophy of Science Association Proceedings 1976*, vol. 2. East Lansing, Mich.: Philosophy of Science Association, pp. 307—334.
———: 1979. *Naturalism and Ontology*. Reseda, Calif.: Ridgeview.
———: 1980. 'Behaviorism, Language, and Meaning,' *Pacific Philosophic Quarterly* **61**, 3—25.
———: 1981a. *The Carus Lectures. The Monist* **64**, 3—90.
———: 1981b. 'Mental Events,' *Philosophical Studies* **39**, 325—345.
Shimony, Abner: 1978. 'Is Observation Theory Laden? A Problem With Naturalistic Epistemology.' In R. T. Colodny (ed.), *Logic, Laws, and Life: Some Philosophical Complications*. Pittsburgh: University of Pittsburgh Press, pp. 185—208.
Skinner, B. F.: 1971. *Beyond Freedom and Dignity*. New York: Bantam.
Sussman, Alan N.: 1975. 'Mental Entities as Theoretical Entities,' *American Philosophical Quarterly* **12**, 277—288.
Young, Julian: 1973. 'Intentionality,' *Review of Metaphysics* **26**, 696—722.

# PART II

# THE USE OF COGNITIVE PSYCHOLOGY IN EPISTEMOLOGY

DONALD T. CAMPBELL

# NEUROLOGICAL EMBODIMENTS OF BELIEF AND THE GAPS IN THE FIT OF PHENOMENA TO NOUMENA [1]

## PREFACE

The traditional terms 'epistemology' and 'philosophy of science' are not quite right for many of the exciting current developments that go by these names. Whereas classical epistemology tried to answer the problem of knowledge without assuming any knowledge in the process, naturalistic epistemologists (Quine, 1969), Copernican epistemologists (Shimony, 1970), and evolutionary epistemologists (Campbell, 1959, 1974a) deliberately beg the question, and undertake epistemological inquiries assuming that present-day physics and astronomy give us approximately valid knowledge of the world to be known, and that evolutionary biology, psychology, and sociology tell us something about man the knower. Such epistemological efforts can be conducted in loyal compatibility to traditional epistemology, as in accepting the negative results of the skeptical tradition: our shared epistemological predicament with Plato's prisoner in the cave, the scandal of induction, the argument from illusion, and the irrefutability of solipsism. To the traditional epistemological question, 'Is knowledge possible?' the logical answer is 'no'. We cannot be sure that we know, or when we know. Insofar as knowledge of the world is concerned, our most valid beliefs will involve presuppositions we cannot prove, and any additions to 'knowledge' (in the practical sense) will be at the expense of additional presuppositions. A 'descriptive epistemology' that takes over from here will be more like science than philosophy, as philosophers have drawn the distinction. It will be hypothetical, contingent, and presumptively descriptive of knowers and the world to be known. Within these limitations, it can still deal with the issue of truth: In what kind of world would what kind of procedures lead a knowing community to improve the validity of its model of the world? But whereas traditional epistemology wanted to do epistemology without ontology, every descriptive epistemology will involve a hypothetical ontology. Longings



*Abner Shimony and Debra Nails (eds.), Naturalistic Epistemology*, 165–192.

for certainty, for a firm anchoring of beliefs, or even for complete explicitness in the evidence supporting beliefs, must be relinquished.

Not only is science today the main arena in which descriptive-epistemological issues get argued, the character of most 'philosophy of science' has also changed. No longer is it abstemiously restricted to a logical analysis of the status of scientific truth. While there is a strong interest in explicating the normative decision rules science *should* use in deciding between theories, this gets mixed up with arguments about which decision rules science *has* used, implicity or explicitly, in presumably valid decisions in the past, and thus can be seen as a hypothetical, contingent search for normative rules. For such endeavors, 'theory of science' becomes technically more appropriate than the term 'philosophy of science'. For such theory of science, the history of science is fundamentally relevant, whereas it is not for the traditional purist designation of the task of the philosopher or logician of science.

Under the general heading of theory of science not only history of science, but also such specialities as sociology of science and psychology of science (Campbell, 1959; Fisch, 1977; Royce, *et al.* 1978) can become relevant to the issue of the validity of science, even though as now practiced, these fields generally avoid this issue. Within this latter field, the present essay makes its conjectures at the level of developmental and evolutionary neurobiology.

This essay is a part of a more general program to mechanize and physicalize all aspects of believing and/or knowing, which in turn follows from a hypothetical realism that examines sensation, perception, cognition, and choice of actions as material and mechanical processes. This extension of the physicalist-realist program, in a reflexive way, supports the analyses of the skeptics and anti-realists by joining in their rejection of a 'direct realism' for perceptual products, or 'real' status for scientific theories. Descartes' evil demon, or modern implanted electrodes, could indeed create neural processes perceived as real. Again with Descartes, we have no analytic or otherwise self-justifying decision rules to decide whether current experience is dream or reality. 'Justification' of beliefs based upon perceptual products must be argued from their general 'reliability', not from their infallibility. Assessments of such reliability beg the very question reliability theory seeks to answer, and turn out to gain most plausibility from Descartes' trust in God's providence or a parallel trust in biological natural selection. (See Paller & Campbell, 1987 for a fuller presentation.)

## VEHICULAR REQUIREMENTS FOR EMBODIED KNOWLEDGE

Even though naturalistic epistemology as a movement announces the relevance of the anatomy and evolutionary biology of eye and brain, most of its discussion, including much of my own (1959, 1966, 1974a, 1974b), employs philosophical concepts and vocabulary. In contrast, the present essay attempts to keep to a language of physical substances, placing 'knowing' in a framework of material things and systems, of physical objects and processes. Pure epistemology may often deal quite profitably with disembodied, unrealized, and abstract belief and knowledge. Not so, however, for the descriptive epistemology I attempt. Instead, the knowledge it studies will be physically embodied in some substance, some vehicle or carrier. This vehicle will have its own physical nature and limitations.

Let us make this more vivid by considering a mosaic mural done in stone fragments and picturing a street scene, as an example of embodied knowledge of the street, buildings, and persons depicted. The size of the stones, the thickness and color of the cement, the range of natural colors available, the restriction to a two-dimensional surface, the required rigidity, etc., all contribute to the substantialized belief or knowledge that is carried, all become a part of the picture, reducing its validity from any ideal of perfection, were such a conceptualization feasible. The end product, knowledge, at its realized best, is some compromise of vehicular characteristics and of referent attributes. Where validity is our goal we of course minimize the vehicular contribution as much as possible, as by using smaller and smaller pieces of stone, and cements that are thinner and more transparent. But we can never completely eliminate vehicular restriction and bias for embodied knowledge. This also holds true for retinas made of rods and cones, for nerve cells, brains, memory processes, visual perceptions, innate reflexes, stimulus-response associations, thought and cognitive structure.

Without having done the logical analyses that might make them compelling, I have leaped to some general principles that will guide my explorations: the vehicular substance that carries knowledge is unavoidably alien to the referents of knowledge — it is a different substance with different structural characteristics. Complete flexibility in depiction, reflection, transmission, or recording, is precluded by the structural requirements of the vehicle. If the vehicle is completely flexible it lacks the rigidity to hold together the picture it carries. These vehicular-structure requirements produce not only restrictions on fineness of detail, but also bias and limitations of aspect. Keeping the vehicle intact becomes a requirement in rivalry with the requirement of validly mapping the referent.

This alien, limited, biasedness I extend to less obviously physical vehicles of knowledge, such as spoken, written, and remembered language, logical symbol systems, and mathematical notations. Their rigid structures of terms and syntax are vehicular requirements distorting the referents to some degree. This analysis can also be extended to the self-perpetuating social systems that are the vehicles for scientific knowledge (Campbell, 1979). The social glue that holds such groups together has structure-maintenance requirements that limit and bias the portrait of the world such social groups sustain.

Descriptive epistemology will need eventually a physical theory of optimal vehicles. Think of plaster-of-paris casts, clay, magnetic tapes, photo-sensitive chemicals, and fixing processes in photography: do these always involve a two-phase process, one phase of maximal flexibility and sensitivity to the shape of the environment (or to reflections of the environment) and a second phase of rigidity? Think of how we choose stone and wood for realistic sculpture: is it required that the physical structure of a good vehicle be fine-grained? Do nervous tissue and genetic codes conform to such principles insofar as they differ from other bodily tissues? How are these physical requirements for stable record related to revising, expanding, and improving embodied knowledge? Under what conditions are partial revisions possible? Is total substitution of a different portrait generally a more mechanically feasible procedure than retouching it?

A similar applied physics of structures is needed for detection and transmission systems. Fritz Heider in his 'Ding und Medium' (1926, 1959) was thinking about such issues. A 'transparent' medium seems to be one that contributes least of its own structure to the knowledge it transmits. But it must have some structure to transmit other patterns at all. If wave forms are being transmitted, must resonances of the transmitter substance always get added, and be removed only by presumptive artificial reconstructions (as by filters that remove the vehicular harmonics, whether or not they were part of the message input)?

Let us pause for a moment in this physicalization. Like most traditional and modern epistemologists, I too regard conscious experience, visual perception, memory of past events, language, and the mathematical formulae of modern physics, as prime exemplars of embodied knowing. I recognize that most epistemologists, descriptive or otherwise, will profitably stay within these bounds. But I also feel that it may be useful for some of us to try placing these prime exemplars in radically different conceptual frameworks. I have started such an exploration in adding the strange examples of mosaic murals and plaster-of-paris castings. My strategy in this paper is next to move to a still wider genus, adding exemplars still more strange, and then to move back toward our common problem area, bringing along some distinctions and processes I find easiest to clarify at these more general levels.

## EXTERNAL AND INTERNAL SELECTORS

For me, embodied knowledge falls within the general class of instances of 'fit', of the fit of one system to another, or of one system to a stable hodgepodge. The fit of animal form to environmental opportunity is an important exemplar of fit. Now I share with William James' contemporary, James Mark Baldwin (1909), and with the modern cyberneticist W. Ross Ashby (1952), and others, the dogma that all instances of fit are achieved, maintained, and improved by profoundly indirect processes involving selection from unforesighted variations. Fit comes from selection rather than mutations that are wise in advance. In a series of past papers I have run this theme into the ground in applying the blind-variation-and-selective-retention recipe to learning (1956a), vision (1956b, 1966), cultural evolution (1965) and scientific development (1974a, 1974b). At the levels of vision, creative thought, and science, my stubborn insistence on the blindness of variations has lost me some otherwise sympathetic readers, perhaps needlessly (Gamble, 1983). Because of this, and because I've expounded the thesis enough in print already, I am merely confessing and not trying to persuade on this issue in this essay.

In what follows, I am substantially extending and complicating a natural-selectionist evolutionary epistemology which I share with 200 to 300 others (Campbell, 1974a; Campbell, Heyes, & Callebaut, 1987), starting with Darwin and extending up to the present day. For all of us, the central insight is that biological natural selection and other selection processes allow the real world to edit and select among variations, providing the fit between belief or knowledge (just as for organic form) and the real world. While this insight is a precious one which I will try to sustain, many presentations of it have naively assumed a directness to nature's selective action which a closer inspection will not support. Thus William James (1880) in a brilliant essay on creative thought, stupidly writes as though all those wild ideas and "absurd conceits" were rejected by the environment through direct encounters with it, an idea in itself so absurd that in an hour's further thought he would have encountered his error, which is corrected in Poincaré's (1913) otherwise similar, but probably quite independent, essay on mathematical creativity. Poincaré suggests that mathematical beauty is a preliminary mental criterion which screens out 99% of the dross produced by the unconscious wild permutation of mathematical ideas. This must then be followed by tedious, conscious, paper-and-pencil sifting in trying to validate the beautiful idea. Positing that 'nature herself' chooses among the variations offered to her in the form of competing scientific theories implies a directness to the real world's selective activity that becomes untenable upon any close examination of the many individual and social discretionary decisions (uncompelled by logic or evidence) which implement changes in the dominance of one scientific theory over others.

In all of my previous essays on evolutionary epistemology (e.g. 1956b, 1959, 1974a), I have provided a partial correction to this oversimplification by stressing *vicarious* blind-variation-and-selective-retention processes, and within these, *vicarious selectors.* Poincaré's mathematical beauty is an example of one, and I will return to this theme later in this essay. More recently (1974b), I have briefly alluded to *structural* selectors, which have already appeared in this essay as *vehicular* selection requirements. Structural selectors, as well as those entitled below 'Decision-Rule Structure' and 'Monitor-Modulate Distortion Correctors', all fall under a class that can be called *Internal Selectors.* Applied to brain processes underlying belief and perception, they are all selection systems that do not involve any selection by the contemporary environment. In the information-processing model of cognition, they do not involve the sensing of 'new information', but rather the neurological restructuring or massaging of an 'information' flow.

I am borrowing the concept of *internal selection* from a line of criticism of the completeness of the neo-Darwinian orthodoxy for biological evolution in general, and am applying it to aspects of neurologically embodied belief or knowledge. Whyte (1965) entitles his book *Internal Factors in Evolution*. Bertalanffy (1967, page 87) says "evolution then appears essentially co-determined by 'internal factors' . . . or 'inner-directed'." Waddington's concept of 'creodes' (e.g. 1969) is used to make a similar point. Thorpe (1969) summarizes this line of criticism, and Blachowicz (1971) extends it to evolutionary epistemology. Except for Waddington, these authors employ the emphasis on internal selection to deny that the awesome fit between animal form and environmental opportunity can be explained solely by selection from undirected mutations. This non-sequitur I, of course, reject. But I find the stress on internal selection descriptively correct, and also a valuable correction to Panglossian adaptationism. In this respect, it is akin to the concepts of punctuated equilibrium and species selection (e.g. Gould, 1982, 1986). The 'fit' between organism and environment is much less than complete. The 'adaptive distance' is great, not only because of historical limitations provided by the accidents of species history, but also because of the requirements of being an 'organism'.

Extended to the evolutionary epistemology of vision (as here), the emphasis on internal selection increases our awareness of the unavoidable gap between belief and the *Ding an Sich*. Extended to science, internal selection would refer to selection for compatibility with the trusted corpus of other scientific beliefs and with the social system requirements of the scientific community. Blachowicz (1971) uses such a concept of internal selection to argue that the awesome fit between scientific theory and physical reality cannot be explained by a blind variation and selective retention of theories. In sharp contrast, I here use it to deny any perfection or direct realism to scientific beliefs. The relativist, social constructionist sociologists of science have described much of this 'internal selection'. (See Campbell, 1986a, for a review.) While I reject their ontological nihilism, I do agree as to the profound indirectness and presumptiveness of any reflection of the external world in scientific theory. My evolutionary epistemology, both at the level of beliefs generated by visual perception (as here) and at the level of scientific theory, has always emphasized *epistemological* relativism: cousins to the amoeba that we are, how could we know for certain?

The specific examples of internal selection which Whyte et al. give all fall under that subtype I have called structural selection. Let me provide my own distinction between this and external selection first with a biological example. Let us accept the metaphor of 'information' or 'biological memory' for the portions of DNA molecules that constitute 'genes'. Consider a DNA molecule being bombarded by cosmic rays that disrupt and rearrange its prior structure. In order for the resulting material to be in the running as a mutant gene which might improve the fit of organism to environment (*external* selection) it must first meet the *structural* selective requirement of being a stable alternate form of DNA molecule capable of duplication. The great bulk of the disruptions produce rearrangements that fail to meet this *structural* requirement, being incomplete or imbalanced. There are also other intra-organismic selection levels involved that could be separated out with profit, but which I will lump for now with the *structural*. The stable DNA molecule must be one around which messenger RNA can form, with this RNA capable of serving as a template for the formation of a stable protein molecule. The proteins produced must form non-lethal composites with the preponderance of the proteins other genes have produced. If, after all of this internal, structural selection, an adult, fertile phenotype is produced, this phenotype is then subject to an *external* natural selection. Of all of these many modes of selection, only this last can involve an improvement in the fit of the organism to the environment, an increase in the 'knowledge' which the genome carries about the external world.

This example fails to go far enough in illustrating Whyte's emphasis on the requirement that organic form *per se* imposes. This requirement, in concert with the punctuated equilibrium model, can serve validly to emphasize the adaptive distance between organism and environment, resulting in or making possible long periods of stasis in the paleontological record concommitant with continual changes in the ecology to which organic form did not adapt.

My goal, of course, is to convey my conviction that we should expect a gap between scientific beliefs and the physical world comparable to that which we find between animal form and ecological niche. My strategy, however, will be to extend my physical and biological illustrations, dramatizing the issue of degrees of dominance of internal selection relative to external selection. Conceptual clarity is achieved best in an 'epistemology of the other one' (Campbell, 1959), avoiding the reflexive incoherence that comes from describing our own describing process as the object of inquiry, and avoiding the need to take current scientific theories as descriptions of the physical world which these same current scientific theories are adapting to. Though I assert that belief/knowledge is an indirectly and incompletely achieved fit to external selectors, strongly constrained by internal organizational requirements and the historical specifics of prior stages, my illustrations come easiest from adaptations in biological structure. (Wimsatt [in preparation], in his marvelous essay on von Baer's law in biology and in the genealogy of scientific belief, provides a better example of what I am attempting to do here.)

The relative balance between internal and external selection can be dramatized by comparing snowflake, diatom, giraffe and flatfish. The six-fold radial symmetry of the classic snowflake shows the dominance of internal, structural selection requirements. But most snowflakes are not in the classic mold, and even in snowfalls in which the classic forms predominate, some external selection is probably involved. Even on a windless day, wafting down through the air disrupts elongated fragile crystals more than the structurally stronger, radially-symmetric ones. (Would it be an irrelevant distraction to call attention to the still stronger selection bias among cognizers favoring the radially symmetric forms?)

In the skeletons of diatoms and radiolarians, the structural convenience of radial symmetry dominates, in contrast with skeletons of giraffes. But the forms of diatom skeletons are also, in part, due to the external selection for survival in their ecological niche. And much of the bilateral symmetry of the giraffe skeleton is due to structural requirements of organismic coherence (including coherent action), plus its species-history in prior niches, rather than due to the 'shape' of its current ecological niche.

The flounder, sole, and other flatfishes show a historicist version of a structurally preferred form being partially crushed and deformed by its current ecological niche. (May I remind those of you who only meet flatfish filleted in the restaurant, that in this family both eyes have migrated to one side of the ancestral model in an 'ugly' compromise.) Now a bioengineer starting from scratch would have no special difficulty in developing a flat bottom hidden predator that was both bilaterally symmetric and had both eyes on its water side rather than one on its mud side. But the flatfishes entered their present niche by a different route, to wit, as a vertically thin fish turning on one side for bottom hiding, with a subsequent migration of the mud-side eye, otherwise useless in that ecology. It is my conviction that evolutionary epistemologists should be prepared to find successful scientific theories bearing just as strong stigmata of their internal structural requirements and historical route as does the flatfish. In spite of the fallibilist, hypothetical, ontological realism of most evolutionary epistemologists, we should have no quarrel in this respect with anti-realists such as van Fraassen (1980, 1985) and Laudan (1981, 1984). (See Paller and Campbell, 1987, for more details.)

'Bean bag genetics' is a phrase that has come to be applied to mathematical models of large numbers of genes each with some net effect on survival, independent of the alleles present at the innumerable other loci, and in disregard of internal organization. While my stress on a nested hierarchy of vicarious blind-variation-and-selective-retention processes (each requiring a separate organizational level, each making the adaptive process less direct, though not less precise) should have precluded it, some of my past presentations of an evolutionary epistemology have come close to being a cognitive equivalent of bean-bag genetics, rightly criticized by Lewontin (1982). To use a physical metaphor, belief is seen as fitting the world as does a *sand casting*, where the numerous minute independent grains of sand collectively 'fit' with high fidelity the model being cast. But these grains of sand (independent of their container), do *not* have the organizational structure that an organism or theory has. A more appropriate model would be the 'cast' of a stone made by the roots of a tree that surround it: the internal structure of the roots, their round cross-sections and branching structure, set limits to the goodness of the fit of the 'mold'.

Associationism in learning theory, whether it be among sense-data elements, among ideas, or between sense-organ neural excitations and muscle twitches, partakes of the sand-casting model. Piaget's (e.g., 1971) distinction between assimilation and accommodation has greater epistemological utility. Once a child has achieved a structured schema for negotiating with environmental events, it tends to hold on to that schema, treating new types of experience as though they were appropriate to it, resulting in schema stability at the expense of poor fit to environmental opportunity. Too-frequent errors in expectations eventually motivate a schema change, in which more of the environmental regularities are 'assimilated' into the schema. In a comparable fashion, Kuhn's (1962) concept of the alternation of paradigm-dominated normal science with extraordinary periods of paradigm change illustrates the importance of cognitive coherence, the stability which paradigm and dominant theory offer, and the looseness of fit (anomalies) which are the price of that coherence and stability.

## DECISION-RULE STRUCTURE

In order to emphasize the gap to be expected between reality and the organic forms (and beliefs) adapted to it, I have oversimplified the competition between internal and external selection. The organic shape and capillary network of the root structure mapping the stone was of course required for survival, and thus externally selected. But conceived of as a 'mapping' of the environment, the fit was much more indirectly achieved, and was much less specifically a fitting to the *immediate* environment — being instead some net fitting to past environments.

Moreover, in my agenda of a preliminary stretching of the concept of belief/knowledge to include all instances of fit to external environment, I have gone too far, even for me. I want seriously to include only the 'fit' found in those mappings which are the product of specialized mapping structures, presumably selected by natural selection because of this adaptive value. I want to include only servo-system networks in organisms which perform some monitoring of the state of the environment. Thus the 'fit' of the snowflake to the ambient air, and the 'fit' of the root shape to the rock are now excluded. (If sand-casting remains, it is only because of the purposes of the artisan, for we may consider the mechanism of fit to include the artisan's tools and manipulations.) So, too, I will exclude most of Cannon's (1963) 'wisdom of the body'. I do not now want to classify the polar bear's fur as 'knowledge' about the arctic climate.

But, within these limits, I still want to expand our concept of knowledge far beyond the central exemplars of traditional epistemology, even though my focal concerns still lie within them. In the course of organic evolution many organisms develop capacities for *transient* anticipatory adaptations to *transient* changes in the environment. With these capacities come primitive versions of decision criteria, action rules, or recipes for behavior (the most primitive versions of the 'intentionality' characterizing beliefs). In my effort to try out radical recenterings of the problem of knowledge, I want the aspects of organic structure which carry such recipes to be included within the category of embodied knowledge. Thus, I want to keep in, as a borderline case, the fruit tree that governs its budding, leafing, fruiting, and shedding by useful but imperfect clues as to the changes in the seasons. I also want to retain the Venus flytrap plant that acts with animal quickness in closing around the insect it has sensed. I want to try out regarding these rule systems as primitive maps of interest-relevant aspects of the environment, and to seek out the implicit assumptions which they make about the nature of the world. If the non-active, fixed aspects of organic form can be said to 'assume' *stability* for corresponding aspects of the environment, then these action-recipes assume and diagnose *regularities among environmental changes*. A primitive hypothetical ontology has been constructed, partially overlapping with ours. For these rooted non-locomotor plants, for sessile animals, and even for primitive locomotor animals (e.g., the water shrew of Lorenz [1982]), these action-recipes are purely temporal maps; they are sequence rules rather than spatial maps. This sequence-rule characteristic persists in higher forms in which spatial maps are also internalized.

At this point we can introduce an internal structural requirement that is externally selected for: *physically implementable action-rule structure*. A highly detailed passive reflection of the environment is, by itself, of no use to the organism, and approximations to it appear very late, if at all, in the evolutionary branching. The prior requirement is the generation of approximately adaptive action. This action requires physically embodied if-then rules. As Ashby (1952) saw, it requires transforming continuous environmental variation into intraorganic step functions, sharp-threshold digitalizing, and, in general, transforming quantitative change into qualitative difference. It also requires physically embodied inference logics which imitate logical entailment in their rigid transmission of decisions.

Here is a very important field for descriptive epistemology. We have beginnings of it in the use of formal logic for developing wiring diagrams in control system engineering. Piaget's (1971) insistence on the intimate linking of the child's logical inference structures to its operations on manipulated objects is of profound importance. I see the need for much more collaboration between logicians and evolutionary neurophysiologists on the problem of how a workable primitive if-then logic can be modified, while still operating effectively, into a more complex logical decision structure involving more numerous detected contingencies, and eventually with superimposed approximate compensations for distortions coming from the initial over-simple rule structures (see Monitor-Modulate Distortion Correctors, below).

This aspect of our physical theory of embodied knowledge will be quasilogical, dealing with physically workable rule-system consistency, i.e., with consistent sets of implementable decision rules. It will be logical in that the intra-organismically workable decision-rule consistency can be evaluated separately from the adequacy with which the rule system as a whole fits the environment. If there are logical truths about inexorable limitations on such inference rules, such logical truths are a part of the selective system editing the mutations which introduce variations into inference and decision-rule anatomy. Thus an abstract analysis of the logically possible can describe part of the environment being biologically adapted to. That this logical 'environment' includes the categories of identity and noncontradiction, as Piaget asserts, seems likely. But Piaget groups other concepts with these, such as object permanence and object discreteness from other objects. These latter seem to be assumptions about the nature of the world, and hence to fall on the synthetic, contingent, descriptive side of the traditional dichotomy. At the same time, permanence and discreteness of objects may be prerequisites for any interpreted logic practically applicable to the physical world in which we evolved, whether or not required by the neurology of logic-embodiment, or the abstract forms of modern logic. Thus here in evolutionarily-selected, physically-implemented decision logics, the analytic-synthetic distinction may again break down. So too, once any logic-machinery has been biologically installed, the installation implies the descriptive assumption that the world is (in part) such as to be so describable.

Recipe-likeness or decision-rule-hood involve aspects of parsimony, simplicity, and generality. If recipes are equally adequate in fitting the environment, the simplest one to engineer is strongly preferred in biological development. At the other extreme, were an organism to be able to record in detail every past environmental impingement and organismic response, this detailed knowledge would be in itself of no use at all, for it would totally lack recipe-likeness or decision-rule-hood. If in addition to step-function digitalizing, discrete terms, and discrete operators, a recipe also has a more continuous form (such as an analogue computer or algebraic formula), then mathematical simplicity in a form embodiable in protoplasm will be essential to some degree. (While number of parameters, order of function, and notational simplicity have failed to give us any precise way of specifying the concept of mathematical simplicity, perhaps the additional constraint of convenience in physical protoplasmic embodiment will improve our ability to formulate our intuitive perceptions of simplicity.) Thus, external selection for decision-rule-efficacy becomes a selection requirement that must be met prior to external selection for adaptive mapping accuracy. Before a mutant on an old rule-structure can be subject to external selection as a potential improvement for the whole organism, it must first meet the prior requirement of being an efficacious, internally consistent, decision-generating rule-structure.

In future essays on natural-selection analogues for the replacement of one scientific theory by another, I hope to be able to treat formal scientific theories as analogous to these embodied decision-rules, and to make a similar point about prerequisite structural requirements. In the trial-and-error competition between rival theories, before any is selected as 'better' through external selection or its surrogates, each theory must first meet the requirements of 'being a theory'. If one were to stubbornly maintain that a 'blind variation' goes on at this level, it would have to be a blind variation of complete theories. This wording would still make the valuable point that one cannot tell which theory will be better in advance of assembling each complete theory and then trying them all out against the trusted facts from the past, including the trusted anomalous observations, comparing each new theory with its predecessors and rivals in this regard. Prescient innovation is still excluded. Haphazard variations on the elements of the prior dominant theory would first have to be edited by the structural requirements of theory-hood, expressed in part as elegance or beauty. Piecemeal variation may not always be a possible route to a new alternate theory. Major shifts in root metaphor (Pepper, 1967), or in central organizational scaffolding, may be occasionally required.

From the standpoint of a completely neutral cognitive ideal of unbiased description, decision-rule-hood can be regarded as a source of distortion. In simple if-then decision structures, the representation of the environment is chunked into several internally homogeneous mutually disjoint terms, which relate to each other through all-or-none operators. The category filters must be broad enough that useful numbers of exemplars are encountered, all of which are treated identically. Such distorted, selective, two-valued, black-white, go-no-go 'images of the world' are subsequently modified by evolutionary development (and also in childhood learning, and in cultural and historical developments within science) in the direction of greater detail, more terms, more complex contingency detection and decision-trees, smoother portrayal of smooth environmental processes and surfaces, and built-in corrections for the distortions that simple decision-rule-hood introduces, as is discussed in the next section. But some important imprint of the structural requirement of decision-rule-hood always remains. In addition, as will be seen, the distortion-correction devices add potentialities for illusion absent in simpler predecessors.

## MONITOR-MODULATE DISTORTION CORRECTORS

Some simplistic evolutionary and comparative neurology may help. Let us begin part way along an evolutionary branch, with a simple nervous system such as a coelenterate's, consisting of a single sensory nerve, one internuncial, and one motor neuron. I would like to distinguish two types of subsequent evolutionary development in our own family tree, both of which lead to brain development or cephalization. In the first type, there is an expansion of the sensory surface, adding more touch receptors, more retinal cells, and their corresponding sensory neurons. In parallel there is an expansion of the number of muscles and motor neurons. One source of thickening or cephalization comes from the corresponding addition of internuncials leading from one sensory nerve to several motor neurons, and internuncials leading into a single motor neuron from several sensory ones, thus making possible contingent responding to the environment.

The second source of cephalization has a quite different character. There emerge supplementary nervous networks whose 'sense organs', if you will, do not monitor any information flow from beyond the organism's skin, but instead monitor the activity of other nerves, as in suppressing or amplifying their conductance. These auxiliary networks play the roles of filters and amplifiers, detecting and responding to rates of change as well as to change. They are the analogues of the devices in radio and television receivers that remove the effects of static and the resonances and squeals intrinsic to the physical structure of the receiver. While they generally improve the validity of the organism's knowledge of the environment, they do not do so by additional sensing of that environment, for they sense only the message structure of other nerve chains. Instead, they monitor and modulate the basic sensory input on the basis of assumptions about the nature of the world.

In our visual system, one such monitor-modulate system inhibits up-stream optic nerve rate of firing in response to high levels of firing monitored downstream, farther from the eye. This system is one of several, including iris adjustment, which improve brightness contrast and color constancy in the perception of objects, by reducing the distracting effects of changes in level of illumination. But these compensatory devices also generate a variety of contrast illusions which a simpler system would not show (e.g., Campbell & Kral, 1958). A parallel evolution has occurred in cameras. Compare the old-fashioned box camera having a fixed aperture, with a modern camera in which a photocell continuously readjusts the aperture. Given the unusual task of identifying the same small square of grey paper placed alternately against a large white background or a large black background, the primitive box camera will yield the most useful prints, and the photocell-iris camera will show the same illusion as does human vision. The assumption, not always correct, that the brightness of the overall visual field is an indicator of the brightness of illumination, rather than the collective brightness of the objects, has built into each mechanism a rigid 'artificial' reconstruction of the probably stable attributes of the probably stable things in the environment. (Let me expand the description of this proposed experiment. With each camera we will photograph our one grey square on each background, white and dark. But when we look at the prints, our interpretation of them will again be confounded by our eye's contrast illusion. To avoid this meta-illusion, let's take scissors and cut out the pictures of the grey squares, thus isolating them from their photo background. Then when these snips are placed on a common background, the two snips from the box camera will be seen to be the same grey, but the two snips from the photocell-aperature-control camera will differ markedly, that from the white background photo looking much darker than the snip from the black background.)

Motion picture perception provides another example. The continuous fluid flow of the moving animals being filmed is chopped up by the camera into a series of still pictures, which alternate with darkness, followed by another still picture that is abruptly different even if in small degree. When this choppy-gappy series is projected on the screen, our human visual system artificially reconstructs from it a smooth flowing continuous motion, usually more validly depicting the reality which the camera photographed than did the intermediate stages in which light sensitive film made occasional rigid transient records of the light reflected from the objects. (The exceptions to this validity in motion pictures come when stagecoach wheel spokes, horses hoofs, and hummingbirds' wings have a real frequency near or above the frequency of the camera shutter speed, and therefore encounter a significant structural limitation.) This *generally* increased validity comes not from sensing the motion directly, but rather from artifically reconstructing it on the basis of the 'assumption' that flexibly stable objects in motion are more likely than a succession of discrete, highly similar, separate objects.

Now it may be thought that this choppy-gappy intermediate process is introduced only because of the mechanical limitation of film and camera. Not so (e.g., Shepard & Zare, 1983; Gregory & Harris, 1984). The human eye and brain have similar mechanical limitations with similar effects. In the normal perception of smooth, continuous motion going on in the world, the small, high-frequency saccadic movements of the eye, the all-or-none response characteristics of retinal cells and optic nerves, the larger-scale scanning movements of eye and head, all provide choppy-gappy sense-organ and neural transmission stages, which are presumptively reconstructed by the central nervous system into a perceived continuous motion (which is usually more valid than what the retina tells the optic nerve, or what the optic nerve tells the brain).

Similar processes can be detailed for the perception of form, binocular vision, color vision, etc. Such a presumptive correction mechanism underlies Locke's (1690) and Berkeley's (1713) temperature illusion. Consider their three basins of water experiment: On the right place a basin of hot water, in the middle a basin of warm, and the left one of cold. Place your two hands in the extreme basins and leave them there a moment to adapt. Then, watching them with your eyes, place both hands in the middle basin. Your hands will disagree about its temperature in a rather unsettling way. To the right hand it will feel cold, to the left hot, yet your watching eyes will support a strong assumption that all the water in that middle bowl is of the same temperature (even though transient temperature inhomogenieties are a possibility). (If done blindfold, one might decide there were two middle bowls.) In this experiment, as in all cases of tactile temperature evidence, the experienced temperature is demonstrably a comparison, a joint product of skin temperature and object temperature. The rigid wiring diagrams and decision trees of the sense organs and brain convert this completely relative evidence into a phenomenally absolute experience of a temperature adhering solely to the object. In this ecologically atypical situation the 'phenomenal absolutism' (Segall, Campbell & Herskovits, 1966; Campbell, 1969) or ontological absolutism of our unreflective experience is wrong, even though the goal implemented by these rigid pre-experiential sense-organ and neural processes is in its most general form correct. That is, they turn relative evidence into presumptive beliefs about objects as they are assumed to exist in independent reality.

Another example of monitor-modulate circuitry is provided by Karl Duncker's (1929) experiments on induced movement. Consider a totally dark room, in which against the front wall there is a luminous frame some 6 feet by 4 feet, on tracks so that it can be moved from right to left; and in the central region a luminous dot, also movable on a horizontal track. When the frame gets moved one foot to the right, everyone sees instead the dot move one foot to the left. Stubborn prephenomenal circuitry in our brains will have taken accurately recorded relative motion and gratuitiously divided it up into perceptions of absolute fixity for the frame and absolute motion for the dot. Now had it been important, biological evolution and childhood learning could probably have produced an experiential process avoiding this illusion when one is sitting in rigid chairs or standing flatfooted. The informational base is available through detection and memory of eye, head, and body movements. But for our ancestral fishes moving in currents the direction and speed of which they had no way of directly knowing, there was no better alternative to the information processing assumption that, in case of doubt, it was the larger bulk of the visualized environment that was stable, the smaller fragment that was moving. So, too, for a moving land animal such as a dog or human chasing a rabbit, a data-reduction decision-tree that does not require precise recording of how one has moved oneself has great net advantages, if the environment is such that miscarriages like the dot and frame illusion are not too frequent.

A final example comes from the computer programs that are used to clean up the pictures of the planets radioed back from the space probes. Inevitably, the increased validity of the picture is achieved at the expense of fine detail that is suppressed as though it were static. In one of the first of the Mars shots, an editing program was being used similar to a two-dimensional variant of the old method of moving averages which we used to use in smoothing graphs. In this program, a single 'pixel' (or transmitted grain or dot) which contrasted in brightness with all of its immediate neighbors was erased, and a value based on its neighbors substituted, a process which is remarkably effective in removing static and in producing sharp contoured pictures that are clear and real-looking. In one of the early runs it was discovered that the setting used in this editing process had erased Deimos, a small but long-known moonlet of Mars. In this instance, the error was corrected by readjusting the correction parameters, but such editing programs obviously remain a threat to the recognition of still smaller satelites not previously known. (I am indebted to my friend William Powers, an astronomical engineer and cybernetic theorist [1973], for this example.) Visual neurology also has mechanical processes which enhance the clarity of dominant contours and suppress the blurring effects of 'static'. Inhibitory dendrites from an excited optic nerve inhibit the excitation of nerves coming from adjacent retinal regions.

To go back to the primitive neural model, all of the monitor-and-modulate circuits improve our picture of reality (where nonillusory), not by being new detectors of aspects of reality, but rather by transforming the otherwise detected reality on the basis of assumptions about the nature of the world which have been built in during the course of biological evolution. These features make our experienced image of the world more vividly real and complete, but do so by an artificial reconstruction of that world, not by directly sensing those vivid added aspects. (Except for the stick bent in water, all of the illusions employed in classic epistemological discussions, including after-images, are the product of monitor-modulate circuitry.)

The general point about these presumptive, corrigible, generally validity-enhancing, artificial, monitor-modulate circuits can be emphasized by considering their relationship to the old issues of primary and secondary qualities, the 'causal theory of perception', the 'argument from illusion', and skepticism more generally. In the times of Leonardo, Kepler, Pascal, and Descartes, it was discovered that light rays communicated pattern information which was focused by the lens to form an image on the back of the inner eyeball, which image was transmitted somehow by the nerves to the brain. Even though there was puzzling about the 'unrealistic' inversion of the image, the discovery of this plausible, dependable process of mechanical transfer might have been thought to support a realist epistemology, or epistemological complacency (as the causal theory of perception seems to have done in some periods [Heffner, 1981, and in this volume]).

Historically, instead, awareness of the arbitrary non-iconic nature of the intermediate links coming from this materialist-realist understanding of the machinery of perception seems to have increased skepticism, and hence to have reduced the acceptability of a realist or critical realist epistemology and ontology, perhaps because of the increased awareness of the possibilities for illusory stimulation of intermediate links. Descartes experimented with the image-forming optics of oxen lenses (Crombie, 1967). Locke was a medical doctor who knew visual anatomy, probably including the many-celled granularity of the retina. Hume can be quoted to show that he was an anatomically and physiologically oriented naturalist (MacCormac, 1980), and was a providentialist with regard to the general dependability of vision only because he lacked a natural-selectionist evolutionary theory (Harré, 1980).

Knowledge of the monitor-modulate circuits adds to the possibilities of illusion, and hence to the skeptical 'argument from illusion'. Consider Locke's skepticism regarding the 'secondary qualities', and his trust regarding the 'primary' ones (the latter perhaps an example of epistemological trust coming from the pattern-preserving power of the lens and image-projection system). Our monitor-modulate circuit analysis has demonstrated that supposedly primary qualities like fixity and motion are 'constructed by the mind' fully as much as the secondary quality, color. So too the perceived three-dimensional form of objects is constructed by the brain from two-dimensional pattern information in a manner that creates the possibility of optical illusions. On the other hand, the classic example of a secondary quality, color, is now understood as mapping an objectively stable attribute of objects, to wit their differential reflection of light of different wave lengths (Yilmaz, 1973).

I am thoroughly sympathetic with the modern revival of a causal theory of perception and/or knowledge (Shimony, 1971; Goldman, 1967). Yet, I would emphasize that it does not provide the epistemological assurance that the classic epistemologists sought. It does not answer skepticism. A physicalist-realist (materialistic) hypothetical ontology for the referents of belief or knowledge, consistently followed through, requires a materializing and mechanizing of the mediational processes involved in the sense organs and brain. Such a physicalism is the most vigorous opponent of attributing clairvoyant competency to vision, or to any sensing-perceiving process. This kind of physicalist realism is clearly as incompatible with modern "direct" realism as it is to naive realism. In this opposition, physicalism/materialism must side with skepticism and *epistemological* relativism. All that such a physical, physiological, anatomical 'causal theory of perception' can do is to provide a plausible mechanism that could sometimes produce valid belief (if everything is working right, and if the external world being perceived at the moment accords with the presuppositions about the general nature of the world built into the mechanism). Such a scenario makes the minimalist point that 'knowledge is *not impossible*'. Augmenting this scenario with a selective-survival model of biological evolution makes it plausible that the visual machinery is apt to be working right, with presuppositions about the world that are generally correct, and thus supports something like 'visually based beliefs are usually correct'. Naturalistic epistemology is thus a modest, presumptive enterprise, and quite consistent with the skeptical analysis of the search for dependable foundations for belief. (See Paller & Campbell, 1987, for a related analysis offering more and different detail.)

## VICARIOUS SELECTIVE SYSTEMS

The concept of "vicarious selectors" has been used in my previous writings to encompass vision (1956b), radar and sonar (1959, 1974a) on the one hand, and creative thought (1960) on the other. The machinery involved is in each case a product of external selection in past environments. Like the monitor-modulate circuits, when the assumptions built into them are locally appropriate, their action improves belief, but in a profoundly presumptive and indirect way. The case of creative thought is most akin to the monitor-modulate circuits in not introducing 'new information' about the environment. In contrast, vision, radar, sonar, echo-location in fish by the lateral-line organ, and the blind-man's exploration of space with a cane, do involve sensing of the current environment, albeit vicariously by presumptive use of imperfect indicators. But while they generally function to improve the practical fit of belief to its immediate referents, they do so in a way that undermines foundationalist longings and claims for 'direct' knowing.

Once again, it seems easiest to illustrate the concept first with some concrete biological examples more primitive than knowledge processes. Let us start with wound healing and limb regeneration. Spiegelman

(1948) has applied a natural-selection analogy to embryological development and regenerative processes. In oversimplified scenario, in each somatic cell all of the genes are continually available and competing as templates for their kind of protein, but most are inhibited by the effects of the surrounding tissues and more specific inhibitory systems. These selectors initially control growth, turning it off facet-by-facet as some template for mature form is achieved. In animals like the salamander, capable of regeneration, the loss of a limb removes this template restraint, and the limb regrows. One can imagine this growth being stopped by an *external* selector, the leg growing down 'until it hits the ground'. This is not how it works, however. Instead, the growth processes are restrained, selected, inhibited, or turned off by an internal vicarious selector when a certain length is reached. The vicarious selector represents already evolved wisdom about optimal leg lengths for locomotion and survival. With regard to length, it contains what can be called a 'reference signal' in control system language (Powers, 1973), or a 'homeostat setting' in Ashby's (1952) terminology, analogous to a thermostat setting. This homeostat setting is continually subject to mutations and natural selection, involving the life, death, and differential procreation of whole salamanders, and will be reset at a different value if it becomes seriously non-optimal. Borrowing further metaphors from control system engineering and statistics, one could speak of the vicarious control system for leg length as being 'nested' within the whole-organism-reproduction control system of natural selection.

It turns out that all of the organismic processes we would call learning involve vicarious selectors rather than a direct encounter with reality itself. The supposedly one-cell paramecium *stentor*, the hero of Herbert Jennings great book of 1906 on the *Behavior of Lower Organisms* that so influenced Ashby (1952), does a primitive kind of problem-solving employing no memory. If it were to collide with objects it couldn't swim through, its trial-and-error of locomotions would indeed be edited by an external selector, by reality itself. But on close inspection it turns out that its fragile waving cilia have instead made the encounter, and this has obviated a full body collision. The cognitive presumption equating impenetrable-to-cilium with impenetrable-to-full-body-locomotion (oversimplified and wrong in some instances) has produced a vicarious selector standing as a screen between the paramecium and a direct encounter with reality. So, too, in its avoidance and escape problem-solving in regard to noxious chemical concentrations in its environment, its locomotions are edited by sense organs that approximately detect very dilute non-lethal concentrations of the likely chemical toxins, leading to a change in direction of locomotion. These chemoreceptors are vicarious selectors, substituting for the direct external selection by environmental realities through the life and death of whole paramecia. In both cases, the homeostat settings of these vicarious selectors — the number of cilia contacts that signal object solidity or the parts per million of the chemical that triggers direction change — are themselves subject to readjustment by the external selection of the life, death and procreation of mutants differing in this regard.

We can now consider some examples in higher organisms. The echoes of the pulses from the fish's own swimming, as received by the lateral-line organ, vicariously explore and select among environmental objects and open paths, like the paramecium's cilia and the blind man's cane, but with greater spatial depth. Echolocation by bats and cave birds, and radar and sonar for man-made ships, have a similar vicariousness. A trial-and-error of radar beam emissions encounters selective reflection, which substitutes for a trial-and-error of whole ship collisions, on the assumption that opaque-to-radar-beam equals impenetrable-to-ship, an identification only approximately true, even though the electromagnetic wave band has been selected just for that equivalence. So, too, for vision (Campbell, 1974a, but see Wachtershauser, 1986). As a final example, higher animal learning is guided by the taste receptors, which are only imperfect vicars for the true nourishingness and noxiousness of the edible environment.

At this point some general descriptive-epistemological conclusions might be tried out. The vicarious selector is only approximately accurate, and has fringes of inappropriateness which can produce illusions. Glass is transparent to light but opaque to locomotion, while fog is vice versa. Sonar echoes come from penetrable schools of fish as well as from objects impenetrable to a ship's locomotion. For radar, flocks of birds and tinfoil strips are sources of illusory perceptions. For taste, illusions are found in non-nourishing saccharin and good tasting ant poison. Even within the ranges of optimal function, the built-in assumptions are approximate truths about past worlds and may no longer hold if the ecology has changed.

A second point is harder to convey but perhaps undermines still more the appeal to Cartesian indubitability as a clue for truth. Let me begin by comparing the selection ratios for natural selection with those of vicarious selection. In biological evolution, the selection ratios for two alternate forms of a given gene (the improvement in life chances) can be very slight and still lead one gene to replace another; a 1% advantage of likelihood of bearing offspring might be enough, 9% versus 8%, for example. If selecting from the range of edible objects those with a given taste provided even such a slight selective advantage over eating those with an alternate taste, this small statistical truth about the environment could become represented as an innate taste preference which would operate on learning and eating with an all-or-none, nearly 100% selection ratio, or at least a greatly enhanced differential. Thus slight and fuzzy environmental truths become exaggerated into clear, all-or-none preferences, and into overclarified conscious perceptions. Similarly, I would argue that physical locomotor collisions (and even tactile fumbling in the dark) provide more nearly direct knowledge of the spatial environment than does vision. Although locomotor exploration also employs assumptions and sensory vicars, it does so much less indirectly than does vision. Yet the experiential clarity of vision far exceeds that based upon locomotor encounters.

## OVERVIEW

An evolutionary perspective on how organisms come to fit their environments is extended to encompass the fit of belief (e.g., in perception or in science) to environmental referents. While employing a hypothetically realistic ontology (Shimony's 'Copernican epistemology'), the analysis more nearly supports skeptical arguments than direct realism or epistemological realism. Several aspects of the physical embodiment of belief are used to illustrate the inevitable gaps in the fit of phenomena to noumena: (1) structural requirements of the vehicles carrying embodied belief/knowledge; (2) 'internal selection' in biological evolution due to the structural requirements of organic form and species history; (3) monitor-modulate distortion correctors in physical information-processing machinery, including neurological ones; and (4) vicarious proxies for external selection requirements.

*Lehigh University*

## NOTE

[1] This paper is built upon Lecture 2 of the author's William James Lectures, delivered at Harvard University, Spring, 1977. It is hoped that the lectures will eventually be published as a whole. In undertaking the present publication, author, editor, and publisher acknowledge this prior plan and agree to place no obstacle to it. The preparation of this paper has been supported in part by NSF Grant BNS 792—5577, by the New York State of Regents Albert Schweitzer Chair at Syracuse University, and by a University Professorship at Lehigh University.

## REFERENCES


Ashby, W. R.: 1952. *Design for a Brain*. New York: Wiley.

Baldwin, J. M.: 1909. *Darwin and the Humanities*. Baltimore: Review Publishing Company.

Bertalanffy, L.: 1967. *Robots, Men, and Minds*. New York: Braziller.

Berkeley, G.: 1927. *Three Dialogues Between Hylas and Philonous*. London: Henry Clements, 1713. (Reprinted, Chicago: Open Court, 1927.)

Blachowicz, J. A.: 1971. 'Systems Theory and Evolutionary Models of the Development of Science,' *Philosophy of Science* **38**, 178—199.

Campbell, D. T.: 1956a. 'Adaptive Behavior from Random Response,' *Behavioral Science* **1**, 105—110.

Campbell, D. T.: 1956b. 'Perception as Substitute Trial and Error,' *Psychological Review* **63**, 330—342.

Campbell, D. T.: 1959. 'Methodological Suggestions from a Comparative Psychology of Knowledge Processes,' *Inquiry* **2**, 152—182.

Campbell, D. T.: 1960. 'Blind Variation and Selective Retention in Creative Thought as in Other Knowledge Processes,' *Psychological Review* **67**, 380—400.

Campbell, D. T.: 1965. 'Variation and Selective Retention in Socio-cultural Evolution.' In H. R. Barringer, G. I. Blanksten and R. W. Mack (eds.), *Social Change in Developing Areas: A Reinterpretation of Evolutionary Theory*. Cambridge, Mass.: Schenkman, pp. 19—49.

Campbell, D. T.: 1966. 'Pattern Matching as an Essential in Distal Knowing.' In K. R. Hammond (ed.), *The Psychology of Egon Brunswik*. New York: Holt, Rinehart, & Winston, pp. 81—106. (Reprinted in H. Kornblith (ed.), *Naturalizing Epistemology*, Cambridge, Mass.: Bradford Books, MIT Press, 1985.)

Campbell, D. T.: 1969. 'A Phenomenology of the Other One: Corrigible, Hypothetical and Critical.' In T. Mischel (ed.), *Human Action: Conceptual and Empirical Issues*. New York: Academic Press, pp. 41—69.

Campbell, D. T.: 1974a. 'Evolutionary Epistemology.' In P. A. Schilpp (ed.), *The Philosophy of Karl R. Popper*, Volume 1. LaSalle, Ill.: Open Court, pp. 413—463.

Campbell, D. T.: 1974b. 'Unjustified Variation and Selective Retention in Scientific Discovery.' In T. Dobzhansky and F. J. Ayala (eds.), *Studies in the Philosophy of Biology*. London: Macmillan, pp. 139—161.

Campbell, D. T.: 1979. 'A Tribal Model of the Social System Vehicle Carrying Scientific Knowledge,' *Knowledge* **2**, 181—201.

Campbell, D. T.: 1986a. 'Science's Social System of Validity-enhancing Collective Belief Change and the Problems of the Social Sciences.' In D. W. Fiske and R. A. Shweder (eds.), *Metatheory in Social Science: Pluralisms and Subjectivities*. Chicago: University of Chicago Press, pp. 108—135.

Campbell, D. T.: 1986b. 'Science Policy from a Naturalistic Sociological Epistemology,' *PSA 1984*, Volume 2. East Lansing, Mich.: Philosophy of Science Association.

Campbell, D. T., C. M. Heyes, and W. Callebaut: 1986. 'Evolutionary Epistemology Bibliography.' In W. Callebaut and R. Pinxten (eds.), *Evolutionary Epistemology: a Multiparadigm Program*. Holland: Nijhoff.

Campbell, D. T. and T. P. Kral: 1958. 'Transposition Away from a Rewarded Stimulus Card to a Nonrewarded One as a Function of a Shift in Background,' *Journal of Comparative and Physiological Psychology* **51**, 592—595.

Cannon, W. B.: 1963. *The Wisdom of the Body*. New York: Norton.

Crombie, A. C.: 1967. 'The Mechanistic Hypothesis and the Scientific Study of Vision.' In S. Bradbury and G. L'E. Turner (eds.), *Historical Aspects of Microscopy*. Cambridge, England: W. Heffer & Sons Ltd., pp. 3—112. (Also issued as *Proceedings of the Royal Microspical Society*, Volume 2, Part 1, 1967, 3—112.)

Duncker, K.: 1929. 'Uerber induzierte Bewegung,' *Psychologishe Forschung* **12**, 180—259.

Fisch, R.: 1977. 'Psychology of science.' In J. Spiegel-Rosing and D. de S. Price (eds.), *Science, Technology, and Society: A Cross-disciplinary Perspective*. London and Beverly Hills: Sage Publications, pp. 227—318.

Gamble, T. J.: 1983. 'The Natural Selection Model of Knowledge Generation: Campbell's Dictum and its Critics,' *Cognition and Brain Theory* **6**, 353—363.

Goldman, A.: 1967. 'A Causal Theory of Knowing,' *The Journal of Philosophy* **44**, 357—372.

Gould, S. J.: 1982. 'Darwinism and the Expansion of Evolutionary Theory,' *Science* **216**, 380—387.

Gould, S. J.: 1986. 'Evolution and the Triumph of Homology, or Why History Matters,' *American Scientist* **74**, 60—69.

Gregory, R. L. and J. P. Harris: 1984. 'Real and Apparent Movement Nulled,' *Nature* **307**, 729—730.

Harré, R.: 1980. 'Knowledge.' In G. S. Rousseau and R. Porter (eds.), *The Ferment of Knowledge*. Cambridge, England: Cambridge University Press, pp. 11—54.

Heffner, J. H.: 1981. 'The Causal Theory of Visual Perception: Its Scientific Basis and Epistemological Implications,' *International Philosophical Quarterly* **21**, 303—332.

Heider, F.: 1926. 'Ding und Medium,' *Symposion* **1**, 109—157. (Translated as 'Thing and Medium' in G. S. Klein (ed.), *Psychological Issues*. New York: International Universities Press, Part III, 1959, pp. 1—34.)

James, W.: 1880. Great Men, Great Thoughts, and the Environment. *The Atlantic Monthly* **46**, 441—59.

Kuhn, T. S.: 1962. *The Structure of Scientific Revolutions*. Chicago: University of Chicago Press. Second edition, 1970.

Laudan, L.: 1981. 'A Confutation of Convergent Realism,' *Philosophy of Science* **48**, 19—49.

Laudan, L.: 1984. *Science and Values*. Berkeley: University of California Press.

Lewontin, R. C.: 1982. 'Organism and Environment.' In H. C. Plotkin (ed.), *Learning, Development and Culture: Essays in Evolutionary Epistemology*. Chichester and New York: Wiley & Sons, pp. 151—170.

Locke, J.: 1690. *An Essay Concerning Human Understanding*. London. (Reprinted by P. H. Niddetch (ed.), Oxford: Clarendon Press, 1975.)

Lorenz, K.: 1982. 'Kant's Doctrine of the A Priori in the Light of Contemporary Biology.' Reprinted in H. C. Plotkin (ed.), *Learning, Development, and Culture: Essays in Evolutionary Epistemology*. Chichester and New York: Wiley & Sons, pp. 121—143.

MacCormac, E. R.: 1980. 'Hume's Embodied Impressions,' *The Southern Journal of Philosophy* **18**, 447—462.

Paller, B. T. and D. T. Campbell: 1987. 'Reconciling Maxwell and van Fraassen Through Sense-organ Evolution, the Ostensive Basis of the Term "Observe", and Optimal Justificatory Practice in Science,' *Synthese* (in press, as a part of Mary Lou Maxwell and C. Wade Savage (eds.), issues on Grover Maxwell).

Pepper, S. C.: 1967. *World Hypotheses*. Berkeley: University of California Press.

Piaget, J.: 1971. *Biology and Knowledge*. Chicago: University of Chicago Press.

Poincaré, H.: 1913. 'Mathematical Creation.' In H. Poincaré, *The Foundations of Science*. New York: Science Press, pp. 383—394.

Powers, W. T.: 1973. *Behavior: The Control of Perception*. Chicago: Aldine.

Quine, W. V.: 1969. 'Epistemology Naturalized.' In W. V. Quine, *Ontological Relativity and Other Essays*. New York: Columbia University Press, pp. 69—90.

Royce, J. R., et al.: 1978. 'Psychological Epistemology: A Critical Review of the Empirical Literature and the Theoretical Issues,' *Genetic Psychology Monographs* **97**, 265—353.

Segall, M. H., D. T. Campbell, and M. J. Herskovits: 1966. *The Influence of Culture on Visual Perception*. Indianapolis: Bobbs-Merrill.

Shepard, R. N. and Zare, S. L.: 1983. 'Path-guided Apparent Motion,' *Science* **220**, 632—634.

Shimony, A.: 1970. 'Scientific Inference.' In R. Colodny (ed.), *Pittsburgh Studies in the Philosophy of Science*, Vol. 4. Pittsburgh: University of Pittsburgh Press, pp. 79—172.

Shimony, A.: 1971. 'Perception from an Evolutionary Point of View,' *Journal of Philosophy* **68**, 571—83.

Spiegelman, S.: 1948. 'Differentiation as the Controlled Production of Unique Enzymatic Patterns.' In *Symposia of the Society for Experimental Biology, II, Growth in Relation to Differentiation and Morphogenesis*. New York: Academic Press.

Thorpe, W. H.: 1969. 'Retrospect.' In A. Koestler and J. R. Smythies (eds.), *Beyond Reductionism*. Boston, MA: Beacon Press, pp. 428—434.

van Fraassen, B.: 1980. *The Scientific Image*. Oxford: Clarendon.

van Fraassen, B.: 1985. 'Empiricism in the Philosophy of Science.' In P. M. Churchland and C. A. Hooker (eds.), *Images of Science*. Chicago: University of Chicago Press, pp. 245—305.

Wachtershauser, G.: 1986. 'On Light and Life.' In G. Radnitzky and W. W. Bartley, III (eds.), *Evolutionary Epistemology, Theory of Rationality and the Sociology of Knowledge*. LaSalle, Ill.: Open Court Publishing Company.

Waddington, C. H.: 1969. 'The Theory of Evolution Today.' In A. Koestler and J. R. Smythies (eds.), *Beyond Reductionism*. Boston: Beacon Press.

Whyte, L. L.: 1965. *Internal Factors in Evolution*. New York: Braziller.

Wimsatt, W. C.: (in preparation). 'Von Baer's Law of Development, Generative Entrenchment, and Scientific Change,' *Philosophy of Science*.

Yilmaz, H.: 1973. 'Perception and Philosophy of Science.' In R. S. Cohen and M. W. Wartofsky (eds.), *Logical and Epistemological Studies in Contemporary Physics*, Volume XIII. Boston Studies in the Philosophy of Science. Dordrecht: D. Reidel, pp. 1–91.

JOHN HEFFNER

# CAUSAL RELATIONS IN VISUAL PERCEPTION[1]

What is the relation of visual perception to the measured order of physical objects? Both vision and measurements are sources of information about physical objects, although they provide this information in different ways and with systematic differences of organization. Although vision is our primary source of information about the external world, it does not supply the whole truth unassisted. Perceptual data do not arrange themselves without ambiguity, nor are they transformed automatically into all the information we desire. The various uses we make of visual data shape the form of knowledge, and to a surprising degree they also shape its content.

I wish to consider at length the possibility that the philosophical problem of perception has the same epistemological status as a scientific problem of high generality. Although there is no one crucial experiment which can solve the epistemological problem, experimental studies of visual perception can contribute greatly to our understanding of it. This approach focuses the epistemological problem on the more specific problem of the veridicality of vision, which, in turn, is largely an empirical matter. I shall thus answer my initial question by taking a frankly descriptive and naturalistic approach.

Far too much epistemology has been written which more or less uncritically assumes a simple analogy between vision and the camera. In such work visual perception tends to be analyzed into a passive sensory component and an active cognitive component. A stratification is assumed, that is to say, which divides sensory data from the cognitive processes of association, inference, etc. by which they are organized into judgments. The more passive the sensory component is considered to be, the more it tends to attract epistemologists with strong foundational leanings, who look to it as a possible source of incorrigible information. The camera analogy leads to a misplaced emphasis on the sensory component of perception as being the most secure part of knowledge. It also leads to a reduced conception of the scope of sensory data, to a minimal or even a sceptical view of the information perception can provide about the physical world, and to an unneces-



*Abner Shimony and Debra Nails (eds.), Naturalistic Epistemology*, 193—214.

sarily suspicious attitude toward the role of cognition — especially when sensations fail to yield obvious correspondences with what is taken to be the true structure of the physical world.[2] In perceptual psychology the analogy has been used as a model for predictions, many of which have been disconfirmed by subsequent experiments. In short, the analogy between vision and cameras has occasioned unnecessary confusion, much of which can be cleared by avoiding the analogy.

For this purpose we shall adopt a cognitive view of perception based partly on information processing. There are two major differences between this view and the camera analogy. First, it allows more readily for alterations in the order of information as it is processed by the visual system. Second, it does not require that the physical world be taken as an ontological standard, against which visual data are to be measured in a hopeless attempt to find correspondences. Except for carefully qualified analytic purposes, it is misleading and artificial to stratify visual perception into physical, sensory, and cognitive layers. What actually occur are functional relations in which all three aspects are combined.[3] We shall illustrate these combinations in perceptual examples which approach the purely sensory component of traditional analysis, and for the most part we shall avoid examples of overt perceptual judgments. Even in the examples we emphasize, we shall see cognitive processing at work.

Visual perception occupies a middle position in a ranking of processes from purely sensory to purely cognitive, and it is affected by both higher and lower processes in this ranking. We need not assume that the perceptual processes occur only in one direction, from a physical stimulus to its eventual registration in the phenomenal field of a passive mind. We shall see, on the contrary, that although somatic processes eventually may be discovered which underlie even the highest cognitive structures, they need not always be known beforehand to construct a reasonably good epistemological account. This account, to be sure, is descriptive and provisional rather than definitive, but it does illustrate that a solution to the mind-body problem need not be a prerequisite to the study of perception. One of the merits of this approach is that causal relations can be examined at various levels, one of which admits cognitive processes as such.

To examine the causal relations in visual perception, we adopt a differential view of causation, according to which causal explanations are sought as regular patterns of modifications among things which already exist. We reject the generative view, according to which causes are sought as explanations of new things created from something else.[4] Early proponents of the causal theory of perception, following Locke and the camera analogy, tended to assume a simple linear and generative view of causation. They regarded sensations as the final events in temporal sequences which begin with physical stimuli. They thus treated the problem of perception as a special case of the mind-body problem. The causal explanations we seek, on the contrary, do not aim at an explanation of why there should *be* phenomenal fields, just as they do not aim at an explanation of why there should *be* other relations sufficiently stable to be referred to as a physical world. However interesting these metaphysical issues may be, and however important they may be in their own right, we aim here at an epistemological rather than an ontological explanation of the causal relations in vision.

We begin by distinguishing the phenomenal from the physical by the different ways in which knowledge about them is obtained. Intersubjective criteria, appropriate for knowledge of the measured spatial order of physical objects, differ from the personal and perspectival criteria appropriate for knowledge of the phenomenal orders of individual perceivers.[5] There is no single epistemologically privileged order of information about the world. There are instead many such orders, having various sources and various standards. We may aim at comparisons among them without lapsing into gullibility on the one hand or dogmatism on the other.

Specific causal relations are discovered empirically, and knowledge about them is attained indirectly and through inference. Usually it is framed as hypotheses which have empirical consequences. In vision these consequences are expressed as the necessary and sufficient conditions for visual *discriminata*, to use R. W. Sellars' (1938) designation for the differentiations of a phenomenal field. Our search for causal relations between the phenomenal orders of visual perception and the measured order of physical objects does not preclude further searches for causal relations within any one order alone, nor does it imply that phenomenal orders are subservient to the physical.

There are many examples of visual perception which are phenomenologically quite similar but which are processed physically at different levels of the visual system. In the waterfall effect, a band of alternate black and white vertical stripes is moved horizontally across the field of view. If it is viewed with the eyes held stationary, the stripes appear to move backwards when the physical motion is stopped. In the induced movement effect, a large frame which encloses a small luminous spot is moved horizontally while the spot itself remains stationary. The spot appears to move, not the frame, and it appears to move in a direction opposite the physical motion of the frame. Despite their phenomenological similarity, the waterfall effect is retinal and the induced movement effect is cognitive (Gregory, 1977, pp. 109–115). Other experiments show that expectation, recognition, and conceptualization, all of which are cognitive factors, can affect the pre-phenomenal processing of perceptual data.[6] Their effects cannot be assigned merely to the subsequent uses made in thought of passively received stimuli.

The perceived visual orders are defined by color and the basic *discriminata* of phenomenal spatial relations. From them all other visual qualities, such as texture gradients and the other distance cues can be derived.[7] Causal explanations of perception generally require that the perceived and the physical orders differ systematically. If they did not, the relations between the two would be described more accurately as analytical. Applied to vision, therefore, the causal theory of perception implies that neither colors nor perceived spatial relations are perfectly veridical, except in unusual or accidental circumstances, and that these circumstances cannot be discovered except empirically by a careful comparison of the phemonenal and physical orders.

What is meant by the claim that neither colors nor perceived spatial relations are veridical? To say that colors are veridical is to say at least that they are related by a one-to-one correspondence and an isomorphic transformation to the frequency distribution of light emitted from luminous objects or reflected from the surfaces of other objects. If there is no such correspondence or no such isomorphic transformation, then colors are not veridical, because they provide no analytical connection between visual phenomena and the physical order. To say that perceived spatial relations are veridical is to say that the *discriminata* in the visual field have the same geometrical properties as the perceived objects are judged to have on the basis of physical measurements. If the *discriminata* of the visual field have geometrical orders which differ systematically from the measured order of physical spatial qualities, then perceived spatial qualities are not veridical.

Color vision provides an important example. The physical characteristics of a beam of light, whether radiated from a source or reflected from the surface of an object, are specified completely by two variables: wavelength and intensity. Any beam of light can be described completely and unambigously in physical terms by its frequency distribution, which is a two-dimensional plot of the intensities of its various component wavelengths. Two beams of light which have identical frequency distributions have identical physical characteristics, but they may be perceived as different colors, as in the phenomenon of simultaneous color contrast. Although two dimensions specify light physically, three dimensions are required to specify color: hue, luminosity, and saturation. None of them correspond simply to either of the physical variables. A pure yellow hue may be obtained either from a monochromatic beam of spectral yellow, with wavelength about 575 nm., or from a mixture of spectral red and spectral green. If the wavelength of a beam is held constant, its color changes with its intensity, its visual context, and the adaptation of the eye which it stimulates. The three dimensions of color vision mean that colors must be plotted in a solid rather than on a two-dimensional graph. In such a solid humans with normal color vision can discriminate about 7.7 *million* different colors, of which only two or three hundred are spectral. Other phenomena of color vision, such as binocular color mixing and the Land effect, complicate colorimetry still more, but as a science it is nevertheless capable of high precision. Even though it is precise, its order differs systematically, even to the number of its dimensions, from the physical order of wavelengths and intensities.[8]

Because the veridicality of vision can be ascertained empirically, it reflects the uncertainties inherent in empirical relations. Such uncertainties characterize all measurements and all other empirical information. They characterize knowledge of both the physical order and visual phenomenal orders. Visual perception could be said to be veridical even if the physical and phenomenal orders were not perfectly identical, but were similar to within a reasonable approximation. In color vision these uncertainties should be distinguished from systematic differences of order. For spectral colors, for example, human vision is most sensitive to differences of wavelength near the blue-green transition (about 480 nm.) and the yellow-orange transition (about 580 nm.). In these parts of the spectrum, differences in wavelength of about 1 nm. correspond to the smallest discriminable differences of color. Elsewhere in the spectrum the smallest differences of color correspond to somewhat larger differences of wavelength. A spectral doublet whose components differ by 0.5 nm., a difference easily measured with a good spectroscope, will not be discriminable by color. A cheap spectroscope, by comparison, might have a precision no greater than 2 nm., in which case some differences of color could not be resolved as different wavelengths by the instrument. Such examples reflect the uncertainties in empirically based information, and in themselves they do not count against the veridicality of color vision. What count against the veridicality of colors are the systematic differences which visual processing introduces between their phenomenal order and the physical order of wavelengths and intensities.

These systematic differences indicate that the physical order of wavelengths and intensities and the phenomenal order of colors form two different epistemological categories. Without a mediating theory, knowledge about the one order does not imply knowledge about the other. This conclusion is established by empirical investigation and not by conceptual analysis, because there is no sufficient analytic reason why colors fail to be an accurate gauge of wavelengths and intensities, or why wavelengths and intensities fail to be related isomorphically to colors. It is not analytically necessary that there be systematic differences between the two orders. If there were none, then knowledge about one would entail knowledge about the other, and no causal explanations would be required to relate the two. The fact that the two orders do differ systematically, and the fact that these differences are introduced by the visual processes themselves, on the other hand, imply that causal relations are required for their adequate explanation. In what follows, a similar conclusion will be argued for spatial vision.

Color vision also illustrates that 'veridicality' designates a possible relation between the two orders and not a characteristic of either order alone. Both orders provide information about physical objects, but they provide different kinds of information, which are not interconnected in any simple way. As was argued above, colors may be predicated quite legitimately of physical objects, as long as they are not taken to indicate physical qualities of wavelength and intensity.

There are at least three different levels in which causal relations can be discovered in vision. Each represents a different way of specifying the functional relations in visual processes as the necessary and sufficient conditions for differentiations of the visual field. They are to be regarded as convenient designations rather than as rigid categories. Just as there is no ontological presupposition behind our use of 'order' in categorizing the physical and the phenomenal, there is no implication here of any final analysis or any specific stratification of visual processes. In describing them I also do not mean to imply any hierarchical relations among them, nor that they are irreducible to one another. They are intended instead to exhibit differences among causal explanations of the various kinds of visual processes.

The first level specifies phenomenal effects as functions of the physical stimulus configurations. It thus provides systematic relations between the physical order and visual *discriminata*, but it does so without specification of the cognitive or somatic factors which mediate the phenomenal effects. Experimental psychologists have successfully produced explanations of many visual effects — indeed, it probably is fair to say that the bulk of experimental work in perceptual psychology has been done at this level. The phenomenon of simultaneous color contrast, which can be explained independently of physiological theories of color vision, is one such example. The work of James Gibson (esp. 1950) and Gestalt psychology are probably the best known examples for spatial vision. At this level the explanations of visual effects are analogous to the explanations of electrodynamics and elasticity in classical physics, in which the relevant phenomena are studied with great experimental and theoretical sophistication, but without reference to the atomic and molecular properties of matter. Gibson, for example, has preferred to speak of light as a stimulus not just to the retina but to the visual system.[9] This level has been explored the most systematically of the three, but because it offers no account of the cognitive or somatic processes which mediate vision, it is also the least complete. Its naive use, moreover, easily produces the misleading analogy between vision and cameras which we rejected above.

This misleading analogy can be avoided by a consideration of the second level of causal explanation, which incorporates a specification of the cognitive processes which mediate vision. Because these processes, as far as we are concerned with them, usually occur pre-phenomenally, they need not be, and indeed usually are not, conscious. Included at this level are memory, expectation, recognition, and in some contexts the effects of classification, strategy, and social and cultural factors. An interesting, though mistaken, early attempt to specify cognitive processes in vision was Berkeley's argument that the third apparent dimension of the visual field is not genuinely sensed, as are the other two dimensions, but is rather a cognitive artifact built of associations between vision and touch. More recently Rock has explained the overall metrical stability of spatial vision as an integrative effect of memory. In contrast to models of visual processing suggested by the camera analogy, the actual processing of visual data does not occur primarily as a point-to-point transmission of retinal images to the brain. It also incorporates inhibitory interactions and much high-level processing of scaling information; gradients of luminosity, color, and texture; transformations derived from movements and scanning; and some kinds of invariants. The processes to be specified at this level are imperfectly understood at present, although recent investigations of them by social and cognitive psychologists have yielded impressive results in studies of the effects of short-term visual memory, perceptual adaptation, and some of the social, cultural, and other contextual factors.[10]

The third level of causal explanation adds to the first two a detailed understanding of the somatic processes which underlie vision in both its sensory and its cognitive aspects. A complete account of these somatic processes would include genetic and evolutionary considerations.[11] Many past treatments of the causal theory emphasized this level to the exclusion of the others. If it is also combined with a metaphysical perspective or a generative view of causation, this emphasis usually leads to an unnecessary stress on the mind-body aspects of the problem of perception. Both the anatomy and the physiology of the basic sensory aspects of vision are reasonably well understood, but apart from some rather tentative studies, the somatic processes which underlie visual memory and most of the other cognitive factors are at present virtually unknown.

Spatial vision illustrates the combination of all three levels. Like color vision, it shows how veridicality can be defined as a relation between the physical and phenomenal orders. For spatial vision the two orders are bridged by geometry. Spatial vision is not veridical if its geometry differs systematically from the geometry of the measured physical order. In both spatial vision and measurements of physical spatial relations, coordinative definitions are required which link the data to the formal concepts and which distinguish systematic order from the uncertainties inherent in any formal ordering of empirical data. The stipulation of appropriate coordinative definitions for spatial vision forms one of the interesting branches of psychophysics. Although we cannot enter into a detailed treatment of psychophysical methods here, we may state that they usually attempt to exploit thresholds of sensitivity in perceived configurations which are designed to minimize the effects of possible interpretations or judgments made by the subjects.[12] Spatial vision can exhibit astonishing precision. Helmholtz showed, for example, that in human vision vernier alignments can be made to within 4" of arc, and that the geometrical centers of the eyes vary so little from their centers of rotation as to make no appreciable perceptual differences (Helmholtz, 1962, 3:165).

Despite its precision, spatial vision is not veridical. There generally is no one-to-one correspondence between visible physical configurations and the perceived results. Many different physical configurations can be visually equivalent, as in the Ames distorted rooms, in which gross, but carefully calculated physical departures from the shapes of normal rooms nevertheless are seen as though they were normal. Such equivalent configurations can be designed for both monocular and binocular vision, and they provide important clues about the nature of visual processing. Binocularly equivalent configurations usually differ from their monocular counterparts, and *vice versa*, so that the physical distortions of the Ames binocular rooms may be seen by closing one eye. The distorted rooms and other experiments show that binocular spatial vision, unlike physical space, is characterized by non-Euclidean metrical relations. When coordinative definitions are stipulated, so that the shapes of the binocular distorted rooms and other configurations may be expressed mathematically, the resulting forms specify the basis for binocular visual mapping of physical spatial relations.[13]

In monocular vision the absence of stereoscopic spatial information is partly compensated cognitively by greater reliance on depth cues such as overlay, texture, and motion parallax.[14] This compensation makes monocular vision more subject than binocular vision to variations of perceptual strategies. In spatial vision perceptual strategies usually are neither conscious nor voluntary. They are cognitive algorithms by which visual data are organized into more or less coherent perceptual mappings of physical configurations. One consequence of the algorithmic flexibility of monocular vision is that there is no simple theory of monocular equivalent configurations. The introduction of motion, whether of the perceiver or of the configuration, greatly enhances the information available to the monocular perceiver, so that statically equivalent monocular configurations generally do not remain equivalent when motion is introduced. In binocular vision, by contrast, many equivalent configurations remain invariant for some kinds of motion, notably for horizontal rotations of the perceiver's own eyes and head.

Monocular equivalent configurations can be designed for specific contexts. Of these, monocular vision with a fixed viewing point has received the greatest attention, because for this condition equivalent configurations are easily derived according to the rules of linear perspective. It is to be emphasized that they represent only one monocular context and that their geometry does not characterize all monocular spatial vision. Such configurations usually photograph well. In the Ames monocular distorted room, the equivalence to a normal room produces incorrect scaling information about the distances from the perceiver of two persons standing along the rear wall of the room, so that one of them is perceived as much shorter than the other, even though they are equally tall. The gross elimination of normal cues for scaling can produce the same perceptual result, as in Gregory's illustration of the Ames-room-without-the-room. In both examples the perceptual algorithm overrides the conscious presupposition or even the knowledge that both persons are equally tall. The cognitive processing which organizes the visual data into a spatial framework also produces a non-veridical result (Gregory, 1977, p. 178; 1970, pp. 26, 28).

The algorithmic flexibility of spatial vision illustrates the importance of avoiding misleading analogies between vision and cameras, and it thus regains the theme of veridicality. Because the rules for linear perspective are derived from the geometry of Euclidean optical projections, linear perspective suggests an appealing, though ultimately mistaken, criterion to establish the veridicality of spatial vision. Following a suggestion of Bertrand Russell's, E. B. McGilvary (1933) proposed that perceived spatial relations should be judged, not from the measurements of physical objects themselves, but from appropriately taken optical projections.[15] This proposal tacitly draws too simple an analogy between spatial vision and cameras. It fails theoretically, because it treats a simple, highly constrained context, namely monocular vision from a fixed viewing point, as though it were a privileged perceptual paradigm.

The proposal also fails empirically. According to it, monocular spatial vision would be said to be veridical if its geometry coincided with the geometry of optical projections. But optical projections do not include the scaling factors which are usually present in vision. Pinhole photographs, which produce Euclidean projections, must be composed carefully to avoid apparent distortions which are really caused by the absence of visual scaling. Objects must not be placed too near the foreground, for example, or their projected sizes appear to be much too large relative to the perspectival sizes of the other objects in the picture. The same is true for the shapes of objects placed at large angles from the forward axis of the cameras. For vistas of large distances, strict Euclidean perspective tends to produce artificial and unconvincing representations. Canaletto, who is unsurpassed in representing visual perspective convincingly, did not rely wholly on the *camera obscura* to delineate his vistas. In his representations of distant buildings, he departed from the projection of the camera by introducing small telephoto effects, which deviate from the overall Euclidean projections of the pictures. These deviations produce more natural visual appearances than would strict optical projections (Gregory, 1977, pp. 163—169).[16] However natural and convincing the resulting pictures may appear to be, their perspective is not strictly veridical if we define veridicality by Euclidean optical projections.

Because it is attained through cognitive processes which vary with context, spatial vision is not characterized uniquely by any one set of geometrical relations. For specific cases, however, these relations can be ascertained experimentally, and they illustrate the underlying perceptual algorithms. Visual geometry is the effect of long-term integrative processes of visual memory; it is not produced directly from retinal images. Some algorithms for spatial vision are learned. Perceivers in Western cultures, for example, tend strongly to see photographs and line drawings as perspectival representations of three-dimensional objects, which requires algorithms derived from monocular vision with a fixed viewpoint. Like other algorithms, they can be exploited by illusions and perceptual paradoxes, which often are representational analogues of the equivalent configurations mentioned above, and which show how persistent well-established perceptual algorithms can be. Escher's engravings and the Penrose triangle are easily seen as representations of three-dimensional objects, even though the objects cannot be constructed physically if the representations are taken naively as plans for building them. The three-pronged figure cannot even be seen as an object, although it is not at all paradoxical if it merely represents a plane figure (Gregory, 1977, pp. 50—59; Segall et al., 1966). Not all illusions are due to the inappropriate processing of visual data by the algorithms for perspectival representations, but many are, and they are as important to our understanding of perception as they are fascinating in their own right. They show that in normal spatial vision, our general sense of a stable co-ordinate framework is a combination of many sophisticated algorithms, which utilize the cognitive flexibilty of visual processing.[17]

These considerations are illustrated most dramatically by perceptual adaptation, in which cognitive processes effect the reorganization of the entire visual field. In these experiments, optical devices are worn to alter the retinal images normally cast by physical objects. Upon first wearing the devices, subjects see physical objects, including their own bodies as distorted. Often they experience general disorientation, nausea, and loss of co-ordination between vision and hearing and touch. After prolonged use of the devices, sometimes within a few minutes and sometimes only after hours or days, they experience an alteration of the *phenomenal* characteristics of the visual field, so as to reduce the distortions. Prismatic spectacles, for example, can cause physically straight edges to cast retinal images normally associated with curves. Upon first wearing the spectacles, the perceiver accordingly sees physically straight edges as curved, but after adaptation they are perceived as straight or at least as less curved than at first. If the spectacles are then removed, the retinal images revert to normal, and converse phenomenal distortions occur. Thus a ruler will appear to be curved when the spectacles are first worn; it will appear to be more nearly straight after adaptation; it will again appear to be curved when the spectacles are first removed; and finally it will appear to straighten as normal vision is regained. Other devices cause alterations of size, distortion of stereoscopic information, or up-down or left-right reversals. Since the retinal alterations persist while the devices are worn, the phenomenal changes are cognitive effects. But because the effects are often incomplete, and because they do not occur at once, they also illustrate that there are limits to the cognitive flexibility of the visual system and that well-established spatial algorithms tend to be persistent.[18]

On evolutionary grounds, the cognitive flexibility of vision hardly surprises us, though we cannot but marvel at how wonderfully well adapted it is to the visual system's biological constraints. As a spatial sense vision enables us to cope rapidly, and on the whole effectively, with a vast array of optically accessible information. It provides spatial information under a great variety of stimulus conditions. We can see objects both near and far, before or when we attend to them, when they are stationary or in motion, when we are stationary or in motion, in full daylight or in near darkness. But despite our resourcefulness and the variety of stimulus conditions under which we can see, our action often is required for ambiguous stimuli. In many such cases our cognitive processes fill the gaps in the stimulus information. The flexibility afforded by cognitive processes in vision thus is attained at a cost: they introduce presumptions of their own which affect the phenomenal results. In some examples, as in the adaptation experiments, they reduce of even eliminate artificially induced distorions, but in other examples, as in the Ames rooms, they occasion mistakes and cause illusions. Even though it is not veridical, vision usually provides reliable spatial information. We thus obtain the interesting result that cognitive flexibility enhances the usefulness of spatial vision by compromising its veridicality (Gregory, 1974, pp. 377—379, 589—601; Campbell, 1966, pp. 81—106).

Perceptual algorithms can order visual *discriminata* without giving them sentential form. The cognitive approach does not necessarily require that perception be the product of implicit processes of reasoning, even though it has roots in British empiricism. Recent work using it is descended primarily from Helmholtz, who argues that perceptions are 'unconscious conclusions'. By this term he did not mean to assert that perception is sententially structured, although so many of his critics interpreted him as having done so that he avoided the term in later statements of the theory. At present the leading proponent of this approach undoubtedly is Richard Gregory, who has developed it with special reference to an analogy between perceptions and hypotheses. In his work the analogy is not taken to indicate an implicit sentential organization of perceptual data, but it is taken to imply flexibility in cognitive processes. Although the details of the analogy cannot be examined here, it provides fruitful ways of studying the phenomenal effects of the various cognitive elements in vision. It is especially useful in exhibiting the implications of illusions and ambiguous configurations and in incorporating the insights — the illustrative failures as well as the successes — of artificial intelligence.[19]

Visual memory, perceptual strategies, and other cognitive processes usually affect vision pre-phenomenally, without molding it into judgments. Neither physical nor phenomenal orders are in themselves either true or false. Veridicality, which designates one possible relation between the two kinds of order, likewise does not itself imply either truth or falsity. Perception, as in spatial and color vision, thus may be distinguished from perceptually based judgments. In the former, veridicality is the significant criterion; whereas in the latter, criteria of evidence dominate. Veridicality is a criterion which is especially appropriate for the aspects of perception which are cognitively prior to their use in the formation of empirical judgments.

Some examples of perception do have implicit judgmental form. These examples occur much nearer a purely cognitive level than spatial and color vision. They include the Ames chair, the duck-rabbit figure, many cases of figure-ground reversal, and symbolic representations. Like spatial and color vision, they usually can be distinguished from explicit empirical judgments. In such examples perceptual strategy tends to override the more limited consideration of veridicality. In the duck-rabbit figure, for example, the literal veridicality of the figure is less essential than the strategy by which the perceiver sees the figure as the representation of a duck or of a rabbit. The point is even clearer if we note that no attempt is made in the black-and-white figure to duplicate the natural coloration of either ducks or rabbits. Yet it is equally clear that sufficient distortion of the figure will prevent its being seen as an appropriate representation. Even at the lower levels of spatial and color vision, cognitive processes structure the phenomenal results. Their effects become more pronounced as we proceed upward toward more purely cognitive examples of perception. These upper levels include perceptual recognitions and identifications, which may have implicit judgmental structure. In their evaluation, criteria of strategy and truth may dominate criteria of veridicality.

At intermediate levels, considerations of veridicality and strategy are mixed. The apparent changes of orientation of the Necker cube and the Schröder staircase seem to depend less on the veridicality of spatial vision than on the interpretation of the figures as representations of three-dimensional objects. The same is true of the Ames chair, which requires both an interpretation of the configuration as an object and the algorithms for monocular equivalent configurations. Examples such as the Müller-Lyer illusion and cognitive contours, on the other hand, seem to require only the non-judgmental processes of ordinary spatial vision. Veridicality applies to the perception of an open figure as closed, just as it does to ordinary spatial vision, but this example also depends explicitly upon perceptual strategy, since in some strategies perceivers can readily detect the gaps in the figures which are ordinarily filled by cognitive processes. Although appropriate perceptual strategies can greatly reduce the tendency to overlook data in context or to misapply them, inappropriate strategies can increase the tendency to err. Thus, although cognitive processes can partially overcome the limitations of our ordinary, non-veridical spatial and color vision, they introduce presumptions and limitations of their own.[20]

With this material as background, we shall consider the important objections to traditional versions of the causal theory of perception and show how they may be answered by the view I have sketched (Hirst, 1965, pp. 18—20; Shimony, 1971).

The objection that the causal theory is contradictory is based on the notion that it uses observations against their own veridicality, and hence against their own truth. Errol Harris (1965, p. 372) claims that "so fatal is this objection that it may be taken as final, and it would be anticlimatic to mention other defects." The objection is based on the idea that if empirical evidence is used to claim that perception is not veridical, then, by virtue of its very non-veridicality, the evidence which supports the claim is itself non-veridical and hence false. Far from being final, the objection is based on two different equivocations, the elimination of either of which eliminates the objection itself.

The first equivocation conflates 'non-veridical' and 'false.' We have seen that the perception of colors, for example, is not veridical in the sense that colors cannot be assigned to the physical order, and indeed, without a mediating theory, they cannot even be taken to indicate physical properties. In this sense, however, the non-veridicality of color vision does not imply that all statements which ascribe colors to objects are false. Similarly, the visual perception of spatial qualities is not veridical in the sense that visually perceived spatial qualities do not have the same geometrical order as do measured spatial qualities. 'Veridicality' connotes not truth but only one possible relation between the two orders. For truth to enter as a consideration, one or both orders must be taken to provide information on which statements are based. The claim that vision is not veridical does not establish that the reading of a length from a ruler (on the basis of its colored marks) is false. Measurements may be legitimate observations, even though the perceived qualities which make it possible to use the instruments are not directly part of the physical order. The non-veridicality of vision does not imply that statements based on visual observations are false. The contrary claim is based on an equivocation between 'non-veridical' as 'false' and 'non-veridical' as 'not coinciding with the physical order.'

The second equivocation on which the objection is based is an equivocation on 'observation.' Observations in the sense of the source of scientific data should not be conflated with obeservations in the sense of seeing something. It is true that in the latter sense, visual observations can lead to false assertions, especially if the observations are taken to be veridical in the sense of supplying directly information about the nature of the physical order. Observations in the scientific sense, however, are not so directly related to the perceived order. Color vision again provides an instructive example. A spectroscopist may obtain the spectrum of wavelengths emitted from a chemical substance which has been heated to incandescence, and for convenience he may speak of the wavelength which he observes as a 'yellow line' in the spectrum. The spectroscopist may or may not realize that there is no isomorphism between color and wavelength, but he need not be understood in any case as claiming that the perceived quality of the line is the direct, veridical observation of a corresponding wavelength. The color is merely a convenient way of identifying a wavelength of radiated energy, and the spectroscopist could perform his work equally well if he were color-blind and forced to use some other indication in place of the color. Indeed, much spectroscopy is done with black and white photographs. Similarly, a person whose vision is astigmatic may see the shape of a ruler change as its position shifts in his visual field. This person, however, can still use the ruler to make observations (in the first sense of the term) of the physical order, provided only that he can read the ruler and has reason to believe that it approximates a rigid body. Scientific observations are based on perception, but the truth of the observations does not depend on an isomorphism between the perceived and the physical orders.

The objection that the causal theory is circular usually is made by epistemologists who are proponents of strongly foundational theories.[21] If such theories are accepted, the objection that the causal theory is circular can be advanced for reasons similar to those advanced for the objection that the causal theory is contradictory. In both cases a presupposition is made that a critical assessment of evidence must be based externally, as it were, from an epistemological standpoint which is accepted as more securely justified than is the evidence to be assessed. A rejection of this presupposition greatly weakens the objection itself. Applied specifically to perception, the objection implies that evidence derived from observation, because it is based on perceptual data, cannot be used as the basis of other evidence which also is attained by observation and hence based on perceptual data. If both kinds of evidence are based on perception, both are subject to the same conditions, and it is circular to use one in assessing the other. The objection seems not to apply to the causal theory as it is developed above, because the objection requires that one order be considered more fundamental than the other. As we have developed it, the causal theory implies merely that there are differences between the perceived and physical orders, and that these differences may be explained systematically in causal terms. No further distinctions of status are required.

The objection that the causal theory is unintelligible applies mainly to metaphysical versions of the theory. Occasionally the objection is stated in terms of the mysterious nature of the causal relations which connect the material world with the mental. In this form the objection is directed mainly against the use of the causal theory as evidence for a dualistic theory of the mind-body relation. In this form the objection also presupposes the generative theory of causation and so does not apply to the causal theory as treated here. Hirst notes that the objection can be stated in a somewhat different form, if the causal theory is taken to imply that sense data rather than objects are perceived immediately. In his words: "If we try to verify the existence of the table by touching it, we simply obtain more sense-data — tactile ones — and if we see our hands touching the table we are only having visual data. Whenever we try to peer around the barrier of ideas, we just get more ideas" (Hirst, 1965, p. 18). Hirst's statement of the objection is well put, if Berkeley's assertion is accepted that we perceive ideas and not objects. On the other hand, there is no reason to deny that we do perceive objects. Physical objects may be said to be perceived directly, but the qualities which they are perceived to have cannot thereby be taken to be identical with whatever qualities may characterize the physical order. To assert that the causal theory is false because objects may be said to be directly perceivable is to play on an equivocation of 'perceive' similar to the equivocations discussed above in the consideration of the objection that the causal theory is contradictory. The relation between the perceived and the physical orders can be discussed with few, if any, overt ontological presuppositions about the need for a knowing subject or about the nature of such a subject. These metaphysical questions may be interesting in their own right, but their consideration is not necessary for an epistemological treatment of perception.

When experimental evidence is applied to philosophical theories, an important implication is obtained for epistemology itself. At least one of its central problems, the problem of perception, may be approached from an empirical perspective as well as by more usual conceptual and analytic routes. Although the adoption of an empirical perspective does not eliminate the analytic and conceptual issues in epistemology, it does make some of them clearer. It further increases the probability that the epistemologist will examine evidence which he might otherwise overlook. This evidence, of course, is relevant to his theoretical aim, even if he does choose to overlook it or dismiss it as irrelevant. The empirical perspective I have advocated envisions the proper relation between scientific and philosophical theories as a dialectical relation of mutual support and criticism. According to this perspective, at least some philosophical theories give rise to testable consequences, and this result supports the conclusion that there is no radical distinction between scientific and philosophical theories.

*Lebanon Valley College*

## NOTES

[1] This work was supported by a Faculty Research Grant, for which I am indebted to Lebanon Valley College. Some of the material is condensed from Heffner (1981).

[2] Even the sequence of visual phenomena need not correspond to the temporal order of the associated stimuli, as in the masking effect, which is treated in many works on vision. An especially good introductory account is found in Haber and Hershenson (1973, pp. 125—131). For a discussion of central and peripheral masking effects, see Uttal (1975, pp. 81—88). The Pulfrich effect is a fascinating and easily demonstrated example. Cover one eye with a neutrally colored filter, such as the lens of a pair of sunglasses, while viewing a pendulum binocularly. The plane of the pendulum should be perpendicular to the forward line of vision. Because the filter delays the processing in the covered eye, the path of the pendulum appears to be elliptical rather than plane. There is an extensive literature on the effect. It is illustrated in Gregory (1977, pp. 82—83).

[3] For a perfect example, see Aristotle, On Dreams 460$^{b}$. An extended development of the position with a surprisingly contemportary tone is given by James (1890). For a recent treatment, see Campbell (1969).

[4] Collingwood (1940, chs. 29, 32); Gasking (1955); Harré and Madden (1975); Taylor (1967). My reasoning on this point is amplified in Heffner (1981).

[5] I assume that the measured spatial properties of physical objects coincide with negligible deviations, by whatever reasonable methods are used, so that there is a single physical spatial order, which, on the terrestrial scale appropriate to spatial vision, may for all experimental purposes be considered Euclidean. The point is treated at greater length in Heffner (1976).

[6] A superbly judicious review of experimental literature on this point is given by Owen (1978). I am astonished that some philosophers still regard this point as controversial. Summaries of relevant experimental evidence may be found in Pitcher (1971, pp. 131—152). See also refs. in Note 10. For an extended and roughly similar treatment of causation, see Hanson (1958, ch, 3).

[7] I do not mean to suggest that they are actually decomposed in the processing of spatial vision. See Gibson (1950).

[8] The color solid is well pictured in Kendler (1963, pp. 126—130). See also Robert M. Boynton, 'Color, Hue, and Wavelength,' in Carterette and Friedman (1974—1979, Vol. 5, pp. 301—347), and Mundle (1971, pp. 140—146). Strictly speaking four variables — the two mentioned above plus polarization and coherence — are needed for a complete physical specification of the light. Polarization and coherence, however, are unrelated to perceived color in humans, although bees can discriminate differences in polarization.

[9] Gibson's term is 'ecological optics'. R. M. Boynton, 'The Visual System: Environmental Information', in Carterette and Friedman (1974—1979, Vol. 1, pp. 285—307); James J. Gibson, 'A Note on Ecological Optics', in Carterette and Friedman (1974—1979, Vol. 1, pp. 309—312); Gibson (1966, p. 155). In philosophical terms, Gibson's is a special form of the causal theory. For a good discussion of epistemological issues related to Gibson's work, see Machamer (1978).

[10] Sources which illustrate the recent advances are Bruner (1973), Haber and Hershenson (1973), Neisser (1967), Rock (1966), and Segall *et al.* (1966).

[11] Three epistemologically oriented treatements, which intersect the perceptual theme of this paper, and which contain further references, are Campbell (1974), Čapek (1971), and Shimony (1971). Campbell's forthcoming William James lectures, *Descriptive Epistemology: Psychological, Sociological, and Evolutionary*, one of which is included in this volume, expand this treatment.

[12] Baird (1979); Tarow Indow, 'Applications of Multidimensional Scaling in Perception,' in Carterette and Friedman (1974—1979, Vol. 2, pp. 493—531), S. S. Stevens, 'Perceptual Magnitude and Its Measurement,' in Carterette and Friedman (1974—1979, Vol. 2, 361—389).

[13] Binocular space perception is treated at much greater length in Heffner (1981). See also Walter C. Gogel, 'Size, Distance, and Depth Perception,' in Carterette and Friedman (1974—1979, Vol. 9, pp. 299—333). Foley has published many excellent experimental studies related to visual geometry (for example, Foley, 1972); his work is reviewed in Suppes (1977).

[14] Helmholtz (1962, Vol. 3, pp. 282—294 and *passim.* in §§ 28—30. Helmholtz's charming essay 'On the Relation of Optics to Painting' is well worth reading in this context; it is reprinted in Warren and Warren (1968, pp. 139—168). For a more recent treatment, see Kaufman (1974, pp. 213—268, 350—366).

[15] Russell's suggestion is found in Russell (1917, Ch. 7).

[16] For a beautiful discussion of the relation between optics and perspective, with illustrations from the work of Canaletto, see Pirenne (1970, esp. pp. 102—103); for an amended treatment see Pirenne's 'Vision and Art' (Carterette and Friedman, 1974—1979, Vol. 5, pp. 433—490). Van Gogh seems to have attempted a perspectival representation of the non-Euclidean space of binocular vision; see Heelan (1972). The relation of photography to spatial vision is treated in Snyder (1980).

[17] The specific analyses of many illusions remain controversial. For a superb philosophical discussion of these points, see Wartofsky (1978). Further examples are treated in Gregory (1977, pp. 132—162) and his 'Illusions and Hallucinations' in Carterette and Friedman (1974—1979, Vol. 9, pp. 337—357); the classic Gombrich (1960); Gregory and Gombrich (1973); Coren and Girgus (1977, 1978).

[18] Research in perceptual adaptation has had a steady and distinguished history, which began with Stratton's work in the 1890s. Binocular vision is somewhat less susceptible to adaptation than monocular vision. The reorganization clearly is phenomenal and not only a re-learning of vision-touch coordination. Whether the reorganization is caused wholly by visual processes or whether it also is mediated by touch is less clear. Rock and Wallach advocate the former view, whereas Kaufman is somewhat more sympathetic to the latter, although he is mistaken in his claim that Rock's theory requires an "*inherent* spatial coordinate system" [my emphasis]. The effects can be explained adequately by the persistence of previously established spatial algorithms. Kaufman (1974, pp. 409—460); Rock (1966); Hans Wallach (1976, Chs. 10—11, esp. pp. 235—297, 353—366).

[19] This subsumes a family of theories, among which there are differences which seem not to affect the broad epistemological implications, but which are important for experimental psychology. The best comparison is Gregory's 'Choosing a Paradigm for Perception' in Carterette and Friedman (1974—1979, Vol. 1, pp. 255—283). Somewhat different views are found in Gogel (Carterette and Friedman (1974—1979, Vol. 9, pp. 299—333); and Rock (1977). For a philosophical assessment, see Kline (1979).

[20] A longer discussion, which focusses on problems of scientific observation, is Shimony (1977).

[21] A closely related objection, which holds that the causal theory leads to an infinite regression, is discussed in Heffner (1981, pp. 327—328), with particular reference to Smythies (1956) and Smythies and Brain (1965).

## REFERENCES

Baird, John C.: 1970. *Psychophysical Analysis of Visual Space*. New York: Pergammon.

Bruner, Jerome: 1973. *Beyond the Information Given*. New York: Norton.

Campbell, Donald T.: 1966. 'Pattern Matching as an Essential in Distal Knowing.' In Kenneth R. Hammond (ed.), *The Psychology of Egon Brunswick*. New York: Holt, Rinehart and Winston, pp. 81—106.

Campbell, Donald T.: 1969. 'A Phenomenology of the Other One: Corrigible, Hypothetical, and Critical.' In Theodore Mischel (ed.), *Human Action: Conceptual and Empirical Issues*. New York: Academic Press, pp. 41—69.

Campbell, Donald T.: 1974. 'Evolutionary Epistemology.' In Paul A. Schilpp (ed.), *The Philosophy of Karl Popper*. LaSalle, Ill.: Open Court, pp. 413—463.

Čapek, Milič: 1971. *Bergson and Modern Physics*. New York: Humanities Press.

Carterette, Edward C., and Morton P. Friedman (eds.): 1974–1979. *Handbook of Perception*. 10 vols. New York: Academic Press.

Collingwood, Robin G.: 1940. *An Essay on Metaphysics*. Oxford: Clarendon Press.

Coren, Stanley, and Joan Girgus: 1977. 'Illusions and Constancies.' In Epstein (1977, pp. 255–283).

Coren, Stanley: 1978. *Seeing is Deceiving*. New York: Wiley.

Epstein, William (ed.): 1977. *Stability and Constancy in Visual Perception*. New York: Wiley.

Foley, John M.: 1972. 'The Size-Distance Relation and Intrinsic Geometry of Visual Space: Implications for Processing,' *Vision Research* **12**, 323–332.

Gasking, Douglas: 1955. 'Causation and Recipes,' *Mind* **64**, 479–487.

Gibson, James J.: 1950. *The Perception of the Visual World*. Boston: Houghton-Mifflin.

Gibson, James J.: 1966. *The Senses Considered as Perceptual Systems*. Boston: Houghton-Mifflin.

Gombrich, E. H.: 1960. *Art and Illusion*. New York: Pantheon.

Gregory, Richard L.: 1970. *The Intelligent Eye*. New York: McGraw-Hill.

Gregory, Richard L.: 1974. *Concepts and Mechanisms of Perception*. New York: Scribner's.

Gregory, Richard L.: 1977. *Eye and Brain*. 3rd ed. London: Weidenfeld and Nicolson.

Gregory, Richard L., and E. H. Gombrich (eds.): 1973. *Illusion in Nature and Art*. London: Duckworth.

Haber, Ralph N., and Maurice Hershenson: 1973. *The Psychology of Visual Perception*. New York: Holt, Rinehart and Winston.

Hanson, Norwood Russell: 1958. *Patterns of Discovery*. Cambridge: Cambridge University Press.

Harré, R., and E. H. Madden: 1975. *Causal Powers*. Oxford: Blackwell.

Harris, Errol: 1965. *The Foundations of Metaphysics in Science*. New York: Humanities.

Heelan, Patrick A.: 1972. 'Towards a New Analysis of the Pictorial Space of Vincent Van Gogh,' *Art Bulletin*, December, pp. 478–492.

Heffner, John: 1976. 'Some Epistemological Aspects of Recent Work in Visual Perception.' In Frederick Suppe and Peter Asquith (eds.), *PSA 1976: Proceedings of the 1976 Biennial Meeting of the Philosophy of Science Association*, Vol. 1. East Lansing, Mich.: Philosophy of Science Association, pp. 165–174.

Heffner, John: 1981. 'The Causal Theory of Visual Perception: Its Scientific Basis and Epistemological Implications,' *International Philosophical Quarterly* **21**, 303–332.

Helmholtz, H. von: 1962. *Treatise on Physiological Optics*. 3rd ed. Ed. James P. C. Southall. 3 vols. New York: Dover.

Hirst, R. J. (ed.): 1965. *Perception and the External World*. New York: Macmillan.

James, William: 1890. *The Principles of Psychology*. New York: Henry Holt & Co.

Kaufman, Lloyd: 1974. *Sight and Mind*. New York: Oxford University Press.

Kendler, Howard H.: 1963. *Basic Psychology*. New York: Appleton-Century-Crofts.

Kline, A. David: 1979. 'Constructivism and the Objects of Perception,' *Nature and System* **1**, 37–45.

Machamer, Peter K.: 1978. 'Gibson and the Conditions for Perception.' In Machamer and Turnbull, 1978, pp. 435–466.

Machamer, Peter K., and Robert G. Turnbull (eds.): 1978. *Studies in Perception: Interrelations in the History and Philosophy of Science*. Columbus: Ohio State University Press.

McGilvary, E. B.: 1933. 'Perceptual and Memory Perspectives,' *Journal of Philosophy* **30**, 309—330. Partly reprinted in Hirst (1965, pp. 199—208).

Mundle, C. W. K.: 1971. *Perception: Facts and Theories*. London: Oxford University Press.

Neisser, Ulric: 1967. *Cognitive Psychology*. New York: Appleton-Century-Crofts.

Owen, Dean H.: 1978. 'The Psychology of Prior Experience.' In Machamer and Turnbull (1978, pp. 267—524).

Pirenne, M. H.: 1970. *Optics, Painting, and Photography*. Cambridge: Cambridge University Press.

Pitcher, George: 1971. *A Theory of Perception*. Princeton: Princeton University Press.

Rock, Irvin: 1966. *The Nature of Perceptual Adaptation*. New York: Basic Books.

Rock, Irvin: 1977. 'In Defense of Unconscious Inference.' In Epstein (1977, pp. 321—373).

Russell, Bertrand: 1917. *Mysticism and Logic*. London: Allen & Unwin.

Segall, Marshall H., Donald T. Campbell, and Melville J. Herskovits: 1966. *The Influence of Culture on Visual Perception*. Indianapolis: Bobbs-Merrill.

Sellars, R. W.: 1938. 'A Statement of Critical Realism,' *Revue Internationale de Philosophie* **1**, 472—496. Partly reprinted in Hirst (1965, pp. 241—242).

Shimony, Abner: 1971. 'Perception from an Evolutionary Point of View,' *Journal of Philosophy* **68**, 571—583.

Shimony, Abner: 1977. 'Is Observation Theory-Laden? A Problem in Naturalistic Epistemology.' In Robert G. Colodny (ed.), *Logic Laws, and Life*. Pittsburgh, University of Pittsburgh Press, pp. 185—208.

Smythies, J. R.: 1956. *Analysis of Perception*. New York: Humanities.

Smythies, J. R., and Russell Brain: 1965. 'The Representative Theory of Perception.' In J. R. Smythies (ed.), *Brain and Mind*. London: Routledge and Kegan Paul, pp. 241—257.

Snyder, Joel: 1980. 'Picturing Vision,' *Critical Inquiry* **6**, 499—526.

Suppes, Patrick: 1977. 'Is Visual Space Euclidean?' *Synthese* **35**, 397—421.

Taylor, Richard: 1967. 'Causation.' In Paul Edwards (ed.), *The Encyclopedia of Philosophy*, vol. 2. New York: Macmillan, pp. 56—66.

Uttal, William R.: 1975. *An Autocorrelation Theory of Form Detection*. New York: John Wiley & Sons.

Wallach, Hans: 1976. *On Perception*. New York: Quadrangle.

Warren, Richard M., and Roslyn P. Warren (eds.): 1968. *Helmholtz on Perception*. New York: Wiley.

Wartofsky, Marx W.: 1978. 'Rules and Representation: The Virtues of Constancy and Fidelity Put in Perspective,' *Erkenntnis* **12**, 17—36.

EDWARD S. REED

# WHY IDEAS ARE NOT IN THE MIND: AN INTRODUCTION TO ECOLOGICAL EPISTEMOLOGY

## INTRODUCTION

Modern epistemology rests on a number of assumptions about ideas, the mind, and the brain. The notions that ideas are in the mind, that the mind is in the brain, and that ideas are somehow caused by whatever causes brain activity, gave birth to both modern psychology and epistemology in the seventeenth century. The successes of this science spurred epistemologists more and more to accept these assumptions. Thus, by the end of the eighteenth century even philosophers who vehemently disbelieved that scientific explanations of mental phenomena could be produced nevertheless went along with the above assumptions.[1] From that time forward epistemologists have rarely debated these fundamental premises, taking them as givens.[2] Yet I contend that we had better begin to query these assumptions, because they have not led to a satisfactory science of cognition. On the contrary, the puzzles concerning human knowledge of the external world have deepened, not lessened, as an experimental psychology based on the above assumptions has evolved. The purpose of this paper is two-fold. First, to give sufficient historical and critical background for readers to appreciate the many scientific anomalies spawned by the hypothesis that ideas are in the mind which is in the brain. Second, I shall sketch some of the fascinating new problems and questions motivated by rejecting these traditional assumptions.

## THE CONTRADICTIONS OF POST-CARTESIAN EPISTEMOLOGY

The problem with post-Cartesian epistemology is that it rests on two inconsistent premises. On the one hand, knowledge is asserted to come through sense perception (despite all the debate over the amount or certainty of this knowledge, or the amount of 'processing' it requires, even the most ardent of anti-empiricists admit that some sense perception is important for cognition). On the other hand, it is asserted that in



*Abner Shimony and Debra Nails (eds.), Naturalistic Epistemology*, 215—229.

sense-perception one is directly aware not of the external objects that cause the sensory stimulation, but only the neural effects of that stimulation. If physics is true, Russell says somewhere, then we are aware not of the world but of our brains; yet, if this be so, it is well-nigh impossible that we could have come to have scientific knowledge of the physical world. As Gibson (1966, p. 47) put it, "If the senses are instruments through which knowledge is obtained they cannot be conveyors of mere sensory quality; if they are conveyors of sensory quality they cannot be instruments for obtaining knowledge." Here is the heart of the dilemma: epistemologists have always wanted to assert that the senses are conveyors of (at least some) knowledge, while simultaneously holding that the senses convey nothing more than mere qualities. I call the attempt to join these two contradictory notions 'Cartesian epistemology' and the second assumption I label 'sensationalism'.[3] While I am aware that there are myriad differences among those who hold these assumptions, in this essay will focus entirely on the consequences of these assumptions, not on the various ways on which one might try to reconcile these irreconcilable premises.

Naturalistic epistemology has not faced up to this inner contradiction. Some naturalistic epistemologists (such as Quine) seem unaware of the contradiction, and act as if a physicalist account of perception had already succeeded. Evolutionary epistemologists have put off facing the problem. The vague idea on which they pin their hopes is that phylogeny might somehow have occurred so that the mind creates accurate hypotheses on the basis of its awareness of sensory qualities. Hidden in this optimism is not one but two profound problems. Sensory qualities are particular and meaningless; knowledge is general and meaningful. The problems of ascending from particular to general, from meaningless to specific meanings have been subsumed together under the 'problem of induction'. It should be clear by now that these problems are insoluble and that even natural selection cannot manage to circumvent logical impossibility.[4] Again, Gibson (1979, p. 253) summarizes the problem well,

> The error lies ... in assuming that either innate ideas or acquired ideas must be applied to bare sensory inputs for perceiving to occur. The fallacy is to assume that because inputs convey no knowledge they can somehow be made to yield knowledge by 'processing' them. Knowledge of the world must come from somewhere; the debate is over whether it comes from stored knowledge, from innate knowledge, or from reason. But all three doctrines beg the question. Knowledge of the world cannot be explained

by supposing that knowledge of the world already exists. [All these theories] imply cognition so as to account for cognition.

## FROM PHYSICS TO PSYCHOLOGY

The hypothesis that ideas are located in the mind which is in the brain received its impetus from the rise of classical physics (Burtt, 1932). The successes of Galilean physics were so impressive that Galileo, Descartes, Boyle and many others began to apply physical reasoning to explain the reasoning of physicists. The fascination of modern theorists of knowledge with the scientific knowing process date from this reflexive move of the seventeenth century. Descartes argued that the new physics itself provided justification for reducing psychology to physics. His original physical treatise (the suppressed *Le Monde*) begins with a discussion of ideas: dividing ideas into two types, those which accurately reflect their external physical causes and those which do not. He contrasted tickles, which manifestly do not represent the external world, with physical ideas which (he claimed) do represent the external causes of sensations. Where the mind is wholly passive, Descartes argued, physical objects cause mere sensations unrepresentative of their objects (ideas in their 'material' aspect, he would say). Where the mind reasons about those those sensations it may produce more accurate representations (ideas in their 'formal' aspect).[5] For Descartes, physics explained why ideas are in the brain and helps to distinguish among kinds of ideas. Given this, psychology's task is to explain whatever else happens to ideas in the mind.

This strategy of using classical physics to detail the causal impact of objects on the brain condemns one's psychology to explaining how the ideas thus formed are representations not of the world but of the physics of the time. Somehow the interaction of so-called 'primary qualities' in the brain gives rise to 'secondary qualities' in the mind. And yet this does not block the mind from knowing the primary qualities via inference from the secondary. In the early 1800s Johannes Müller modified this philosophical assumption in his 'law of specific nerve energies'. Müller stated that the sensory quality of an awareness is a function only of the sensory nerve excited, regardless of the source of the stimulation. Hence a blow to the ears produces a ringing sound whereas a percussion of the eyes yields a sharp colored spot.[6] What is specified in awareness is which cranial nerve is stimulated, not the environmental source of the stimulation.

Following Müller most psychologists and physiologists of the last century and a half would hold that in sensory awareness one *cannot* be aware of the external world. Sensory awareness is of one's nervous system's state of arousal, and it is only from this, by dint of mental work, that one can infer the causes of that specific stimulation and awareness, Helmholtz codified this research strategy in the three volumes of his monumental *Treatise of Physiological Optics*.[7] The first volume covers the physics of light in the eye; the second discusses the sensations of sight aroused by the optic nerve; the third discusses perception, which is 'the theory of the interpretation of visual sensations'.

The philosophical contradictions outlined above thus became embodied in concrete scientific research strategies. Sensations are knowledge of stimulation of the brain, but — one assumes — knowledge of the external world is somehow based on these sensations. Scientific psychologists who make these assumptions analyze the physical parameters of stimulation in order to discover the specific causes of those sensations. The perceiver is then faced with a problem not unlike the scientist's: how to discover the causes of sensations. The tendency to reduce psychology to physics in turn leads to a tendency to reduce all observers' minds to the hypothetical scientists described in post-seventeenth century epistemology.

Müller's law gave scientific legitimacy to what would otherwise be seen as a relic of Aristotelian physics. Why should a particular *locus* of matter (or even a particular *form* or *organization* of matter) such as the brain be so special as to give rise to absolutely unique phenomena — mental states? Leibniz pointed out the bizarreness of this hypothesis in the *Monadology* with his fantasy of enlarging a brain to the size of a factory and walking around inside it looking for perceptions. Müller's elegant hypothesis simply asserted that it was a law of nature that different cranial nerves produce different sensory qualities. Just as Newton had no explanation of the 'occult force' of gravity other than that it was a natural law leading to useful experiments, so Müller's law allowed for experiments on the quantitative relations between physical stimuli and mental results without raising embarassing questions about the nature of this odd relationship.

What is it about the nervous system that gives rise to mental phenomena, and what is it about the sensory nerves that gives rise to specific sensations? What are the possible specific sensations and mental phenomena? Despite centuries of research, no answers to these questions have been forthcoming. Even placement in the skull (Müller's favorite hypothesis) cannot explain all the sensory differences, if smell, touch and kinesthesia are included in the list of specific sense modes.[8] Half a century ago a number of philosophers showed how classical physics and the psychology to which it led are inconsistent.[9] A naturalistic epistemology must face up to this contradiction. How is a naturalistic explanation of knowing possible? Is it possible on the basis of the physical view of the world? If so, why has classical psychophysics not led to a scientific understanding of cognition (as it demonstrably has not)? If not, then we need a new approach to both ontology and epistemology. After showing the muddles and confusions that arise as a consequence of applying classical physics to the explanation of perception via an analysis of Müller's law, I will argue the need for a new, ecological, basis for the theory of knowledge.

## CRITIQUE OF MÜLLER'S LAW

Although Müller's law (or variants of it) is widely accepted as a fundamental premise of neurophysiology, it is by no means the case that the details of Müller's theory are universally accepted. In fact, there is considerable disagreement over how to explain a number of anomalous facts. What is *not* disputed by most writers is that some specific factors in the process of sensing constitute the objects of awareness. It is my contention that the following anomalies should suffice for us to question whether *any* neural factors of the sensory process are the specific basis of our awareness of external objects.

The most important difficulty facing proponents of Müllerian sensationalism is that there has never been an independent way of distinguishing the sense modalities. Aside from placement in the skull, there are no physiological differences among the sensory nerves significant enough to account for the remarkable differences among seeing, smelling, tasting, feeling, and all the various objects so sensed.[10] Until this mystery is clarified it would be justified to say that a study of visual sensations was a study of auditory information processing — for there is no non-subjective way of distinguishing the two. Those who respond to this criticism by appealing to the different regions of the sensory cortex do no more than restate the problem. Two centuries ago Thomas Reid recommended this sort of theory as it led to the conclusion that God had simply ordained specific correspondences between areas of the body and sensations. Whether Reid's conclusion is acceptable to modern psychologists and epistemologists remains to be seen.

A second anomaly of sensationalism is 'sensationless perception' of which there are at least three varieties. First, there are perceptual systems whose normal functioning involves no specific sensations, and in which sensations arise only in pathology. This is true of the basic orienting system. Balance and orientation is maintained by a combination of articular (joint and muscle), visual and vestibular proprioception, without any sensation of balance at all. Only where the normal coordinated functioning of this system (especially the visual-vestibular loop) is disrupted are there unpleasant sensations of dizziness, vertigo, motion sickness, etc. A second variety of sensationless perception is what Michotte dubbed 'amodal' perception, as when one sees an object that has moved out of sight behind another surface.[11] A third form of sensationless perception occurs when the perceptual system actually involved in information detection is itself not detected. This is the case in the "obstacle sense" of the blind, in which it is reported that one "feels" obstacles on the face when in fact one is detecting them by means of echo-location. Visual proprioception is another example, as it is often sensationless or accompanied by non-visual sensations, especially in cases of disorientation. As Gibson (1966, p. 3) says, the "organs of perception are sometimes stimulated in such a way that they are not specified in consciousness" — directly contradicting Müller's law.

Another serious anomaly arising from sensationalism is the discovery that changes occur in psychophysical correspondence, even without feedback or reinforcement.[12] Discrimination thresholds are actually laboratory artifacts. Even the two point limen (the 'minimum' separation capable of being felt as two distinct points) decreases with mere repetition — and many of these threshold changes transfer and generalize as well! Thus there are whole ranges of changes in psychophysical correspondence that occur without specific physiological transformations of the sensory nerves involved and without psychological mechanisms (such as reinforcement) as well. Why should repeating a double jabbing of the skin make it easier to separate stimulation on one's skin? Or, to take another related line of evidence, why should staring intently at one curved line 'cause' it to 'straighten out' and also lead to systematic changes in how curvature is seen?[13] The hypothesis that there are subjective sensory atoms caused by specific kinds of stimulation cannot be true. At best there are basic sensory differences, related to dimensional properties of stimulation. But even this hypothesis is problematic, especially when the variability and adaptability of so-called sensory thresholds is taken into account.

Wolfgang Köhler (1913) long ago articulated yet a further confusion in the theory of specific nerve energies. He argued that, if sensations are the consequences of specific neural activity, then perceptions could never be unified brain events. In perceiving my typewriter I am seeing, hearing, feeling and acting, creating activity in many areas of my nervous system. If my perception of the typewriter as an object is a combination of all these specific sensations then it is happening almost everywhere in my nervous system and body. If simple sensations are localized in the brain, then perceptions are not, and vice versa. Which localization shall we choose — if any?[14]

Sensationalism also requires what might be called a 'two environments' theory, the idea that there is a mental (or phenomenal) environment as well as a physical one. This two environments hypothesis has produced a considerable amount of research ever since writers like Mach (1913) began speculating about the shape of visual and tactile space. Do all the sensory spaces have a similar structure, are they at least homeomorphs, how do they relate to physical space? These are testable questions. Yet once one tries to explicate how phenomenal space is caused by physical objects, or to differentiate the phenomenal manifold into 'objective' and 'subjective' components, one runs into deep difficulties and contradictions, experimental and theoretical. The experimental contradiction is the most telling: to do experiments on one sensory space, it is necessary to isolate it. So there are experiments on feeling in which the body is held motionless, on seeing in which the eyes head and body are still. These are anti-naturalistic experiments and (as shown in a century of mystifying reuslts) they do not inform us about the processes whereby human beings come to know the world around them. Theoretically, it is impossible to prove the two environments hypothesis while holding a sensationalist theory that denies observers have perceptual access to external reality. If we are condemned to judge the external causes of our sensations then the world of physics and perception alike are mental constructs, constructs made for different purposes, but neither World is a place to inhabit. Sensationalism condemns us to a life of 'as if': it is always 'as if' there were a physical world or a perceived world or some connection between our various judged worlds, but our apprehension of these various worlds is equally indirect and unlikely.[15]

Self perception is another difficulty for Müllerian theorizing. The mental self is not a physical object that can cause sensations, so awareness of it must be non-perceptual.[16] The bodily self, being non-mental, can cause sensations (through the so-called proprioceptors of its muscles, joints and viscera) but these cannot be specific to desires and goals. That is, proprioceptions specify muscle contractions and motions of the skeleton, not mental states. Sensationalist observers must infer their own existence and purposes every bit as much as they infer the existence of other people and objects. No wonder sensationalist epistemologists are the only people who take solipsism seriously!

There are further anomalies deriving from Müller's law and scientific sensationalism, but these cannot be treated here as to do so would require close analysis of experimental practices. The most important of these anomalies are the predilection for treating observers as passive and non-purposeful, as mere objects receiving physical radiation. For example, in the field of visual perception we know a great deal about how observers react to photons, frequencies, flashes of light and shade, snapshot images of geometrical figures, and so on, but we know precious little about how people use vision to get around in everyday life. It is about time that epistemologists faced up to the critical fact that sensationalist hypotheses have created more confusion than understanding.

## ECOLOGICAL EPISTEMOLOGY

The questions raised by the ecological approach to epistemology are radically different from those motivated by traditional epistemology, even by naturalistic or evolutionary epistemology. Because of the great contrast in motivation, it is difficult to compare ecological epistemology with other approaches to the theory of knowledge. It is not at all clear, for example, what sorts of arguments are needed to persuade epistemologists to favor one approach over another. The relevant considerations are many and their implications widespread and often vague. For this reason I will focus here on the motivations and potential for raising new problems of an ecological epistemology.

Why do things appear as they do to evolved, historical, purposeful observers? Not why is the sky blue, nor why do certain frequencies of light look yellow sometimes, but why does the earth below us seem to afford support for locomotion, why does this look good to eat, and here a place to shelter, there an event of comforting ritual? Ecological epistemology starts from the assumption that it is not the physical world, but the uses or values of things that are perceived. In other words, ecological epistemology starts from the hypothesis that what there is to be perceived is a meaningful environment, an environment full of affordances for acting and apprehending, affordances that are perceived, not judged or inferred. As Gibson (1979, p. 1) put it: "How do we see where we are in the environment? How do we see whether or not we are moving and, if we are, where we are going? How do we see what things are good for. How do we see how to do things?" If we perceive the use-values of things, their affordances, then many of the traditional questions of epistemology must be revised or simple discarded.

Ecological epistemology is naturalistic in that it takes as its starting point the role played by perception, knowledge and action for animals situated in their environment.[17] The 'basic orienting system' is fundamental; there is, in Marjorie Grene's words, a 'primacy of orientation'. The fundamental epistemological ability of all animals is to maintain orientation to the more persistent features of their environment, while dealing with the changing events surrounding them. To locomote is not only to make those movements which take one from here to there, but to do so without a breakdown in those postures that orient one (in the terrestrial case) to the sky above and earth below, and horizon out there. One of the most profound effects of blocking orientational information (as is done in so-called sensory deprivation and also in various kinds of torture) is the breakdown of general psychological well-being and the destruction of coherent cognition. Observers in these situations become desperately confused, frequently hallucinate, lose track of time, place, bodily functions and emotions.

There are some similar if less intense forms of disorientation which affect our awareness and thinking. Sudden movements across continents, or between cultural and linguistic groups often produce physical and mental discomfort and even vertigo. The first contact between cultural groups is often very disorienting at 'higher' levels of cognition, as witness the European discovery of alien cultures in such books as *The Persian Letters*. Restrictions of autonomous movements and adjustments, as with imprisonment, have serious psychological consequences. Even social restrictions of autonomy, such as the degradation of modern jobs, or contemporary housework, have pronounced mental effects.[18] It is time we recognized just how true the image of the philosopher in an armchair really is: to be able to spend one's time just thinking about abstract issues, with little concern to the pressing orientational demands of daily life, is to be in a very special ecological situation. In addition to having one's physical needs taken care of, one has to live in a society where the division of labor has led to specialization in 'intellectual labor'.[19] Georgraphic, linguistic, and historical forces are all intertwined creating that to which philosophers orient. Standard epistemology treats ideas and belief formation as more or less a natural process that is impinged upon by social factors. This has been codified in the standard distinction in philosophy of science between 'internal' and 'external' factors in the life of an idea. This distinction separates the inseparable: to be a thinker is to be in a unique historical situation. To look for both 'internal' and 'external' causes of an idea's formation is to combine two factors that were not separated in the first place. The ecological concept of orientation as a background to all forms of thinking and acting offers a way of treating human ideas as a naturally social phenomenon.

However, orienting is only a background skill. To perceive, think or act, it is necessary to orient, but awareness is much more than this. One of the virtues of the ecological approach is that is places strong emphasis on analyzing what it is that we are aware of. The objects, events, places, and persons surrounding us are studied along with the processes by which we gain awareness of them. One of the primary tasks of ecological epistemology is to discover what there is to be perceived and thought about in our environment: its meaningful layout of objects, events, persistences and changes, and our human historical additions of languages, customs, forms of exchange and so on. The post-Cartesian division of labor of philosophers into epistemologists, ontologists and axiologists, is called into question by the ecological approach.

In somewhat technical terms, the ecological approach holds that the objects of perception (here meant in the broad sense of 'entities that can be perceived') are intentional objects. That is, they are the real objects of animate desires, awarenesses, actions and beliefs (see Reed, 1983 and Searle, 1983). Cartesian epistemology has always tried to reduce intentional objects to some aspect of the activity of awareness. This is the motive behind Müller's law, where the passive body receives some stimulation on which the mind acts to form a representation — a mental object. The mind is then said to be aware not of the external object but rather of that which it has caused, the mental object. This immediately solves the difficult problem of non-existent intentional objects (as when we desire something that does not exist, such as a cure for cancer) as the mental object/representation always exists on this theory. However, this reduction of intentional objects to mental objects — the doctrine that ideas are in the mind — saves non-existent objects only at the price of losing the entire external environment, whose existence is at best inferential on this theory.

The alternative to this forced choice between external and mental objects, non-existent objects and solipsism — the alternative to Descartes' uneasy truce between mentalism and physicalism — is to abandon sensationalism. If information exists specifying the affordances of the environment then one can consistently maintain that environmental objects are the objects of intentional states and that not all intentional objects exist. When information specifying something is picked up despite the absence of that entity (as in a *trompe l'oeil* painting), or when misinformation is picked up, observers become aware of something not actually present.[20] Otherwise, the pickup of information entails the awareness of the meaningful objects that gave rise to that specific information.

One of the most powerful arguments for an ecological epistemology is that it can explain the distinction between being visible and invisible, between going out of sight, and going out of existence. When an object is out of sight does it still exist? It depends. If it goes out of sight in such a way as to progressively occlude one or more of its surfaces, then it still exists; if not, then it has undergone at least a change of substance if not a change of existence. If one questions whether an object now visible in fact exists, progressive occulsion provides an excellent test of existence: Does the object in question occlude the background? Is this occlusion progressive as one moves toward the object, regressive as one moves away from it? Do other objects progressively reveal and conceal the object in question as they are interposed along the line of sight? If yes, the object exists and has a definite location or path of motion, if not it either does not exist or exists in an indefinite location (as with rainbows, whose 'position' is a function of the line of sight and line of light).[21] No alternative epistemology has anywhere near as practicable a means for distinguishing existent from non-existent objects.

Some of the most important intentional objects of the environment are other animate beings — other intentional agents. Mutual awareness, the awareness of others as aware of things and of what they are aware of, is a fundamental fact of social life. The manipulation of information so as to make others aware of something (via gestures, signs, languages) enriches and transforms the whole process of awareness. The traditional epistemological problem of other minds is a by-product of the insistence that meaningful objects must be judged, not perceived. Ecological epistemology replaces this entire problematic. Instead it asks what information is available that allows one to perceive another animal as an intentional agent? What is the basis for perceiving the specific goals and desires of others? What is the effect of communication by scans of transforming information on the whole process of mutual awareness?

In sum, ecological epistemology raises a host of new questions. Especially, it raises the ontological questions of what are the meaningful aspects of the environment that can be perceived? How do these meanings and values change with historical and social change? What is the information available that is the basis of specific awarenesses? How is this information used in perceiving? In learning to perceive? How is information and the values that it specifies transformed by social processes of mutual awareness and communication?

## CONCLUSION

By calling for a naturalistic epistemology I take it we are calling for the development of a theory of knowledge that has more to say about how sentient beings come to be aware of the surroundings they inhabit than about the rather abstract forms of judgment engaged in by those very few sentient beings known as scientists. It this be so we should give some serious consideration to abandoning the Cartesian strategy of starting with physics to account for the physicist's knowledge. In addition to yielding centuries of puzzles this strategy has caused us to focus on the unusual case of an organism separated from most forms of normal purposeful activity, the use and enjoyment of the affordances of things. No doubt we shall have to explain how nature begat such creatures, but I imagine a naturalistic epistemolology will start with more humble cases of habitation and awareness. If this is so, then the ecological approach to epistemology has much to recommend it for the task at hand.[22]

*Drexel University*

## NOTES

[1] For a sketch of this history see Reed (1982a and refs. therein).

[2] The major exceptions here are some of the phenomenologists and, to a lesser degree, Brentano. Husserl (1970) tries repeatedly to discover just what it is that philosophers have assumed that has created so much confusion. A small group of 'pragmatists' and 'neo-realists' in the United States around 1900 — especially E. B. Holt and H. Heath Bawden — also are important, if virtually unknown, exceptions to this generalization.

[3] My termininology thus differs form that of Agassi (1977) although my analysis owes much to that work.

[4] For an explication of the problems of induction see Popper (1972, Ch. 1). For a more detailed analysis of why evolution cannot rescue Cartesian epistemology, see Turvey *et al.* (1981).

[5] See Aquila (2977, pp. 6—7) for a good discussion of Descartes' 'material' versus 'formal' ideas.

[6] In our post-Darwinian era, most psychophysiologists would add that animals have evolved so that their sensory nerves are protected from extraneous stimuli. Thus ears and eyes are structured so as to channel light rays and sound waves respectively, blocking other sources of stimulation. The question is, nevertheless, how one could know the source of any form of stimulation if that source is not specified in the stimulation.

[7] Helmholtz's masterpiece was first published in 1866—67. A similar division of topics can be found in his *On the Sensations of Tone*.

[8] See the general critique of Gibson (1966, Chs. 1—3). See also Puccetti and Dykes (1978).

[9] See Burtt (1932, Ch. 8). Also: Husserl (9170, *passim*); Whitehead (1926, Chs. 1—4).

[10] See Puccetti and Dykes (1978). For some suggestions on how to understand the functional differentiation of the brain without appealing to the concept of specific nerve energies, see Reed (1982b and 1984).

[11] Those who are skeptical that this should be called 'seeing' should perform the experiments exhibited in Kanizsa (1979, Ch. 1).

[12] E. J. Gibson (1953) reviews the literature concerning this issue.

[13] Gibson (1937) reviews this and other related phenomena.

[14] See Reed (1984) for more on this.

[15] Vaihinger (1925) is an overlooked but briliant exposition of the impossibility of reconciling the two environments hypothesis by any means but 'as if' judgments.
[16] The controversy over whether 'sensations of innervation' (volition) existed was about precisely this. These hypothetical sensations were nothing more than the assumption that there 'had to be' a sense channel for motor commands in order to explain the awareness of volition. See James' (1890) chapter on the will for an overivew of the arguments.
[17] See Gibson (1966, Ch. 4) and Reed (1982b) for accounts of the basic orienting system.
[18] Terkel (1975) and Olsen (1978) are richly frightening analyses of the plight of modern daily life: lack of control over place, purpose, and general orientation in the human world leads many people to identify more with machinery than with other humans.
[19] See Bahro (1978, parts 1—2) for a thorough analysis of the division of 'intellectual' labor away from everyday life of ordinary people in modern society.
[20] The details of this theory of illusion require to be worked out, and Givner (1982) has begun to do so. See also Gibson (1966, Chapter 11).
[21] The experimental evidence for these assertions can be found in Reed and Jones (1982, Ch. 2.7). See also chapter 2.9 for more on misperception and occlusion.
[22] For a more historical analysis of these issues, see Reed (forthcoming).

## REFERENCES


Agassi, Joseph: 1977. *Towards a Rational Philophical Anthropology.* The Hague: Nijhoff.

Aquila, R.: 1977. *Intentionality: A Study in Mental Acts.* University Park: Pennsylvania State University Press.

Bahro, Rudolf: 1978. *The Alternative in Eastern Europe.* London: New Left Books.

Burtt, E. A.: 1932. *The Metaphysical Foundations of Modern Physical Science.* 2nd ed. Garden City: Doubleday.

Gibson, E. J.: 1953. 'Improvement in Perceptual Judgments as a Function of Controlled Practice or Training,' *Psychological Bulletin* **50**, 401—431.

Gibson, James J.: 1927. 'Adaptation With Negative After-Effect,' *Psychological Review* **44**, 222—244.

———: 1966 *The Senses Considered as Perceptual Systems.* Boston: Houghton-Mifflin.

———: 1979. *The Ecological Approach to Visual Perception.* Boston: Houghton-Mifflin.

Givner, D.: 1982. 'Direct Perception, Misperception, and Perceptual Systems in J. J. Gibson and the Problem of Illusion,' *Nature and System* **4**, 131—142.

Helmholtz, H. L. F. Von: 1962. *Treatise of Physiological Optics.* English translation of 3rd ed. (1909—12). New York: Dover.

Husserl, Edmund: 1970. *The Crisis of European Science and Transcendental Philosophy.* Evanston: Northwestern University Press.

James, William: 1890. *Principles of Psychology.* New York: Henry Holt & Co.

Kanizsa, Gaetano: 1979. *Organization in Perception.* New York: Prager.

Köhler, Wolfgang: 1913. 'On Unnoticed Sensations and Errors of Judgment.' Reprinted in Mary Henle (ed.), *Selected Papers of Wolfgang Köhler*. New York: Liveright, 1971, pp.

Mach, Ernst: 1913. *Space and Geometry*. New York: Open Court.

Olsen, Tillie: 1978. *Silences*. New York: Dell.

Popper, Karl R.: 1972. *Objective Knowledge*. London: Oxford University Press.

Puccetti, R., and R. Dykes: 1978. 'The Sensory Cortex and the Mind-Body Problem,' *Behavioral and Brain Sciences* **1** 337—376.

Reed, Edward S.: 1982a. 'Descartes' Corporeal Ideas Hypothesis and the Origin of Scientific Psychology,' *Review of Metaphysics* **35**, 731—752.

———: 1982b. 'An Outline of a Theory of Action Systems,' *Journal of Motor Behavior* **14**, 98—137.

———: 1983. 'Two Theories of the Intentionality of Perceiving,' *Synthese* **54**, 85—94.

———: 1984. 'From Action Gestalts to Direct Action.' In H. T. A . Whiting (ed.), *Bernstein Re-Assessed*. Amsterdam: North Holland.

———: 1986. 'Seeing Through History,' *Philosophy of Social Science*.

Reed, E. S., and R. K. Jones (eds.) 1982. *Reasons for Realism: Selected Essays of James J. Gibson*. Hillsdale: Erlbaum.

Searle, John: 1983. *Intentionality: A Study in the Philosophy of Mind*. London: Cambridge University Press.

Terkel, Studs: 1975. *Working*. New York: Avon.

Turvey, M. T., R. E. Shaw, E. S. Reed, and W. M. Mace: 1981. 'Ecological Laws of Perceiving and Acting,' *Cognition* **9**, 137—304.

Vaihinger, Hans: 1925. *The Philosophy of 'As If'*. London: R. & K. Paul.

Whitehead, A. N.: 1926. *Science and the Modern World*. New York: Basic Books.

ABNER SHIMONY

# COMMENT ON REED

Ecological epistemology has a strong affinity to phenomenology, particularly the version of Merleau-Ponty (as Reed, 1983, points out). There is a common emphasis upon the richness of experience, the irreducibility of perception to sensation, the importance of proprioception, and the inseparability of valuations from presentations. Ecological epistemology, however, is more dedicated than phenomenology to a controlled, experimental study of perception, and pays closer attention to the physics of the interplay between the perceiving subject and the environment. James Gibson's demonstration of the intricacy of this interplay constitutes a permanent contribution to experimental psychology, even if he was not as successful and as revolutionary in solving traditional epistemological problems as Reed has claimed.

No matter how rich a phenomenological description may be, it is not a surrogate for a fundamental theory of sensory and cognitive functioning — unless all theories are understood only instrumentally, as devices for organizing and predicting experience. Since Reed does not seem to espouse such an instrumental view of theories, he has as great a need as any heir of Descartes for an ontology which accommodates both the physical aspects of the percipient subject and the subject's consciousness. The evidence is overwhelming that sense organs and a nervous system are necessary for perception, and much is known about the way they work physically: the build-up of electrical potential in neurons, the transport of ions across neural membranes, the excitation of photosensitive molecules in the retina, the existence of specialized receptors for specialized enzymes in the brain, etc. Our ignorance of a bridge — by causality or by duality of aspects or otherwise — between this physical functioning and the perceiver's experience is certainly not diminished by acknowledging the immense richness of the latter, and not even by insisting that this experience is characterized by intentionality. Even if intentionality has primacy in the order of knowing its status in the order of being is mysterious, and it is in need of theoretical explanation.

Among the philosophers whom Reed cites as pointing out "how classical physics and the psychology to which it led are inconsistent" is



*Abner Shimony and Debra Nails (eds.), Naturalistic Epistemology*, 230—234.

Whitehead (Note 9), who deplored the 'bifurcation of nature'. Like the phenomenologists, Whitehead stresses the epistemological importance of openminded, uncensored empiricism (see esp. 1933, Ch. 15). Moreover, he explores in some detail "perception in the mode of causal efficacy" (1929, *passim*), which is clearly cognate to the intentionality of the phenomenologists. But Whitehead's philosophical method consists not only of an appeal to direct experience but also of an application of the hypothetico-deductive method, e.g.,

> The true method of discovery is like the flight of an aeroplane. It starts from the ground of particular observation; it makes a flight in the thin air of imaginative generalization; and it lands again for renewed observation rendered acute by rational interpretation. (1929, p. 7)

A central one of his hypotheses is that the ultimate, concrete entities of the universe are 'occasions' of experience, of which human experience provides extremely special cases. The characteristics which Whitehead attributes to all occasions are suggested by human phenomenology, but the immense extrapolation which he proposes could be warranted only by the explanatory power of his theory. It is indeed impressive how much is explained, at least in the sense of 'falling into place': for example, the existence of physical laws, the geometry of space-time, the occurrence of both discreteness and continuity in physical phenomena, wave-particle duality, particle creation and annihilation, the possibility of composite bodies with hierarchical structure, the possibility of the evolution of organisms endowed with high level mentality from systems of primitive occasions, the role of the nervous system in high level mentality, the interplay of consciousness and unconscious mentality, the specious present, short and long term memory, proprioception, and the feeling of the immediate presence of an external world. The rigor of Whitehead's explanations leaves much to be desired, unfortunately, for the meanings of some of his basic terms ('feeling', 'subjective form', 'satisfaction', 'prehension', etc.) are quite obscure (see Rorty, 1963), and the predictions of his theory are imprecise and only partially confirmed (Shimony, 1965). Nevertheless, Whitehead deserves credit for his understanding of the depth of the problems with which he grapples, for his sustained effort to work out a metaphysics and an epistemology which mesh with each other, and for his daring in proposing a radical unification of physics and psychology.

Finally, a few remarks about Whitehead's treatment of perception in the mode of causal efficacy are relevant to Reed's theses.

(1) The non-anthropocentric character of Whitehead's philosophy is evident in the way he subsumes perception in the mode of causal efficacy under his general categories:

> . . . Perception in its primary form is consciousness of the causal efficacy of the external world by reason of which the percipient is a concrescence from a definitely constituted datum. The vector character of the datum is this causal efficacy. (1929, p. 184)

Presupposed in this passage are the fundamental Whiteheadian hypotheses that a new occasion derives 'data' for its experience from the occasions of the past, that those occasions have 'objective immortality' by entering into new occasions, and that the 'prehension' of the past occasions by the new one is the essence of efficient causation. Whitehead proposes that these hypotheses have universal validity. Only consciousness sets off perception in the mode of causal efficacy as a special case, since in Whitehead's view only high-level experience rises to the level of consciousness.

(2) Even though Whitehead holds that perception in the mode of causal efficacy provides a direct warranty of some propositions about the external world, he stops far short of Direct Realism and, I think, of Reed's position. The activity and autonomy of the concrescent occasion have the consequence that its experience is a complex mixture of appearance and reality, with appearance entering when prehensions of past occasions are played up, played down, abstracted, transmuted, and synthesized, and when they are infused with conceptual prehensions. The chapter 'Symbolic Reference' in *Process and Reality* (which was important enough for Whitehead to have published it separately, with variations, as *Symbolism: Its Meaning and Effect*, 1928) insists upon a large element of contingency in the connection between perception in the mode of presentational immediacy (the presentation of sense data) and perception in the mode of causal efficacy, and in many respects his doctrine resembles Critical Realism and Helmholtz's theory of unconscious inference. Whitehead would probably be sympathetic to Reed's statement

> ecological epistemology starts from the hypothesis that what there is to be perceived is a meaningful environment, an environment full of affordances for acting and apprehending, affordances that are perceived, not judged or inferred.

But he would surely have important reservations, as indicated by the statement

> What is directly perceived, certainly and without shadow of doubt, is a grey region of the presented locus. Any further interpretation, instinctive or by intellectual judgment, must be put down to symbolic reference. (1929, p. 261)

His reservations should not be attributed to epistemological conservatism, for they are implied by his subsumption of the perceptual process to his general metaphysical principles.

(3) Whitehead's Philosophy of Organism need not be treated as an indissoluble package. I do not see in principle why his thesis that the fundamental actual entities are occasions of experience cannot be separated from the doctrine of prehension. The latter might be replaced by a postulate of the dynamical evolution (preferably stochastic rather than deterministic, in view of the physical evidence) of the state of the world. With this replacement, the similarity of a concrescent occasion to earlier occasions, the retention of a pattern or a type of feeling tone, would be consequences of the fundamental dynamical law, rather than manifestations of 'objective immortality' of a past occasion in its successors. Such a variant of Whitehead's philosophy would be even farther removed from phenomenology and Reed's theory than is his own formulation. There would, however, be reason to prefer this variant if it should turn out to have greater explanatory power than Whitehead's own, and then perception in the mode of causal efficacy (and hence intentionality) would have to be interpreted as largely an inferential process. This is not an issue that ought to be prejudged. We have abundant evidence that other apparently direct aspects of our perception are derivative, inferential, and subjective (see, for example, Heffner's paper in this volume). It is often necessary to assess indirectly the legitimacy of a claim to direct knowledge.

*Boston University*

## REFERENCES

Reed, Edward S.: 1983. Two Theories of the Intentionality of Perceiving, *Synthese* **54**, 85–94.

Rorty, Richard M.: 1963. 'The Subjectivist Principle and the Linguistic Turn.' In George L. Kline (ed.), *Alfred North Whitehead: Essays on His Philosophy*. Englewood Cliffs, N. J.: Prentice-Hall.

Shimony, Abner: 1965. 'Quantum Physics and the Philosophy of Whitehead,' In M. Black (ed.), *Philosophy in America*. London: Allen and Unwin. (Reprinted in R. S. Cohen and M. W. Wartofsky (eds.), *Boston Studies in the Philosophy of Science*, vol. 2. New York: Humanities Press, 1965.)

Whitehead, A. N.: 1928. *Symbolism: Its Meaning and Effect*. London: Macmillan.

———: 1929. *Process and Reality*. London: Macmillan.

———: 1933. *Adventures of Ideas*. New York: Macmillan.

LOUISE M. ANTONY

# NATURALIZED EPISTEMOLOGY AND THE STUDY OF LANGUAGE

Quine (1969) in 'Epistemology Naturalized', attacks an entire conception of how philosophy ought to approach the topic of human knowledge. This conception, which we could call *external* epistemology, relies on a number of distinctions which are unsupportable by Quine's lights: *a priori* vs. *a posteriori* knowledge, necessary vs. contingent truth, matters of meaning vs. matters of fact. Lying at the basis of these distinctions are the assumptions that there are such things as linguistic rules, and that these rules have a crucial role to play in a philosophical account of knowledge. Quine's attack on external epistemology is ultimately an attack on these fundamental assumptions.

Quine wants to show that the notion of 'rules of language' presupposed by the analytic/synthetic distinction has no role to play in a scientifically respectable epistemology, or, to put it another way, that any epistemological project which requires appeal to such a notion is illegitimate. The internal or 'naturalized' epistemology Quine advocates makes no essential use of any linguistic notions, syntactic or semantic. Quine believes that both the syntactic and semantic structures of language are indeterminate, and thus not proper objects of scientific investigation. In accounts of human knowledge, language effectively drops out of Quine's explanatory scheme.

I think it's undeniable that much of Quine's criticism of external epistemology is correct. What I want to consider in this paper is whether Quine is correct about the relation between epistemology and language — whether a naturalistic approach to knowledge really can or must make do without appeal to rules of language. My view is that such an approach to knowledge is not only compatible with the view that human language use has determinate, objective structure, but that a naturalized epistemology requires that the empirical study of linguistic structure be taken seriously. I will argue that Quine, by his own lights, is unjustified in his view of linguistic theory, and that recent developments in this field not only re-instate linguistic rules as important elements in an account of human knowledge, but bear in interesting ways on the original epistemological project of logical positivism.



*Abner Shimony and Debra Nails (eds.), Naturalistic Epistemology,* 235—257.

The first step in understanding the relation between language and epistemology is to understand how the analytic-synthetic distinction functioned for Quine's predecessors.[1] The notion of a 'rule of language' only made its way into positivistic philosophy via epistemological issues. The analytic-synthetic distinction was appealed to initially to account for the peculiar epistemic features of the 'synthetic *a priori*': the propositions of mathematics, and the most general, quasi-logical principles of science. The difficulty was how to explain, within an empiricist framework, how propositions so apparently substantive and manifestly useful could be known, as they seemed to be, independently of experience. It was not just that people seemed to accept such propositions in advance of experience. The problem was more in seeing how experience even could serve as a basis for belief in these propositions: how could any amount of agreeable data warrant the extremely high degree of confidence invested in claims so wide in scope?

The positivists attempted to solve this problem by distinguishing the *grounding* of the problematic propositions from that of legitimate empirical claims. The latter, it was claimed, reported the actual content of experience; indeed, the meaning of such claims could be identified with the set of experiences one would expect to have if they were true. The former, however, precisely because their truth was compatible with every imaginable experience, had to be true in virtue of something else. The something else, it was decided, was the structure of the language used to make empirical reports. Kant's synthetic *a priori* propositions turned out to be merely analytic.

In order for this move to work, though, the positivists needed to make some very strong assumptions about the nature of language, assumptions that went beyond the already controversial identification of meaning with empirical consequences. Since the whole problem was explaining how there could be *a priori* knowledge at all, the classification of statements according to their grounds was only the first step. To complete the job, it had to be assumed that our *access* to the grounds of analytic propositions was fundamentally different from our access to the grounds of synthetic truths — in short, that our knowledge of *language* was non-empirical.

There is, on the surface, no difficulty about such an assumption. The positivists observed that language was an artifact of human cognitive activity, and inferred from this fact: (1) that the structure of language was immediately available by introspection to anyone who successfully used the language, and (2) that linguistic facts were not really objective, since they simply reflected a formal structure imposed on the empirical world by human beings. Language, then, was both transparent and conventional.

In all this, the positivists were of course greatly influenced by the spectacular development of formal logic. Drawing on treatments of syntax and semantics in formal languages, Carnap produced the apparatus for rendering a precise account of the *a priori* (now the analytic) in terms of linguistic rules. In formal languages, there is a sharp division between the language in which things are 'said' — the *object-language* ('*L*') — and the language used to talk about the object-language — the *meta-language* ('*M*'). Both the syntax and the semantics of *L* can be laid out explicitly in *M* — the syntax, through a set of formation and transformation rules for sentences in *L*, and the semantics through a schematic specification of their 'truth-conditions'. This division between *L* and *M* permits a precise delineation of the class of sentences of *L* which serve as the formal analogues of analytic truths in natural language, namely, the set of *valid* sentences of *L*. It's worth looking closely at the way valid sentences are picked out in formal languages, because it's in the formal case that one can get the clearest sense of what it might mean to say that a sentence is 'true in virtue of language'.

Validity has both a syntactic and a semantic face: both the syntactic and semantic rules serve to pick out a privileged class of sentences which are called valid. The syntactic rules pick out this class either by explicit list, or by the specification of procedures which generate valid sentences, or by both means. The formal semantics introduces the notions of 'interpretation' and 'truth under an interpretation'. A valid sentence is then defined to be one which is true under all interpretations.

Syntactic validity on its own would be of no use in explicating analyticity in natural language. We could certainly regard syntactically valid sentences as both true and true in virtue of linguistic rules, but the notion of 'truth' in use here would be purely formal. Syntactic validity becomes more illuminating when it is wedded to semantic validity.

A formal semantic theory can be thought of as having two components: a part which stipulates the semantic values for the terms of *L*, and a second part which specifies how the truth-value of an entire sentence is to be determined, given the semantic values assigned to the sentence's parts. These two components together yield a set of conditional evaluations of the sentences of *L*, of this form: 'if the components of sentence *S* are assigned values such-and-such, then *S* will have this-or-that truth-value'. It will turn out that truth for some sentences will be manifestly contingent upon the assignments made to their components: different assignments will yield different truth-values. But the truth of valid sentences will not depend upon the details of interpretation — valid sentences are those which come out true under *any* assignment of values to their component parts. What, then, does determine a valid sentence to be true? If not the particulars of interpretation, then it must be the general rules of the semantic meta-theory. The valid sentences, it seems clear, are true *because of* the way the rules work — they are true *in virtue of* those rules.

The application to natural language seems obvious: if we take 'possible experiences' as the analogue of 'possible interpretation', then synthetic statements are those whose truth-values depend upon experience, and analytic statements are those whose truth (or falsity) is guaranteed by the very rules which relate statements to experience.

There is a second way in which semantic validity seems to afford insight about analyticity. The semantic meta-theory does not make assignments to every vocabulary item in *L* — only to (what is tendentiously called) the non-logical vocabulary. The so-called logical constants are not interpreted directly, as are, for example, the predicates. Rather, they receive their interpretation indirectly through that part of the semantics which stipulates how a sentence is determined to be true or false in accordance with its composition. Thus it could be said that the meanings of the logical vocabulary are determined by the rules set down in the meta-language. Since the truth of valid sentences does not depend upon the meanings of their *non*-logical components, it might be held that their truth depends upon the meanings of their logical components, and thus, once more, on the rules of the language.

Either way, the notion of a sentence's being true in virtue of language seems quite precise and intelligible in the case of formal languages. But this is only because of the availability of explicit rules governing the behavior of items in the object language, and this availability is due to the wholly conventional and artificial nature of the language. The notion of a 'truth of language' therefore becomes highly problematic when transferred to a natural language, for despite the seductive parallels between validity and analyticity, it is the *dissimilarities* between formal and natural languages that are telling. Natural language does not arise through deliberate acts of legislation, nor does it come neatly divided into object- and meta-language. If natural languages were our artifacts in the same sense that formal languages are, if we did consciously set out syntactic rules and stipulate meanings, then it would be no puzzle whence our commitments to analytic truths arose — we engendered them when we constructed the language — nor why we clung to them with such tenacity — to abandon them would be to discard the very language we'd created. But as it is, natural languages lack the crucial aspects of artifice, and so the positivists' account of analyticity lay fully open to the coming challenges from Quine and others. But more on this below.

While Carnap and his followers certainly realized that natural languages differed from formal languages, they believed that the *implicit* structure of natural language (and theories expressed in them) could be formally represented. And, not only would the possibility of such formalizations underwrite the positivists' account of *a priori* knowledge, it could reform and clarify the practice of philosophy itself. Philosophy would cease its production of meaningless metaphysical systems and transform itself into a purely analytical discipline. The goal of philosophy, Carnap argued, should become the production of a 'rational reconstruction' of science — a formalization of scientific theory that would lay bare both its structure and its empirical foundations.

A rational reconstruction of a scientific theory would include an explicit meta-linguistic specification of the logical and semantic features of that theory: rules of deductive and inductive logic, and semantic 'bridge laws' that would reduce (at least partially) highly theoretical statements to statements about actual or possible observations. It is extremely significant, in light of what was to come, that Carnap viewed procedures of justification as part of the formal structure of theoretical language, even as part of the semantics of the language. Since a sentence's meaning was identified with the set of experiences which would confirm it, the bridge laws, or reduction sentences, served both to define theoretical vocabulary, and to specify the relationship between theory and evidence.[2]

By distinguishing clearly formal principles from substantive, empirical claims, a rational reconstruction also served to differentiate the kinds of epistemic commitments a rational being could have. On the one hand, there were the quasi-voluntary commitments engendered by one's 'decision' to adopt a particular language with a particular set of linguistic rules. (Notice that these rules include inductive rules.) On the other hand, however, were those commitments entailed by the first: once one had 'decided' to embrace a language, one was bound to reason in certain ways, and to accept certain sentences as true in certain circumstances.

Demonstrating the existence of this dichotomy between kinds of epistemic commitment was necessary, not only to account for the *a priori*, but to explain and legitimate Carnap's conception of philosophy as a *normative and non-empirical* enterprise. In fact, the two problems were the same: accounting for the utility and acceptability of *a priori* principles was precisely what was needed to preserve a role for philosophy within the positivists' framework. Philosophy would hereafter be *analysis* — the discipline which would separate empirical wheat from metaphysical chaff. In producing rational reconstructions, philosophy would not only clarify the empirical commitments of given scientific theories, it would reveal the emptiness of the pseudo-sciences:

> In our 'Vienna Circle' . . . the conviction has grown, and is steadily increasing, that metaphysics can make no claim to possessing a scientific character. That part of the work of philosophers which may be held to be scientific in its nature — excluding the empirical questions which can be referred to empirical science — consists of logical analysis. The aim of logical syntax is to provide a system of concepts, a language, by the help of which the results of logical analysis will be exactly formulable. *Philosophy is to be replaced by the logic of science* . . . . (Carnap, 1937, p. xiii)

This passage reveals how deeply important it was to Carnap that philosophy play a *normative* role in the investigation of human cognition. This envisaged role had two facets: (1) to ensure that scientific theorizing had a secure, empirical basis (through the analysis of the concepts of science), and (2) to legitimate the procedures according to which science developed from that basis (by displaying the formal structure of scientific language).

The latter goal underwent some evolution. For a short while it had been possible to believe that there was a single, universal logic underlying all theoretical systems; if this had been the case, then science could have been legitimated absolutely. But with the recognition of alternative logics, it had to be admitted that science could at most be shown to be *self*-legitimating, that is, able to achieve its self-ascribed objectives within self-imposed constraints.

Still, Carnap believed that even if science's legitimation had to be relativized to choice of 'language', invidious comparisons between science and pseudo-science could be supported. Though it could not be shown that one was rationally obliged to adopt one language rather than another, it was still the case that a language, once adopted, carried with it a set of epistemological norms. Metaphysical and pseudo-scientific systems could still be criticized for not being rationally reconstructible — for not having a coherent set of epistemological norms, or for failing to conform to them.

All this highlights, once again, the fundamental importance of the analytic-synthetic distinction for positivistic epistemology. Science could be rationally legitimated if and only if normative analysis, i.e., philosophy, was possible. And both philosophy and the legitimation of science depended upon there being a sharp distinction between matters of meaning and matters of fact. If philosophy was to characterize and somehow validate the norms of science, then philosophy itself had to be *external* to science.

The need for an absolute division between an internal perspective and an external perspective — a standpoint from which science is done, and one from which the doing of science is observed — was in fact the basis of the doctrine of analyticity. The idea that genuinely empirical matters could be distinguished from mere semantic matters was elaborated into the program of dividing issues, through rational reconstruction, into internal matters, resolvable by appeal to experience in accordance with the rules of the language, and external matters, resolvable only by the 'pragmatic' considerations that govern choice of language. Even if they were not separable in practice, 'material-mode' questions had to be *conceptually* distinct from questions in the 'formal mode' (Carnap, 1937, p. 331). But the program of rational reconstruction itself required the possibility of an external standpoint — one free of empirical commitments and epistemically neutral. To deny the possibility of such a standpoint would be to deny not simply the distinction between semantic and empirical matters, but the distinction between rational and non-rational bases of choice, and thus to undermine logical positivism itself. As Peter Hylton explains:

> Any attempt to ask, in a normative spirit, about the justification of a scientific theory as a whole demands that we have some standard of justification which is independent of that theory. For Carnap it is the choice of a language which fixes this standard. If this choice is made on the same basis as that of a statement within science, then we have no

> standpoint from which we can ask the normative question. From Carnap's point of view, therefore, the analytic-synthetic distinction is essential to the kind of account of science which philosophy aims to provide, and so essential to philosophy itself. (Hylton, 1982, pp. 9—10)

The doctrine of analyticity was thus not a linguistic doctrine at all. I've already noted the ways in which the differences between formal and natural languages threatened the strategy of treating *a priori* truths as truths grounded by linguistic rules; once we appreciate the necessity of the analytic-synthetic distinction for Carnap's epistemology, we can see why it was impossible for him to give it up, regardless of the difficulties in applying it to natural language. The positivists did not approach the issues of meaning and syntax from an antecedent interest in language — their theories of language were really epistemological theories. Neither the verificationist criterion of meaning, nor the notion of linguistic rules, developed under the pressure of otherwise inexplicable *linguistic* data. Rules of language were posited as a kind of inference to the best explanation — but of epistemological 'facts' rather than linguistic ones.

All this explains the nature of Quine's attack on the doctrine of analyticity in particular and on philosophical appeals to language in general. His skepticism about the empirical legitimacy and explanatory utility of the notion of meaning is precisely his skepticism about the division between internal and external standpoints. His complaint against the analytic-synthetic distinction in 'Two Dogmas of Empiricism' (1953) is not merely that its explication is circular (as has been widely presumed).[3] Rather, his charge is that the doctrine is, for epistemological purposes, completely vacuous. He points out what's already been remarked on, namely, the fact that the features of *formal* languages that make comprehensible the notion of language's grounding truth are features utterly absent in natural language. As Quine (1976a, for example) makes clear, natural language does not naturally support a neat division into object- and meta-language, so that the standpoint from which one might distinguish truths *of* language from rules *governing* language is not available. Appeals to conventions or rules of language cannot, therefore, explain the origin of *a priori* knowledge.

Nor can such appeals explain the tenacity of certain beliefs. The positivists had wanted to say that belief in analytic propositions is justified by different kinds of norms from those which justify empirical belief. The norms which justify analytic claims are those bound up with the choice of one's language, so that one's commitment to those claims is bound to be extremely stable and impervious to *empirical* disconfirmation.

Quine cuts very deep, therefore, in insisting that the kinds of factors that might prompt rejection of, say, a 'logical law', are of exactly the same kind as the factors that affect belief in the case of 'synthetic' sentences. According to Quine, it is not a strict difference in epistemological norms that distinguishes belief in so-called analytic truths from belief in synthetic truths, but rather only habit, custom, or biologically-based predilection, and even so the differences are matters of degree only. (Ironically, in acknowledging, as he had to, that pragmatic concerns could enter into one's decisions about empirical matters, Carnap actually set the stage for Quine's critique.)

Quine agrees with Carnap that there can be no genuine account of science, in the sense of a rational legitimation of scientific practice, unless science can be viewed from the outside, from a standpoint where norms and rules can be distinguished from mere empirical beliefs. But since he thinks no such distinction is possible, he rejects the viability of an internal-external dichotomy, and thereby, the idea of a normative epistemology. He urges us to give up the attempt to say *why* science works, and to content ourselves with describing *how* it works: "We may not be able to explain why we arrive at theories which make successful predictions, but we do arrive at such theories" (Quine, 1969, p. 79). We should recognize, *contra* Carnap, that there is no distinction between the rational and the pragmatic, the decisions *governed by* the rules of a language and the decision *to be governed by* the rules of a language.

For Quine, therefore, any philosophical appeal to language is bound up with the discredited epistemological project of attempting to 'validate' science. But is there no empirically legitimate approach to language? Quine thinks not. He considers that his attack on the analytic-synthetic distinction constitutes an attack on the very existence of a domain of linguistic fact — an attack on meaning itself, not just some particular way of conceiving it.

Quine has two strategies for showing the non-objectivity of linguistic 'fact'. The first is his defense of semantic holism, which he takes to follow directly from his attack on verificationism. The second is the complex and controversial defense of the indeterminacy thesis. Neither strategy, I contend, is successful. The first fails precisely because of its close connection with the attack on positivistic epistemology, and the second fails, ironically, because it offends against the *naturalized* epistemology Quine offers as a replacement. I will discuss each strategy in turn.

Quine's defense of semantic holism begins with his critique of analyticity in "Two Dogmas ..." (1953). Having rejected various attempts to provide a direct account of the analytic-synthetic distinction, he allows that the verificationist theory of meaning offers a last hope, since its general criterion of empirical significance yields an operational test of analyticity: an analytic statement is one compatible with all possible experiences (Quine, 1953, p. 37). Thus, the two 'dogmas' — analyticity and verificationist reductionism — are "at root identical" (p. 41), and stand or fall together.

Quine proceeds to argue against reductionism with his familiar arguments for *confirmational* holism (also known as the Quine-Duhem thesis): sentences do not have empirical consequences on their own, but only in conjunction with the sentences of entire theories; any sentence can be held immune from revision provided one is willing to suitably distort the rest of one's theory; and so on. In short, because no sentence has a unique range of confirming experiences all its own, the verificationist criterion of analyticity is useless.

Confirmational holism leads to *semantic* holism by an interesting route. Notice that the doctrine of reductionism is detachable from the verificationist criterion of significance itself — that is, one could grant, as required by confirmational holism, that sentences have no identifiable empirical import of their own, and still maintain that meaning is somehow to be understood in terms of empirical import. This is precisely Quine's position — he rejects the need for individual sentence meanings, but not the general criterion of empirical significance proposed by the positivists. Thus, confirmational holism, taken together with essentially positivistic assumptions about meaningfulness, leads to *semantic* holism — the doctrine that the smallest significant unit of language is the language itself:

> The Vienna Circle espoused a verification theory of meaning but did not take it seriously enough. If we recognize with Peirce that the meaning of a sentence turns purely on what would count as evidence for its truth, and if we recognize with Duhem that theoretical sentences have their evidence not as single sentences but only as larger blocks of theory, then the indeterminacy of translation of theoretical sentences is the natural conclusion. And most sentences, apart from observation sentences, are theoretical. This conclusion, conversely, once it is embraced, seals the fate of any general notion of propositional meaning. (Quine, 1969, pp. 80—81)

Semantic holism, though it embodies a version of the verificationist criterion of meaningfulness, still constitutes a radical attack on the notion of meaning. Once one disposes of sentence-meanings as isolable factors, talk of meaning within an empiricist framework is obviated; one might just as well speak of a theory's being empirically confirmable as being empirically meaningful. Semantic holism is thus not a substantive account of meaning, but rather a form of eliminationism.

A genuine theory of meaning would provide, among other things, the basis for addressing questions about semantic knowledge — e.g., what does a speaker know in knowing the meaning of a sentence? The verificationist theory did offer answers to such questions, even though the answers were not empirically motivated. Quine's semantic holism, on the other hand, rejects all such questions. Since meaning is identified with empirical import, and since to know what confirms a single sentence is just to know what confirms an entire theory, semantic knowledge collapses into empirical knowledge.

Quine thus believes that in attacking the distinction between internal and external standpoints, he is ultimately removing the grounds for any *possible* separation of matters of language from matters of fact, and thus eliminating the potential subject matter of a theory of language. But this pessimism is quite unwarranted. All along, Quine simply identifies linguistics and semantics with positivistic epistemology. He presumes that the aim of any theory of linguistic meaning or structure could only be the provision of an empirical grounding for the analytic-synthetic distinction. Once the issue is framed in these terms, the outcome — that there is no real linguistic domain — is a foregone conclusion.

The first move in any attempt to operationalize the thesis of analyticity must be to separate sentences which command prompt and universal assent from those which will be accepted only in certain circumstances. But then, Quine hastens to point out, one must distinguish analytic truths from merely obvious ones. And to accomplish *this* separation, one needs to invoke precisely what the whole operation is supposed to yield — a way of separating semantic fact from widely-shared empirical belief.

> My rejection of the analyticity notion just means drawing no line between what goes into the mere understanding of a language and what else the community sees eye-to-eye on. I doubt that an objective distinction can be made between meaning and such collateral information as is community — wide. (Quine, 1969, p. 86)

Quine grants the existence of 'clear cases' which give the doctrine of analyticity an initial plausibility, but sees no reason to think that these can be made the basis of an empirical taxonomy of truths:

> One quickly identifies certain seemingly transparent cases of synonymy, such as 'bachelor' and 'man not married', and senses the triviality of associated sentences such as 'No bachelor is married.' Conceivably, the mechanism of such recognition, when better understood, might be made the basis of a definition of synonymy and analyticity in terms of linguistic behavior. On the other hand, such an approach might make sense only of something like degrees of synonymy and analyticity. I see no reason to expect that the full-width analyticity which Carnap and others make such heavy demands upon can be fitted to such a foundation in even an approximate way. (Quine, 1976b, p. 129)

I believe Quine is right in thinking that there is no point in trying to construct an empirical shoring-up of a notion that itself has no empirical origins. But Quine's crucial assumption, and the one clearly open to challenge, is that the study of meaning is inevitably motivated by, and bound up with, a concern for the *a priori*.

As I noted above, the positivists' views on language were distorted by the requirements of their epistemology. Specifically, the doctrine of analyticity could not do its assigned epistemological work unless it was assumed that the workings of natural language are *transparent* to its users. But the assumption that the semantic and syntactic structure of language was directly introspectible made the idea of an *empirical* theory of linguistic structure otiose. This, perhaps more than anything else in the positivists' program, connects meaning with *a priorism*, suggesting that the investigation of linguistic structure will inevitably be an exercise in speculative psychology. Quine follows this suggestion where it leads, and concludes that the desired output of any such investigation must be a specification of beliefs distinguished with respect to origins and status from empirical beliefs.

But a genuinely empirical approach to language would be constrained neither by the supposition that there are *a priori* beliefs, nor by the presumption that if there are such beliefs that it falls to the theory of language to account for them. An empirical theory of language is above all concerned with explaining *linguistic* phenomena. This is not to say that a linguistic theory will have no implications for matters outside its immediate domain, but only that such a theory develops primarily in response to data from within its domain.

Here we reach the crux of the matter. While it may be arguable whether or not there are linguistic facts, no one, including Quine, will deny that there are linguistic phenomena, standing in need of explanation. The issue that separates Quine from contemporary defenders of linguistic theory is whether these phenomena can be or need to be given specifically linguistic explanations. This issue (as well as the distinct issue of whether such explanations will make appeal to anything like linguistic rules) is surely an empirical matter. And yet, Quine's 'in-principle' case against a theory of language — the set of arguments for the indeterminacy of translation — is by no means empirical. Which brings me to my second point: the indeterminacy thesis is in serious conflict with the notion of a *naturalized epistemology*.[4]

How, precisely, one ought to interpret the arguments for indeterminacy is a matter which has generated great debate. I need assume only the following, which I trust is not too controversial: in order to demonstrate that translation is genuinely *indeterminate*, Quine must show that the underdetermination that afflicts theories of meaning is qualitatively different from the general underdetermination of theory by data. My claim (which I will be unable to defend adequately here),[5] is that all of Quine's attempts to demonstrate this involve assumptions about the status of the physical sciences which are insupportable from the point of view of naturalized epistemology, taken together with facts about language and cognition.

Quine has basically two responses to the charge that lingusitics and the other behavioral sciences are no worse off than physics: he points either to the 'analytical hypotheses' necessary for the completion of any translation, or to an alleged 'extra level' of underdetermination affecting linguistics. But neither response is successful.

One supposes, reading what Quine has to say about the analytical hypotheses, that nothing like them can be found in the physical sciences. He suggests that translation is non-objective because one must appeal to principles *other than* observational adequacy to decide among competing translations.

> . . . one has only to reflect on the nature of possible data and methods to appreciate the indeterminacy. Sentences translatable outright, translatable by independent evidence of stimulatory occasions, are sparse and must woefully under-determine the analytical hypotheses on which the translation of all further sentences depends. (Quine, 1960, p. 72)

Besides being extra-empirical in the sense of going beyond observational adequacy, the analytical hypotheses are subjective, even conventional, and recommend themselves largely on pragmatic grounds.

But it is Quine who taught us that *any* principle that contributes to the progress of empirical enquiry ought for that reason to be counted as empirical. And it is Quine who called our attention to the role of similarly extra-empirical principles of theory selection in the physical sciences. (In *The Web of Belief* (Quine and Ullian, 1978, p. 66), such principles are honorifically called the 'virtues' of a hypothesis.) Most tellingly, Quine's own characterization of principles like simplicity and modesty reveal that they possess the same features that seemed to de-legitimate the analytical hypotheses (subjectivity, conventionality, pragmatic grounding, etc.).[6]

In fact, there is nothing about the analytical hypotheses that distinguishes them from analogous tie-breaking principles in physics, and thus nothing that distinguishes linguistics *methodologically* from physics. The real reason why a translation constructed with the aid of analytical hypotheses is non-objective is, it turns out, that the *domain* of translation theory is non-objective.

> . . . the analytical hypotheses, and the grand synthetic one they add up to, are only in an incomplete sense hypotheses . . . . The point is not that we cannot be sure whether the analytical hypothesis is right, but that there is not even, as there was in the case of [occasion sentences], an objective matter to be right or wrong about. (Quine, 1960, p. 73)

There is, in short, no argument for indeterminacy based on the character of the analytical hypotheses. The methodological irregularity of linguistics must follow from its non-objectivity, and not the other way around.

The second argument, that linguistics is subject to an additional level of underdetermination, is even more clearly question-begging. In response to Chomsky's charge that the underdetermination of linguistic theory is no different from the underdetermination of physical theory, Quine argues:

> Though linguistics is of course a part of the theory of nature, the indeterminacy of translation is not just inherited as a special case of the underdetermination of our theory of nature. It is parallel but additional. Thus, adopt for now my fully realistic attitude toward electrons and muons and curved space-time, thus falling in with the current theory of the world despite knowing that it is in principle methodologically under-determined. Consider, from this realistic point of view, the totality of truths of

> nature, known and unknown, observable and unobservable, past and future. The point about indeterminacy of translation is that it withstands even all this truth, the whole truth about nature. (Quine, 1975, p. 303)

Quine is arguing that linguistics is indeterminate — non-objective — because it is underdetermined *with respect to physics.* The crucial premise, revealed here in phrases like "the totality of truths of nature", and "the whole truth about nature", is that reality is circumscribed by physics, that the only real facts are facts describable in the language of physics.

Quine is not coy about making this assumption: he says forthrightly: ". . . theory in physics is an ultimate parameter. There is no legitimate first philosophy, higher or firmer than physics, to which to appeal over physicists' heads", suggesting even that, for him, physics *is* a 'first philosophy'. This in itself is not contrary to the spirit of naturalized epistemology: the replacement of metaphysics, or 'logical syntax', with empirical science is precisely what naturalized epistemology mandates. The problem lies in Quine's singling physics out from all other empirical enterprises: from what standpoint, other than an external one, could such an assignment of priority be justified? Is Quine's first philosophy physics, or *physicalism*?

Remember that Quine's attack on the analytic/synthetic distinction was fundamentally an attack on the notion of a privileged standpoint, a perspective free of empirical commitments. Philosophy could no longer view itself, he argued, as a discipline apart from and neutral with respect to the empirical sciences. Philosophy must be conceived as simply the most general form of empirical enquiry. Traditional philosophical questions, if they survive at all, survive as general questions about the results of scientific investigation. Ontological questions are 'naturalized' into questions about the kinds of entities going sciences commit us to. Epistemology, which in its old form addressed questions about the existence of knowledge, now takes knowledge for granted, and becomes the empirical study of knowers.

But all this suggests a corresponding naturalization of the most general metaphysical questions about 'the ultimate parameters of reality', *viz.*, see what kinds of theories we need in order to adequately describe our world. A truly naturalized epistemology must, it would seem, take it as an open and *empirical* question what constitutes an objective domain. Specifically, it would seem that Quine's own naturalism obliges him to admit the possibility that 'special sciences', not reducible to physics, might be required to adequately account for human behavior. Even those with a taste for desert landscapes must be prepared to let a hundred flowers bloom.

Thus, I contend, Quine must present an *empirical* justification for the priority he assigns physics, or be convicted, by his own standards, of *a priorism*. I will argue shortly that Quine is sensitive to this difficulty, and that he believes he has an internalist defense of physicalism. (I'll also argue that defense is inadequate.) But Thomas Ricketts (1982), in his paper, 'Rationality, Translation, and Epistemology Naturalized', offers Quine a different kind of defense against this charge.

Ricketts agrees that the argument for indeterminacy rests on the assumption that physics defines the objective, and he appreciates the question one could raise about this assumption. As he puts it,

> Quine claims that translation is underdetermined relative to any exhaustive description of physical fact deliverable by natural science. But why does he draw the boundaries of natural science just where he does? . . . If Quine is doing more than just offering a list, then he must have in mind some distinguishing feature, presumably some methodological mark, which mathematics, physics, plant physiology, and behaviorist psychology all share but which Chomskian linguistics, anthropology, and literary studies lack. (Ricketts, 1982, p. 130)

But instead of going on to explain what methodological mark Quine offers, Ricketts tells us that

> In the end, Quine does rest content with producing a list: he thus appears to offer a doctrinal rather than a methodological characterization of science. (Ricketts, 1982, p. 131)

The reason why Quine's list is "something other than an expression of prejudice", according to Ricketts, is that Quine's rejection of analyticity, and with it, Carnap's project of rational reconstruction constituted a rejection of any *normative* questions about science. Carnap wanted to validate and explain our preference for science; to justify treating physics as a paradigm for empirical enquiry.

> Quine, on the other hand, denies that there is a place outside of science from which to provide any understanding of rationality which legitimates acceptance of his and Carnap's common paradigm. Our acceptance of natural science as the paradigm of knowledge has to be taken at face value. (Ricketts, 1982, p. 133)

On Ricketts's reading, then, Quine sees no need for, nor any possibility of, a defense of his physicalistic conception of the objective. His preference for physics is just that — a preference. One could stigmatize this preference as a 'bias', but to do so would be to presuppose precisely the kind of epistemic norms that Quine is at pains to reject.[7]

Ricketts's interpretation raises precisely the issues about normativity within naturalized epistemology that I want to take up. While his reading of Quine is certainly defensible, I think it is ultimately incorrect. Or, if Ricketts is right about what Quine thinks, then Quine thinks something he ought not to think.

It's true enough that Quine rejects the possibility of giving an 'account' of science, in Carnap's sense. But the problems he sees with producing such an account all have to do with the problems about an external standpoint: we have no non-arbitrary way of distinguishing our reflections about language and scientific practice from our 'substantive' theories about the world. The canons of rational practice, as well as the rules of logic and language, are simply part of our system of empirical belief. They can enjoy no privileged status, but must be acknowledged to be subject to the same processes of vindication and disconfirmation as any other beliefs.

Indeed, what was once the philosophical study of epistemic norms — epistemology — becomes, on Quine's view, the *empirical* study of normative reasoning — psychology. Epistemology, fully naturalized, becomes descriptive — aiming not at a validation of scientific practice, but seeking instead to uncover the processes by which human beings arrive at good empirical theories.

But in eschewing the normative goals of old-fashioned epistemology, Quine certainly cannot mean to be advocating the *naive* descriptivism that Ricketts attributes to him. Part of Quine's assault on the internal/external distinction involves emphasizing the way in which theory and judgment intrude at every level of theorizing — the notion of a pure, normatively neutral description is as antithetical to Quine's epistemology as an analytic truth.

This all suggests that Quine is not so much abandoning the notion of rational norms, as he is transforming it. It may be that, after studying the mechanisms of knowing, we are able to conclude that some cognitive strategies (e.g., reasoning) are to be privileged with respect to others (consulting a Ouija Board). But if so, it will only be because the favored strategies yield the most satisfying results, relative to our *de facto* standards of satisfaction. The empirical elucidation of those standards is, indeed, one of the tasks of a naturalized epistemology, and is one to which Quine devotes a great deal of attention in *Word and Object, The Web of Belief*, and elsewhere.

The standards and norms uncovered by scrutinizing the actual conduct of science are no less normative, on that account, than if they had been recognized by introspective reflection. Quine's philosophy of science is thus continuous in important ways with traditional metaphysics and epistemology. Quine's investigation of (as it were) our epistemological values in fact serves the same objective as earlier disputations about the nature of reality and the relation of truth to reason: to produce a characterization of the *objective* — of what we are aiming at in our search for knowledge. Earlier philosophers erred only in trying to distinguish this highly general, 'philosophical' project from corresponding 'empirical' efforts to fill in the detail.

*Pace* Ricketts, I see nothing in Quine's work that suggests he is hostile to the idea of internal, *de facto* norms. Indeed, I don't see how else to construe his frequent discussions of scientific values. (What is *The Web of Belief* about?)[8] And if we are to take seriously Quine's characterization of the philosopher as scientist, we must assume that he espouses these values himself.

At any rate, Quine cannot get away, even if he wanted to, with the defense that he's aiming at flat description. The fact is that *not* everyone shares Quine's view of physics. Linguistics, like cognitive science generally, enjoys a high degree of legitimacy, even within the limited community Quine would recognize as 'scientific'. Quine saying that physics is "our" preferred paradigm is like Burt Parks crooning "there she is, your ideal". If these are factual assertions, they're false. If they're normative recommendations, we have a right to know why we ought to accept them.

All Quine's talk of 'communities' is idealized, right from the start. He does not feel bound to represent the epistemic values of mystics, hippies, lunatics, or freshmen in his 'naturalized' characterization of rationality, nor should he. One is not obliged, just because one is doing science, to regard all phenomena as equal. But idealizations ought to be motivated; one must be prepared to defend one's decisions as to which details are relevant and which can safely be ignored. Quine's physicalism is only one possible idealization of empirical enquiry, and he is culpable by his own lights if he thinks that nothing more need be said about it.

Indeed, as I've been arguing, Quine does not think this. He has much to say about why he regards physics as the premier science, focusing mainly on its generality. The defense most congenial to a naturalized epistemology, because it is the most 'internal', is this one: Quine (1978, esp. pp. 162ff.) argues that it is a claim *of* physical theory that physics should enjoy a fundamental and preemptive position among the other sciences.

It's difficult to evaluate this argument,[9] for it's not clear exactly what Quine thinks physics commits us to. Sometimes it looks like physicalism is simply materialism — "there is no change without a difference in the positions or states of bodies" (Quine, 1978, p. 162). But cognitive psychology is eminently compatible with materialism, and yet Quine's physicalism is supposed to yield the consequence that any quantification over mental states would be a "pointless ontological extravagance" (p. 163).

But even if physical theory entails such a consequence, naturalized epistemology requires any world-view, no matter what its origin, to stand up against the data, which in this case include a growing number of increasingly successful theories that quantify over, not only mental states, but even *linguistic rules*. The cognitive sciences argue for their posits in exactly the same way the physical sciences argue for theirs: by demonstrating that they're needed to explain the phenomena.

Quine's psychology of choice is behaviorism. Its inadequacy as an account of animal cognition and human linguistic behavior is now generally recognized.[10] Its theoretical shortcomings spring from the austerity of its methodology, which prohibits any reference to mediating mental variables. Such a prohibition rules out in advance a whole range of possible theoretical models.

This *a priori* restriction — analogous to a restriction on physical theories to posit only observable objects — offends most grievously against the naturalized spirit when promising theoretical models are, by its means, ruled out. This is, of course, Chomsky's (1959, 1975) main complaint against behaviorism, and against Quine's treatment of linguistics.

Chomsky argues that the facts about the acquisition and use of language are most effectively explained by the hypothesis that human beings are computational devices who output systems of linguistic rules — grammars — in response to environmental stimulations, constrained by a set of innate, species-specific principles — universal grammar.

No doubt Chomsky's appeal to linguistic rules suffers, in Quine's mind, from guilt by association. But in fact, the difference between Chomsky's rules of language and Carnap's could not be more striking. Chomsky introduces the notion of a structured body of linguistic knowledge as part of the solution to the empirical problem of explaining human verbal behavior. These grammatical rules are not, on Chomsky's view, available to direct introspection (even though they are 'tacitly known' by every competent speaker).[11] Rather, they must be inferred from speakers' patterns of use and judgment, and must eventually square with what becomes known about the physiological bases of language. As was not the case for Carnap, for Chomsky the structure of language must be discovered empirically, and the motivation for searching for such structure must be empirical as well.

Nothing in Quine's attack on the positivists' appeal to language, then, has any bearing on contemporary appeals to linguistic rules within linguistic theory. Indeed, if my construing of naturalistic epistemology is correct, and if the facts about verbal behavior are as they appear to be, then the substance of Quine's attack on the analytic/synthetic distinction actually vindicates rules of language. In conclusion, I'd like to suggest that these 'naturalistic' linguistic rules may even be able to address some of the epistemological concerns that motivated Carnap's appeal to language in the first place.

I've already argued that epistemology naturalized is not an epistemology completely stripped of normative import. Carnap's mistake, I argued, was not in wanting philosophy to be prescriptive, but in misconceiving the standpoint from which prescriptions could legitimately be made. I pointed out that Quine's own discussions of scientific practice are normative in character, and argued that they can be viewed as empirical (*albeit* mistaken) elucidations of the notions of rationality and objectivity.

I think that the notion of linguistic rules as it has been elaborated by contemporary linguistics enables us to go even further in the reconstruction of epistemological normativity. Contemporary linguistics — indeed all of cognitive science — employs a specific form of idealization, known as the competence/performance distinction.[12] The empirical claim is that one's actual observable behavior — one's *performance* — cannot be adequately explained without postulating an underlying system of knowledge — a *competence* — that functions as one contributing cause. A theory of syntactic — or deductive, or inductive — competence will not give us a theory of *a priori* belief of quite the sort demanded by the positivists. But such theories might well explain, for example, why certain learning tasks are easier than others for humans, and hence contribute to an account of why certain statements command prompt and almost universal consent, apparently in advance of experience. And such theories, while they will not enable us to 'validate' our scientific practices, may well serve to justify intuitions that rational practice has a certain core — that there's a system of rules underlying all our attempts to know the empirical world.

The empirical study of the mind, in short, can be conceived as the search for a theory of rational competence, of cognitive design. A theory of the well-functioning rational agent is an empirical theory of what constitutes rationality, just as an empirical theory of the syntactic knowledge of a competent speaker constitutes a theory of grammaticality. The bet is that such theories will legitimate a good number of our pre-theoretic intuitions about rational strategy, just as a theory of physiological function legitimates our hunch that it's not good to have a hole in one's liver. A naturalized epistemology, in telling us what we *do* do, does not thereby abjure telling us what we ought to do.

*Bates College*

## NOTES

[1] My interpretation of Carnap's views on analyticity was greatly influenced by discussions with Peter Hylton, who in addition to spending many hours talking with me, was kind enough to make available to me in pre-print his 'Analyticity and the Indeterminacy of Translation' (Hylton, 1982). Naturally, he is not responsible for any errors I may have made in interpreting Carnap.

[2] See Carnap (1936, Secs. 7 and 8) for an explanation of bridge laws, or 'reduction sentences'.

[3] See, for example, Grice and Strawson (1956).

[4] Ricketts (1982) argues exactly the opposite. I want to thank Professor Ricketts for stimulating discussion on this issue.

[5] The claim has been adequately defended by others, however. The arguments I find persuasive are made, for example, by Rorty (1972), Chomsky (1975), and Friedman (1975).

[6] I will not quote the relevant passages here, but the reader is invited to compare Quine's discussions of simplicity in Quine and Ullian (1978, pp. 71—73) and Quine (1960, pp. 20 ff.) with his disparaging comments about the non-objectivity of analytical hypotheses in, for example, Quine (1960, p. 72).

[7] This way of summarizing Ricketts's conclusion was suggested to me by Warren Goldfarb.

[8] Significantly, Ricketts (1982, p. 132) finds Quine's apparent attempts to account for our conception of simplicity 'uncharacteristic'.

[9] Joseph Levine (in 'Quine on Psychology', this volume), argues that Quine's defense of physicalism is successful.

[10] See Gleitman (1981) for a summary of the relevant reseach. Fodor (1968) reviews the philosophical critiques of behaviorism. The classic critique of behaviorist accounts of language is Chomsky (1959).

[11] Chomsky's appeal to speakers' 'grammatical intuitions' must not be confused with Carnap's reliance on introspection. It was vital to Carnap's epistemological purposes that rules of language be transparent — immediately accessible to introspection. Chomsky, who has no such purposes (at least none that he looks to the theory of grammar to fulfill), need make no such assumption. He must assume that grammars are in some sense known by every competent speaker, but this knowledge may be only implicit, displayed through linguistic behavior ('sub-personal' in the jargon of some computationalists).

[12] For an explanation of the competence/performance distinction, see Chomsky (1965).

## REFERENCES

Carnap, Rudolf: 1936. 'Testability and Meaning,' *Philosophy of Science* **3**, 419—471.

Carnap, Rudolf: 1937. *The Logical Syntax of Language*. New York: Harcourt, Brace and Company.

Chomsky, Noam: 1959. 'A Review of Skinner's *Verbal Behavior*,' *Language* **35**:1, 26—58.

Chomsky, Noam: 1965. *Aspects of a Theory of Syntax*. Cambridge, Mass.: MIT Press.

Chomsky, Noam: 1975. 'Quine's Empirical Assumptions.' In Davidson and Hintikka (1975, pp. 53—68).

Davidson, D., and J. Hintikka: 1975. *Words and Objections*. Dordrecht and Boston: Reidel Publishing Company.

Fodor, J. A.: 1968. *Psychological Explanation*. New York: Random House.

Friedman, Michael: 1975. 'Physicalism and the Indeterminacy of Translation,' *Noûs* **9**, 353—373.

Gleitman, Henry: 1981. *Psychology*. New York: W. W. Norton and Company.

Grice, H. P., and P. F. Strawson: 1956. 'In Defense of a Dogma,' *The Philosophical Review* **65**. Reprinted in J. F. Harris and R. H. Severens (eds.), *Analyticity*, Chicago: Quadrangle Books, 1970.

Hylton, Peter: 1982. 'Analyticity and the Indeterminacy of Translation,' *Synthese*.

Quine, W. V. O.: 1953. 'Two Dogmas of Empiricism.' In *From a Logical Point of View*. New York: Harper and Row.

Quine, W. V. O.: 1960. *Word and Object*. Cambridge, Mass.: MIT Press.

Quine, W. V. O.: 1969. 'Epistemology Naturalized.' In *Ontological Relativity and Other Essays*. New York: Columbia University Press, pp. 69—90.

Quine, W. V. O.: 1973. *The Roots of Reference*. La Salle, Ill.: Open Court.

Quine, W. V. O.: 1975. 'Reply to Chomsky.' In Davidson and Hintikka (1975).

Quine, W. V. O.: 1976a. 'Truth by Convention.' In *The Ways of Paradox and Other Essays*. Cambridge, Mass.: Harvard University Press.

Quine, W. V. O.: 1976b. 'Carnap and Logical Truth.' In *The Ways of Paradox and Other Essays*. Cambridge, Mass.: Harvard University Press.

Quine, W. V. O.: 1978. 'Facts of the Matter,' *The Southwestern Journal of Philosophy* **9**:2.

Quine, W. V. O., and J. S. Ullian: 1978. *The Web of Belief*. Second edition. New York: Random House.

Ricketts, Thomas G.: 1982. 'Rationality, Translation, and Epistemology Naturalized,' *The Journal of Philosophy* **79**:3, 117—136.

Rorty, Richard: 1972. 'Indeterminacy of Translation and of Truth,' *Synthese* **23**, 443—462.

JOSEPH LEVINE

# QUINE ON PSYCHOLOGY

## I

"Epistemology," Quine (1969, p. 69) says at the beginning of 'Epistemology Naturalized' (EN), "is concerned with the foundations of science". That is, the fundamental epistemological questions ask: 'What is the nature of (scientific) evidence?' and 'How are (scientific) claims confirmed?' There are two approaches one can take in seeking answers to these questions: the *a priori* approach and the empirical approach. The former Quine refers to as the project of 'rational reconstruction'. The latter is just that part of psychology which explains the behavior, exhibited by so-called 'rational' creatures, of theory construction and confirmation. The burden of EN is to argue for the second approach. Thus, on Quine's view, epistemology is a field of natural science, itself the activity epistemology is supposed to explain.

Quine's main argument for the empirical approach to epistemology is that the only plausible attempt at rational reconstruction — the attempt to translate all scientific talk into the language of observation, set theory, and logic — has definitively failed. This project had initial plausibility since all the evidence, at least ultimately, for any scientific theory must be expressible in these term (assuming one can justify an observation/theoretic distinction among terms, that is). Therefore, if one could carry out the translation, the answer to the second fundamental epistemological question falls out immediately. Confirmation involves only deduction and simple induction. No explanation of an apparent epistemic leap from the immediately detectable to the unobservable is necessary.

However, it is now a philosophical truism that no such translation is possible. Individual sentences, though surely established ultimately on the basis of sensory evidence, do not maintain their own unique set of confirming/disconfirming experiences; and thus they are not translatable into terms that describe such experiences. If the justification for the rational reconstruction approach no longer exists, then, Quine (1969, p. 78) argues, "it would seem more sensible to settle for



*Abner Shimony and Debra Nails (eds.), Naturalistic Epistemology*, 259—290.

psychology. Better to discover how science is in fact developed and learned than to fabricate a fictitious structure to a similar effect".

For present purposes, I will assume Quine is right about this. Epistemology becomes psychology. But what is the nature of the psychology with which Quine wants to identify epistemology? Quine has a very definite view about this, expressed in 'Epistemology Naturalized' and throughout his work. He believes that the psychological theory which will explain how we build theories from evidence will be a non-mentalist, behaviorist theory. As he puts it, "Any realistic theory of evidence must be inseparable from the psychology of stimulus and response, applied to sentences" (Quine, 1960, p. 17). Many philosophers who are generally sympathetic to the aims of 'Epistemology Naturalized', find his position on the nature of psychology untenable. My purpose here is to clarify both the nature of Quine's claims about psychology and his arguments for them. What I wish to show is that, while his critics have accurately pointed out important weaknesses in his position, there remains an argument which must be taken seriously by the advocates of mentalistic psychology.

Let me begin by describing a general view of the form an epistemological psychology might take; one that Quine definitely opposes. We see the human organism as a device that takes sensory stimulation as input and gives theories as output. (So far no difference from Quine.) Our theory of this device involves ascribing to it an internal cognitive network: a systematic representation of rules and principles, the mediation of which provides the link between the sensory input and the theoretical output, as well as a representation of the inputs and outputs themselves. By calling this a *cognitive* network I mean that the mediating action of the internal rules and principles is performed by *rational* transformations. That is, unlike a biochemical description of the brain's changes from state to state, this is an *intentional* account, on which changes of state inside the organism are seen to obey the laws of logic (broadly construed to include induction). This entails that the objects of the mediating processes be entities with propositional structure and content. As some have pointed out, this amounts to positing a 'language of thought' (cf. Fodor, 1975).

How is the sort of theory described above an *epistemological* psychology? That is, how would the fundamental epistemological questions be addressed by a mature cognitive psychology? First of all, to the question, 'What is the nature of (scientific) evidence?' we will turn to the theory of perception. So, for instance, take the traditional epistemological puzzle over the nature of perceptual primitives: are they organized perceptual judgments or simple sense data? A naturalized approach to this question involves developing a theory of the reception of physical stimulation by the sense organs and the stages of processing the information contained therein. Assuming we have a criterion for the ascription of intentional state, we can determine at what stage in the process primitive sensations become perceptual judgments, and at what stage the information is accessible to consciousness.

To answer the second question, 'How are (scientific) claims confirmed?' we need a theory of the rules and principles we unconsciously employ in assessing these claims in the light of the outputs of our perceptual systems (along with, of course, previously confirmed claims). In particular, two notions are crucial in this regard, for they serve to restore some normative content to epistemology: these are the competence/performance distinction and the notion of modularity.[1] Their relevance to epistemology is this. The mind as a whole is viewed as a system of interconnected, but separable cognitive capacities — modules. Each such capacity, or module, functions according to its own set of rules and principles, which constitutes the competence of that module. A description of the competence of a module will normally be an idealization, and the discrepancies found between its actual and idealized performance will be treated as 'noise'. Among others, one source of such noise could be located in the interactions among the modules. This could be an explanation of someone's irrationally holding onto a belief, say. By the rules of her inductive capacity the belief is unfounded. This is what makes it 'irrational'. But the action of her 'wish-fulfillment' capacity intervenes in various ways to cause her to affirm the belief. Thus, a theory of the competence of the human belief-confirmation capacity would serve as a normative standard for the evaluation of the reasonableness of beliefs; or, the closest thing to one that a naturalized epistemology can provide.

*Prima facie*, the above is just the sort of theory (or better, framework for a theory) that Quine should welcome. Yet, as mentioned above, he does not. He believes that the sort of psychology which will explain how we build theories from evidence will not involve any intentional attributions at all. At first blush, this is astonishing. What we expect from an epistemology, even a naturalized one, is some account of rational inquiry: the standards we appeal to in adjudicating conflicting claims as to the state of things around us. One would think such an account must refer to processes, rules, and the like, which operate upon entities with propositional content; i.e. internal representations of the claims being adjudicated, and of the relevant evidence. Furthermore, these processes are expected to be sensitive to the logical structure of the entities the processes are defined over. How is one to capture all this in an account which limits its explanatory principles to those of a stimulus-response psychology? Of course these are just the arguments Chomsky, Fodor, and others have made for years. (Cf. Chomsky, 1959; Fodor, 1978.)

Why does Quine object to cognitive psychology playing the role of naturalized epistemology? One might see his position as merely a preference for behaviorist theory; he just thinks it is right. Though it is probably true that he does believe behaviorism is the best bet for psychology, he has more 'philosophical' reasons for his opposition to cognitive psychology. His position rests on a claim involving the ontological status of the mental states — the propositional attitudes — such a psychology attributes to human beings, an attribution that is essential to its explanation of the theory construction/confirmation process. Quine believes that one can distinguish 'hard' facts from 'soft' facts: the former are those that 'really' obtain while the latter do not; the former are objective while the latter are not. Of course this is not Quine's terminology. He makes the distinction by saying that about some issues there is a 'fact of the matter' while about others there is not. Perhaps the best way to put the distinction is this: soft facts obtain, to a crucial extent, in virtue of the interpretation given them, the way they are described, by a rational, language-using subject. Hard facts are independent of, or invariant across, the ways they are interpreted and described by a rational subject. Quine claims that facts about what propositional attitude — in particular what beliefs — a person has, are soft facts.

Quine's thesis concerning the ontological status of states of belief is known as the thesis of the 'indeterminacy of translation'. The relation between translation and belief is straightforward. If a French speaker asserts "Il pleut" I take her to believe what she is saying. Now, if I translate her assertion into English one way, say as "It's raining," I will be attributing a particular belief to her: namely, that it's raining. If, on the other hand, there is no determinate answer whether to translate her utterance as "It's raining" or "It's snowing," there will be a corresponding indeterminacy as to what she believes. Conversely, if there is a fact of the matter concerning what she believes, there is a fact of the matter concerning the correct translation of her utterance.

Quine has two main arguments for his thesis, though it is not clear he would distinguish them himself. First, as elaborated especially in EN and Quine (1960, ch. 2), he ties the indeterminacy thesis to his arguments against the analytic-synthetic distinction. With regard to this argument, I will try to show that to connect the two theses he must rely upon a residual, unjustified 'dogma of empiricism'. The second argument, most clearly stated in Quine (1975b, 1978), derives the indeterminacy thesis from the general doctrine of physicalism. This argument, I believe, carries great force, and poses a real challenge to intentional psychology. In Section II I take up the first argument, in Section III the second, and state my conclusions in Section IV.

## II

The analytic-synthetic distinction exists only if there is an absolute distinction between matters of meaning and matters of fact. For such a distinction to exist it must be possible to separate the linguistic contribution from the factual contribution to the truth or falsity of a sentence. One way to do this, motivated by the empiricist doctrine that all knowledge, and hence, meaning, derives from sense experience, is to identify the meaning — the semantic content — of a sentence with the difference its truth would make to sense experiences. As we mentioned, Quine claims such a sentence-by-sentence assignment of observational consequences is impossible. Furthermore, he seems to believe this vitiates the attempt to assign individual sentences any separable semantic content. Thus the basis for an analytic-synthetic distinction vanishes.

Quine (1969) takes the untranslatability of single (non-observational) sentences into observational terms to entail that semantic content can only be applied to theories as units. He says:

If we can aspire to a sort of *logischer Aufbau der Welt* at all, it must be to one in which the texts slated for translation into observational and logico-mathematical terms are mostly broad theories taken as wholes, rather than just terms or short sentences. . . . We might better speak in such a case not of translation but simply of observational evidence for theories; and we may, following Peirce, *still fairly call this the empirical meaning of theories.* (p. 79, my emphasis)

He then goes on to argue that this consideration leads to a philosophical problem about translation from one language into another; say from English into Arunta. Again:

> If the English sentences of a theory have their meaning only together as a body, then we can justify their translation into Arunta only together as a body. There will be no justification for pairing off the component English sentences with component Arunta sentences, except as these correlations make the translation of the theory as a whole come out right. Any translation of the English sentences into Arunta sentences will be as correct as any other, so long as the net empirical implications of the theory as a whole are preserved in translation. But it is to be expected that many different ways of translating the component sentences, essentially different individually, would deliver the same empirical implications for the theory as a whole; deviations in the translation of one component sentence could be compensated for in the translation of another component sentence. Insofar, *there can be no ground for saying which of two glaringly unlike translations of individual sentences is right.* (p. 80, my emphasis)

He seems to be saying that since individual sentences (save, perhaps, observation sentences) cannot be assigned a set of confirming/disconfirming experiences on their own, their translation from one language into another is indeterminate. He is even more explicit in the next paragraph, where he says:

> If we recognize with Peirce that the meaning of a sentence turns purely on what would count as evidence for its truth, and if we recognize with Duhem that theoretical sentences have their evidence not as single sentences but only as larger blocks of theory, then the indeterminacy of translation of theoretical sentences is the natural conclusion. And most sentences, apart from observation sentences, are theoretical. (pp. 80—81)

Quine's argument for the indeterminacy thesis (1960, ch. 2) is well known. He describes the situation facing a field linguist who is attempting to translate a radically foreign language into her own — say English. He claims that the only hard evidence the linguist has to go on is the occurrences of native utterances in particular stimulus conditions. This evidence will include, after a while, the native informant's assents and dissents to queries posed by the linguist in various situations. Eventually, the linguist will be able to compile 'stimulus meanings' for all the native informant's sentences. A sentence's stimulus meaning is the ordered pair composed of the class of stimulations which would prompt assent to the sentence and the class of stimulations which would prompt dissent from the sentence. It is to be understood as relativized to a speaker, a time, and a given 'modulus' of stimulation.

It is crucial to note that despite the linguist's ability to directly observe whether or not, for a given sentence and a given situation, the native informant assents, a finite number of such observations will not determine stimulus meanings for the informant's sentences. For stimulus meaning involves the informant's dispositions to assent or dissent in all possible stimulus conditions. Nevertheless, Quine sees no problem in claiming that stimulus meanings constitute genuine facts about the language of the informant. The leap from data to hypotheses in this case only involves the standard uncertainties of inductive reasoning. It's really no different from determining that all emeralds are green.

So, despite the normal inductive uncertainties attending the attribution of stimulus meanings to the native utterances, Quine believes such attribution constitutes a description of genuine facts. No one disputes this, as far as I can tell. But Quine goes further. He states that, ". . . stimulus meaning, by whatever name, may be properly looked upon still as *the objective reality* that the linguist has to probe when he undertakes radical translation" (1960, p. 39, my emphasis). He claims, then, that stimulus meanings *exhaust* the genuine facts about the native language — indeed, about language in general.

As Quine demonstrates, a determination of stimulus meanings for the sentences of a language leaves quite a lot undetermined about the semantics of that language. This indeterminacy can be brought out in two ways, corresponding to what Quine (1970) calls the 'argument from below' and the 'argument from above'. First, consider the sentence, 'There goes a rabbit'. While one might plausibly argue that its stimulus meaning constitutes a rough approximation to its truth conditions, it cannot be plausibly argued that it determines satisfaction conditions for its component terms. For instance, as Quine's well-known example goes, we could assign to the term 'rabbit' the class of rabbit-stages (temporal slices of a rabbit's history) as its extension, instead of the class of rabbits. This might be done without altering the sentence's stimulus meaning at all, since to confront a rabbit is to confront a rabbit-stage. This is the argument from below.

The argument from above is to demonstrate that stimulus meanings do not even constitute a rough approximation to truth conditions for many, many sentences. While stimulus meanings provide a basis for translating observation sentences from one language into another, they do not for non-observational occasion sentences and standing sentences. ('Occasion sentences', for Quine, are those to which one assents only upon repeated promptings by the appropriate stimuli; as opposed to 'standing sentences', for which one such prompting is enough. This is slightly oversimplified, but for a full discussion see Quine, 1960, ch. 2.) Observation sentences are those occasion sentences which have substantially the same stimulus meaning for everyone in the speech community.) This is so in the one case because of the intrusion of collateral information, and in the other because of the relative sparseness of their stimulus meanings in the first place. Thus, if we are constrained in translation only by stimulus meanings, many sentences will be susceptible to significantly conflicting translations.

Now, of course, Quine admits, we are not so constrained, and we tend to come up with fairly determinate translations; and so would the linguist faced with radical translation. Quine calls the group of methodological principles needed to provide a determinate translation 'analytical hypotheses'. Analytical hypotheses, when added to the constraint of capturing stimulus meanings, are sufficient to pin down the truth conditions of sentences and the satisfaction conditions of terms. The problem, Quine claims, is that these analytical hypotheses have no basis in 'objective reality'. They do not determine genuine facts. Thus the indeterminacy of semantic content — and thereby of the content of propositional attitudes — that exists without the intervention of analytical hypotheses is a genuine, ontological indeterminacy.

Why should only the process of pinning down stimulus meanings, but not the rest of the process of translation, involve the discovery of real facts? Quine's (1960) answer is this:

> Yet *one has only to reflect on the nature of possible data and methods to appreciate the indeterminacy.* Sentences translatable outright, translatable by independent evidence of stimulatory occasions, are sparse and must woefully under-determine the analytical hypotheses on which the translation of all further sentences depends. To project such hypotheses beyond the independently translatable sentences at all is in effect to impute our sense of linguistic analogy unverifiably to the native mind. (p. 72, my emphasis)

His argument seems to be that since the only concrete data available to the linguist pertain to the native's dispositions to assent and dissent in various stimulus conditions, the whole truth about the semantics of the native's language must consist of what is embodied in the stimulus meanings of her utterances.

Let me now connect his argument (in Quine, 1960) with the argument presented above (from Quine, 1969). He says that if, as was once believed, every meaningful sentence could be associated with its own set of confirming/disconfirming experiences, there would be a sound basis for an analytic/synthetic distinction, and no indeterminacy of translation. Since observation sentences can legitimately be said to have individual sensory meanings — because, as he puts it (1960), their meanings can be identified with their stimulus meanings — there is no significant indeterminacy about their translation. However, when it comes to those sentences which cannot individually be associated with a set of confirming/disconfirming experiences — that is, their meanings are not captured by their stimulus meanings — a significant indeterminacy about their translation, their semantic content, sets in. The arguments thus appear to be substantially the same.

Quine's critics (cf. Chomsky, 1975; Rorty, 1972; Friedman, 1975) have pointed out an apparent inconsistency in Quine's treatment of the relation between data and theory. When it comes to physical theory, or even the determination of stimulus meanings themselves, the fact that the hypotheses are under-determined by the data does not cause Quine to disclaim their ontological rigidity. But when it comes to the part of translation theory left over after stimulus meanings are determined, he seems to take the under-determination of theory by data in this case to have ontological consequences for the theory. The critics claim there is no basis for this distinction. Either the facts of physics are ontologically soft in the way that the facts of semantics are alleged to be soft, or both sets of facts constitute genuine facts. He can't have it both ways.

Why does Quine believe the data exhaust the content of a theory of semantic content, but not the content of any other empirical theory? I think it is because his view of language is constrained by a 'residual dogma of empiricism'. What I am calling an empiricist 'dogma' is the principle that semantic content is empirical content, and empirical content consists of the stimuli that impinge on the language user.[2] If this were true, if the semantic content of a language user's entire corpus of utterances, actual and potential, consisted of the stimulus conditions that prompt them, then indeed there would be no genuine semantic facts other than those that determine stimulus meanings. The question is, though, why believe this principle?

That Quine appeals to some such principle in his argument for the indeterminacy of translation is evident in the passages quoted earlier (Quine, 1969, pp. 79—81; note especially the part about his agreement with Peirce). He takes this principle together with the irreducibility of non-observation sentences to observation sentences to entail the indeterminacy of translation. Only if one accepted this residual dogma would one find it plausible that the linguist's analytical hypotheses are different in any way from the methodological maxims of natural science. But we haven't seen an argument for the principle. The key to Quine's view on this matter is contained in the following passage:

> Should the unwelcomeness of the conclusion [the indeterminacy of translation] persuade us to abandon the verification theory of meaning? Certainly not. The sort of meaning that is basic to translation, and to the learning of one's own language, is necessarily empirical meaning and nothing more. . . . Language is socially inculcated and controlled; the inculcation and control turn strictly on the keying of sentences to shared stimulation. Internal factors may vary *ad libitum* without prejudice to communication as long as the keying of language to external stimuli is undisturbed. *Surely one has no choice but to be an empiricist so far as one's theory of linguistic meaning is concerned.* (1969, p. 81, my emphasis)

The fundamental tenet of empiricism is that all knowledge derives from sensory stimulation. Add to this the plausible premise that you can't speak about what you can't know about, and also the premise that language is learned from others in a public setting, and one gets the corollary that all semantic content derives from sensory stimulation. This seems to be Quine's argument in the passage just quoted, as in an earlier passage, where he states that an unassailable cardinal tenet of empiricism is that "all inculcation of meanings of words must rest ultimately on sensory evidence" (1969, p. 75).

Surely in some sense this cardinal tenet is true. Whatever we may have going on inside our heads, we connect to the world through our sense organs. But it is one thing to say that information, and therefore our semantic attachment to the world, comes to us *through* our sense organs, and another to say that it is therefore somehow *about* what impinges on our sense organs. Furthermore, the fact that language is learned from others only shows that one must infer the meanings of terms from publicly accessible data, not, again, that the terms can only refer to the data. It is the slide from the two former claims to the two latter claims that constitutes the residual dogma of empiricism.

To see what I mean, consider how we can avoid this slide. We might agree with Quine that theories are confirmed as wholes, with no single theoretical statement capable of confirmation or disconfirmation on its own. However, we might still maintain that individual theoretical sentences have semantic contents of their own. 'Electrons have negative charge' has truth conditions, is in fact true, in virtue of the fact that the requisite terms refer, and the entities in the extension of the subject term constitute a subset of the entities in the extension of the predicate term. But, one objects, isn't it only in virtue of this sentence's role within physical theory as a whole, and that theory's connection to sensory experience, that it is even meaningful? True enough, but why couldn't its role within the theory, and the theory's connection to sense experience, endow it with its own semantic content? The argument from the fact that theories are confirmed as a block to the claim that they get semantic content as a block just isn't there.[3]

Of course I haven't said yet *how* the role a theoretical sentence plays, together with the containing theory's connection to sense experience, manages to endow the sentence with its own, individualized semantic content. The reason is that I don't know. The problem of reference is a big one, and no theory of reference that I know of has been both worked out in detail and generally accepted. The point is, this doesn't validate a radical empiricist view by default. There is no reason to believe that our referential ability stops at the surface of our sense organs. So long as we refuse to accept this limitation on our ability to refer, believing instead that we somehow refer *through* our sense experience *to* what accounts for it, indeterminacy does not result from the irreducibility of theoretical sentences to observation sentences. The residual dogma of empiricism is the crucial, missing link.

So far I have tried to show that Quine's only reason for distinguishing the case of physical theory and semantic theory is an unjustified, radical empiricist premise. But let's suppose this premise were justified. It turns out Quine still couldn't distinguish the two cases, because this premise would rebound to affect the case of physical theory as well.

I explained Quine's differential treatment of physics and semantics by reference to a thesis concerning the nature of semantic facts. If the language which is the subject of semantics were distinguishable from the language within which physics is expressed, there might be no problem. But of course, this is not the case. So, given Quine's thesis concerning the nature of semantic facts, he must admit that there is no fact of the matter as to which of two apparently conflicting theories with the same observational consequences is true. In fact, there is no sense to there even being a conflict in the first place, since both theories would have, on this view, the same semantic content. If from holism about confirmation we can derive holism about semantics, this conclusion is inescapable. But then there is no longer any difference between semantics and physics concerning the ontological status of their claims.

Quine himself explicitly notes the parallel between physics and semantics. In the following passage he argues that nonetheless there is a difference:

> The indefinability of synonymy by reference to the methodology of analytical hypotheses is formally the same as the indefinability of truth by reference to scientific method. Also the consequences are parallel. Just as we may meaningfully speak of the truth of a sentence only within the terms of some theory or conceptual scheme, so on the whole we may meaningfully speak of interlinguistic synonymy only within the terms of some particular system of analytical hypotheses.
>
> May we conclude that translational synonymy at its worst is no worse off than truth in physics? To be thus reassured is to misjudge the parallel. In being able to speak of the truth of a sentence only within a more inclusive theory, one is not much hampered; for one is always working within some comfortably inclusive theory, however tentative. . . . Not so the analytical hypotheses that constitute the parameter of translation. We are always ready to wonder about the meaning of a foreigner's remark without reference to any one set of analytical hypotheses, indeed even in the absence of any; yet two sets of analytical hypotheses equally compatible with all linguistic behavior can give contrary answers. . . . (1960, pp. 75—76)

He seems to be saying that there really is no sense to be made of the claim that two physical theories, both of which have the same observational consequences, can conflict. Nevertheless, we normally pick one theory and use it as if all its sentences were on a par. Now, when we do fix physical theory in this way, it turns out that semantic theory, for the reasons adduced in the discussion of radical translation, remains unfixed. Given a physical theory, it is still possible to develop significantly different translation manuals consistent with that theory. Thus the indeterminacy of translation is really over and above the underdetermination of physics.

We are getting closer now to what I referred to in Section I as

Quine's 'second main argument'. However, from this passage it is still possible to extract different arguments, only one of which is the strong one I will discuss later. First, here is one interpretation. If we look at science as a whole, there is no real sense in speaking of the truth or falsity of its sentences, only its adequacy in predicting and regularizing our experience. To compare, therefore, two apparently conflicting total sciences, asking which one is true, is to misunderstand the nature of the theories in the first place. However, since we are always working within some total theory, we can normally speak of truth and falsity without bothering to relativize it; this is just understood. Now, add to this story the assumption that our total science can be identified with physical theory, and the additional indeterminacy of semantics over and above physics takes on special importance.

If this is his view, one can immediately object that physical theory is not even close to our total science. Given this obvious fact, what can possibly be the basis for choosing to call genuinely factual only that which is determined by physics? Rorty in particular makes this objection. (See ahead, quotation on p. 282) Now, if in fact Quine is not a realist when it comes to total science, I don't think he has any adequate reply to this charge of mere scientific bias in favor of the field we call 'physics'. That is, if the truth of 'Electrons have negative charge' is a matter, not of its correspondence with a state of affairs (i.e. that of electrons having negative charge), but merely of its contribution to the systematization of our sense experience, as Quine's agreement with Peirce and Duhem together suggests, then the fact allegedly described by this sentence is pretty 'soft' itself. To single out semantic facts because they are left undetermined after one settles on a physical theory seems like merely a way of expressing one's preference for one theory over the other. If there are no really genuine matters of fact (except, perhaps, those described by observation sentences), then why distinguish among various kinds of non-genuine ones?

On the other hand, suppose we interpret Quine as believing that theoretical sentences generally do correspond to genuine facts. If so, it must be possible for theoretical sentences to have individualized semantic contents even though these be under-determined by a speaker's dispositions to verbal behavior.[4] Suppose, that is, we ignore the aspect of Quine's doctrine that is in agreement with Peirce — we ignore his residual dogma of empiricism. But then, though we now have some hard facts around to oppose to soft ones, we would seem to have lost the basis for claiming semantic facts are soft to begin with. So, it looks as if, either way, whether Quine is a realist or not, he loses.

However, there is yet another way to see Quine's entire argument, one that depends on treating him as a thoroughgoing realist. The problems we then face are these: One, why does the under-determination of meaning by stimulus meaning constitute an ontological indeterminacy? Second, why believe physics under-determines semantics, and even if it does, why care? Taking the realist in Quine as a basis, I believe that an interesting and important argument for the indeterminacy thesis can be developed; and, therefore, that these two questions can be answered. In Section III I will try to make good on this claim.

## III

In Section II I emphasized one side of Quine's attitude toward the relation between theory and observation. On this interpretation, observation sentences play a role quite like the 'protocol sentences' referred to by Carnap and other positivists. Along the same lines, his view of the semantic content of theoretical sentences — that they only have semantic content as a unitary block, and it is constituted by the observational consequences, the 'empirical meaning', of the entire theory — was interpreted to be closely akin to, if in fact not identical with, some positivist views. It's clear from the several quotes above that Quine sometimes espouses such views. But Quine, it seems, is of two minds about the role of observation sentences, and therefore of the entire doctrine of semantic content built upon them.

For one thing, note that he explicitly claims that observationality is a matter of degree (Quine, 1960, p. 42). The wider the agreement among speakers of the language concerning the stimulus meaning of a given sentence, the more observational it is. Furthermore, despite the fact that observation sentences allegedly wear their semantic contents on their sleeves, it is supposed to be possible to deny an observation sentence even when faced with stimulus conditions which are included in its affirmative stimulus meaning. This is so because it may be demanded by one's belief in a deeply entrenched theoretical sentence (together with logical consistency). Now, there may be a philosophical use for this somewhat diluted conception of observation. For instance, it may help to clarify the nature of actual scientific practice — as is the job of a naturalized epistemology. But I don't see how it can provide a basis for the radically empiricist semantics Quine suggests. For if observation sentences can be revised due to pressure from 'above' — i.e. conflict with well-entrenched theoretical sentences — then they can hardly be constitutive of the semantic content of those very same theoretical sentences.

These considerations suggest that Quine ought to distinguish between the claim that theories are confirmed as wholes from the claim that they have semantic content only as wholes. Ignoring the textual evidence cited above, then, let's assume that he does make this distinction. Does this automatically entail that the sentences of a theory are endowed with their own individualized semantic contents? Perhaps not. But it does now allow room for a realistic semantics, one that relates individual sentences with individual truth conditions. So, since a realist attitude is consistent with Quine's Duhemian argument against the analytic/synthetic distinction, since it is a popular position within recent philosophy of science, and since much of what Quine says about physics is interpretable in this way, I will for the rest of this section assume that Quine's argument begins with the doctrine of scientific realism. I will then attempt to develop his argument for the indeterminacy of translation from there.

A good starting point would be to clarify what role the facts of neurophysiology play in the argument for indeterminacy. That is, we've seen that a crucial part of Quine's argument is his contention that the physical facts do not determine the semantic facts. But so far we've really only seen reason to believe that *behavioral* facts don't determine semantic facts. What about all that goes on inside the nervous system the result of which is the behavior the psychologist or linguist observes? In one sense Quine explicitly involves neurophysiological facts in his argument. He has to because of his claim that stimulus meanings are dispositions, and his further claim that dispositions are identifiable with manifest structural properties. As he puts it:

> Dispositions are . . . conceived as built-in, enduring structural traits. An example was the 'would prompt assent' of section 8. For there again a disposition was concerned, albeit unnamed: *some subtle neural condition*, induced by language-learning, that disposes the subject to assent to or dissent from a certain sentence in response to certain supporting stimulations. (1960, p. 223, my emphasis)

So, at least to a certain degree, he does explicitly take neurophysiological facts into account in his argument.

But, of course, this isn't enough. If a cognitive theory, of the sort sketched in section I, is indeed the best explanation of rational behavior, then a person's neurophysiological structure must embody much more than merely 'behavioral dispositions'. For one's behavior is to be explained by, among other things, the interaction of internal cognitive states, which, on this sort of theory, have a syntactic structure and a semantic interpretation. One's neurophysiological structure must be such that it can support such structure and its interpretation. Perhaps if behaviorism were right, our neurophysiological structure, as Quine believes, would not suffice to determine a unique best semantic interpretation among a host of conflicting alternatives. But since I don't want to interpret Quine's argument in such a way as to beg the issue against cognitive psychology, his argument must be able to run even under the assumption that cognitive psychology, not behaviorism, is in fact the best bet for theoretical progress in psychology. So now we need to know, is it still plausible that the neurophysiological facts, complex as these must be to underlie such a rich cognitive network, do not determine, among significantly conflicting alternatives, a correct semantic interpretation?

We have now changed gears completely. Within an attitude that is realist generally, and takes the posits of psychology seriously in particular — as Quine the naturalized epistemologist should — we are no longer specially concerned with behavioral dispositions (except insofar as they contribute to the constitution of whatever cognitive structures psychologists might posit). Instead, our first question is this: Do *all* of the relevant physical facts determine the semantic facts? This is in fact exactly how Quine puts it in his reply to Chomsky:

> Consider, from this realistic point of view, the totality of truths of nature, known and unknown, observable and unobservable, past and future. The point about indeterminacy of translation is that it withstands even this truth, the whole truth about nature.[5] (Quine, 1975b, p. 303)

Before we can see whether or not Quine's assessment of the situation is justified, it is important to briefly consider just how cognitive states are supposed to relate to physical states. The crucial notion here is that of 'realization'. Being materialists, most proponents of cognitive psychology believe that a person's cognitive states are somehow realized in her brain states (or, states of her central nervous system). (See Fodor, 1975, ch. 1.) What they mean by 'realization' is exemplified by the internal workings of computers — which in general serve as the basic metaphor of cognitive models of the mind. When we are discussing the computer in terms of its programming states — from the point of view of its software — we attribute to it states that bear informational, semantic contents. We say things like, 'It takes the value stored in memory location X and puts it in Y'. When we say this, we mean it. That is, we literally attribute to the machine states like having memory locations with particular values stored in them. On the other hand, it is also clear that there are physical mechanisms, describable without recourse to intentional vocabulary, by virtue of which the information states obtain in the computer. This is the machine's hardware, and we say that it realizes the software.

So, we can put the indeterminacy thesis this way. It is possible for there to be conflicting cognitive models of a human being such that all the physical facts do not determine which of them is realized in the person's nervous system, to the exclusion of the others. Formulated this way, it is clear that the under-determination of the theoretical models by the data is not the issue. Rather it turns on a metaphysical view concerning the privileged role accorded physical facts in constituting what is and is not the case.

So far we have dealt with indeterminacy in an abstract way. We haven't actually seen an instance of indeterminacy. In fact, for anything like an interesting case, it is virtually impossible to exhibit indeterminacy in a full-blown, convincing manner. So, what expositors of the thesis do is present mini-instances just to give the idea (cf. Harmon, 1975). This is what I propose to do. First, I will describe a simple calculating device, for which a convincing case of indeterminacy can be made. Then, I will describe, in a much sketchier manner, a device which could be actually considered an instance of a full-blown cognitive model. I will then consider whether the sort of indeterminacy that could be convincingly exhibited with the first device, plausibly attaches to the states of the second device as well.

First, the simple calculating machine. Suppose it is constructed as follows. It consists of a number of flat, rectangular strips, each suspended in a line. Attached to this structure are tracks which lead to each strip, along which marbles of various sizes can roll down toward the strips. The strips are weighted in such a way that they can flipped over by the impact of a marble of sufficient size. Once flipped over, they stay that way until struck again with sufficient force. Also, the pivot point of the strips divides them into unequal lengths. When the longer side is up, flipping a strip will cause it to flip the strip behind it. When the short side is up, it doesn't affect the other strips. Finally, we have two sizes of marble. One is of sufficient mass to flip over the strips, and the other is not.

What we have here is a device that will perform binary addition. It is quite primitive indeed (I clearly have no expertise in computer design), and probably more trouble than it is worth. Nevertheless, a binary adder it is. If we call the big marble '1', the little marble '0', similarly for the long and short sides of the strips respectively, and designate the strips in turn as successive powers of 2 (starting from $2^0$), we're ready for action. The device is an accumulator which adds by having the first number loaded in and then representing the sum of this number with each successive number loaded into it. Loading is accomplished by rolling marbles along the tracks. Depending upon the power of two a particular digit stands for (or, to put it more mechanically, the position the digit in question occupies in the numeral), one chooses the appropriate track to roll the marble down.

Binary addition is a rather natural interpretation for the machinations of this device. In fact, that's what it was 'intended to be'. The function it computes can be represented as $f(x) = x + c$, where $c$ is the number already loaded in the machine. It is only in the context of this interpretation that it makes sense to call the marbles '0' and '1', the short and long strips the same, and a particular place in the line of strips as the 'ones place' or the 'twos place'.

Suppose someone came along and said we had it all wrong. She contends that the place a strip occupied in the line represented the power of two that results if you reverse the order of the exponents. (The values of the inputs remain the same.) She also maintains that the marbles are to be rolled down the tracks in the same order as before. If we protest that, read this way, the machine yields nonsense, not sums, she responds that the device computes a different function, $f'$, which is defined as follows:

$f'(x) = g(h(x))$, where $g$ and $h$ are defined as follows: $g(x) = (n/d)p$, where $n$, $d$, and $p$ are determined as follows:

(1) If $k$ is the number of strips, $p = 2^{k-1}$;

(2) $d = 2^i$, where $i$ is such that $2^i \leqslant x < 2^{i+1}$;

(3) Take the odd integers from 1 to $(2^{i+1} - 1)$ in order and let them constitute a $1xL$ matrix, where $L$ is the number of these odd integers:
  (a) label these columns $C_1 - C_L$,
  (b) take the columns $C_q - C_L$, where $q = (L/2) + 1$, and transpose them so as to make them a second row in the $2x(L/2)$ matrix that results,
  (c) continue this process of creating new rows until there remains a $Lx1$ matrix, $L$ rows and 1 column,
  (d) label these rows $r_1 - r_L$,
  (e) for $x = 2^i + m$, $n = r_{(m+1)}$ (i.e. the value that appears on that row).

$h(x) = x + 1(c)$, where $c$, again, is the number already in the machine and 1 is the inverse of $g$.

Compared to $f(x)$, $f'(x)$ is terribly complex and unnatural. Our imagined friend clearly has a bizarre sense of interpretation. On the other hand, this function is not really *ad hoc*, in the sense of being merely a finite list of arguments and values; rather, it has true generality. Finally, note that the two functions, $f$ and $f'$, yield incompatible values for the same arguments.

I take this to be a case of indeterminacy. For, ask yourself, which function is the device *really* computing? Is there a fact of the matter? I think most people would agree that there isn't. Of course, this may be for reasons peculiar to this rather artificial case. For there is very little structure to this machine, and very little it does. Also, there actually being an *interpretation* of its states — privileged or otherwise — might seem to some an artificial way of looking at it. So it's clear the example won't establish indeterminacy for the hard cases (i.e. beings, like us, who have real psychological lives). But, it does exemplify the phenomenon under consideration. That is, on my interpretation of Quine's indeterminacy thesis, the indeterminacy attaching to the contents of our psychological states is quite like the indeterminacy attaching to the states of the accumulator (which follows from the indeterminacy concerning which of the two functions, $f$ or $f'$, it's computing).

But now let me describe a more complicated device. I call it the 'block-head'. It is a robot which looks and acts very much like human beings. However, its 'brain' is constructed quite differently. It contains zillions of tiny block letters, somewhat like the ones children play with. They are in fact English letters, and they combine to spell English words and English sentences. Also, inside the block-head's 'head', there are various boxes labelled 'BELIEF', 'DESIRE', and names of other common psychological attitudes. The boxes at all times contain groups of block letters that spell sentences (suitably regimented, so as to display their logical form). The blocks interact with each other by way of variously shaped hooks and crevices on their surfaces. If one were to watch the blocks' movements they would appear to follow the transformation rules of deductive logic, as well as those of some system of inductive logic. Finally, to cap the picture, it turns out that one can also find connections between these internal goings-on and the block-head's behavior, such that one can explain the latter in terms of the former.

On the computational model of mind, the model briefly sketched back in Section I, our brains are essentially like the block-head's. Certain neurophysiological structures supposedly spell out sentences in the 'language of thought'. Now, since there will certainly be subtle problems arising when we try to parse these neurophysiological structures into syntactic types, the case for the determinacy of psychological attitudes can only be better if one deals with a case like the block-head. Since this is easier to discuss anyway, let's just assume that the little block configurations in the blockhead play the same role as do neurophysiological structures in us. Now, does the indeterminacy we found with the accumulator carry over to the case of the block-head?

At first blush, there do seem to be crucial differences between the two cases. For one, it doesn't really make sense to consider any of the accumulator's states to be sentences. One could of course interpret its having a strip in a certain position as its believing-true the sentence 'The number $2^2$ is stored here'. But nothing about the way the strips interact makes use of the syntactic structure thereby attributed to the state in question. On the other hand, the workings of the strips constitute a nonsentential syntactic structure of a sort. For while the strips and marbles represent numbers, and the changes of state follow the rules of addition, their interactions are syntactic operations in the sense that it is by virtue of the 'shapes' of the representations that they affect each other (cf. Fodor, 1980). Still, one might argue that the syntax in this case is too simple, and too simply related to the world, to determine which of $f$ and $f'$ it is computing.

Even more important, however, there is nothing about its 'behavior' which exhibits sensitivity to intentional properties. If called upon to account systematically for its behavior, the requisite generalizations could be couched in terms of marbles, masses, and the like. Whereas, in the case of the block-head, assuming it does really act like a human being, it probably exhibits sensitivities to the sorts of properties which require an intentional explanation. Thus one might think this alone endows the posits of the explanations with the same ontologically genuine character as the posits of physics. (Cf. Fodor, unpublished).

Agreed, then, that the two cases are importantly different. Still, we can ask, isn't it possible, in the case of the block-head, just as in the case of the accumulator, that significantly conflicting interpretations could be given to its internal states, all of which are consistent with the relevant physical facts? As soon as one considers this possibility for the block-head, an ambiguity arises. For there are really two ways the psychological model of the block-head could be indeterminate. It might be that the identity conditions for the block letters, words, and sentences, are the same as those for English letters, words, and sentences. Therefore, the logical transformations are the same. However, this still leaves open the possibility that the *semantic* properties of the block words and sentences are indeterminate (cf. Rey, 1980). The second way the psychological model of the block-head can be indeterminate is for there to be conflicting ways to parse the block sentences, yielding different interpretations as to which logical/computational operations were realized in a given block interaction.[6] If one likes, this could be seen as a new form of the difference between the 'argument from below' and the 'argument from above' (though, of course, they do not literally correspond to Quine's distinction). At any rate, I will assume that both semantic indeterminacy and syntactic indeterminacy (which, of course, also gives rise to semantic indeterminacy in many cases) are at issue in our discussion. Both should be troubling to the advocates of cognitive psychology.

With the accumulator we could exhibit explicitly a case of indeterminacy. This clearly cannot be done for the block-head. For one thing, we don't even have a single actual theory of the block-head's inner workings, much less two, the way we had the functions $f$ and $f'$ in the accumulator case. So we'll have to settle for indirect considerations to judge whether indeterminacy applies to the block-head's psychological states. Well, why wouldn't there be more than one psychological model that fits the block-head? To answer this, the cognitive psychologist can provide two sorts of considerations which will address both syntactic and semantic indeterminacy. Basically, the claim is that internal computional role, together with the network of causal relations that obtain between the block-head and its environment, will suffice to determine a correct psychological model from among the initial alternatives.

However, at this point in time I don't see how we can know whether the internal causal mechanisms together with the causal relations holding between the block-head and the world can determine a unique, best fit between psychological theory and physical theory — or, better, between the alleged psychological facts and the physical facts. I said above that the accumulator case does not provide anything like decisive evidence that, when dealing with a full-blown psychological theory, the physical facts under-determine the psychological facts. Nevertheless, reflection upon that example reveals something. It reveals just how large, how open, the range of possible interpretations is for a given physical system. True, the physical goings-on within the block-head are structured much more finely, and complexly, than are those within the accumulator. True, we didn't have the kinds of systematic causal relations between the accumulator and its environment that could help to determine which of the two functions was the right one. But still, how many people would have known of the possibility that it was computing $f'$ had this not been explicitly pointed out? It makes me think that, maybe, there is always another way of interpreting a given physical system, even if nobody is clever enough to figure it out. But, of course, this is only an expression of confidence, not an argument. In the absence of a finished cognitive psychology and neurophysiology, I don't see how an argument can be given one way or the other.

This doesn't mean, however, that there is now nothing left to discuss. For not everyone agrees with Quine that if indeed the physical facts under-determine the psychological facts, then we can conclude that there is no fact of the matter concerning the psychological facts. One reason to argue the point, even though we can't as yet decide on the truth of the antecedent of this conditional, is that there is, as I tried to evoke, some intuitive plausibility to the belief that the antecedent is true. In fact many of Quine's strongest critics take issue more with his claim that anything follows from the hypothesis that the physical facts under-determine the psychological facts, than they do with his claim that it is true. For the remainder of this section, then, I will take up the arguments surrounding the inference that goes from the under-determination of psychological facts by physical facts to their ontological indeterminacy.

Immediately, however, we have to deal with the following objection. One might wonder whether the claim that there could be two conflicting psychological theories each consistent with all the physical facts is even coherent. That is, one might say that whatever difference there appears to be between the two theories is *merely* apparent, and we can tell this by the fact that there are no physical facts that could help us choose between them. To borrow a phrase of Quine's from another context, rather than two conflicting theories, perhaps what we have are one theory and two formulations of that theory (see Note 4). Quine, in one passage in Chapter 2 of *Word and Object*, explicitly acknowledges the legitimacy of this move. As he puts it:

> It may be protested that when two theories agree thus in point of all possible sensory determinants they are in an important sense not two but one. . . . Similarly one may protest that two systems of analytical hypotheses are, as wholes, equivalent so long as no verbal behavior makes any difference between them; and if they offer seemingly discrepant English translations, one may again argue that the apparent conflict is a conflict only of parts seen out of context. Now this account is fair enough . . . . the conflict is precisely a conflict of parts seen without the wholes. (1960, p. 78)

The reason Quine feels he can acknowledge the legitimacy of this move, and yet maintain his ontological thesis, is that this position actually gives him what he wants — namely, that only the physical facts are the 'hard' facts.

After all, what is one saying when one claims that an apparent conflict in one domain is only apparent because there is agreement in another domain? That sounds to me like a claim that what *really counts* is what goes on in the other domain. If not, why should agreement there dissolve the conflict in the first domain? Thus the no-conflict response to the indeterminacy thesis is actually a way of giving in to it; i.e. to the view that the domain of intentional ascriptions is one of 'soft' facts. This is why Quine is not bothered by that response. For the whole point of the indeterminacy thesis is to demonstrate the ontologically soft nature of intentional attributions. What Quine's opponents need to argue is that despite agreement on input/output relations and co-realizability as interpretations of the nervous system, formal incompatibility can reflect a real conflict; where 'real' implies that there is a fact of the matter concerning which interpretation is the right one.

Rorty, in his 'Indeterminacy of Translation and of Truth', argues just this point. He notes that Quine himself admits that the choice of translation manual made by the linguist from among those consistent with speech behavior, is not an arbitrary one. So, Rorty asks, why is it that these nonarbitrary canons of linguistic methodology should carry any less ontological import than those canons of scientific methodology which narrow the field of admissible physical theories? If in fact there is no epistemological difference between linguistics and physics, why, argues Rorty, should there be an ontological difference? He put the matter forcefully in this passage, where he comments on Quine's reply to Chomsky (quoted above, p. 274).

> As for 'the totality of truths of nature,' Quine seems to be saying that these range from statements about the distribution of particles to statements saying that so-and-so made the following sounds, but that when the man from Cassell's comes along and helpfully adds 'and those sounds meant . . .' he is not tacking on another 'truth of nature' but doing something quite different — not stating a fact at all. But this claim does not give an argument for an 'additional' indeterminacy; it just restates the point at issue. (Rorty, 1972, p. 452)

As I have been interpreting Quine's argument in this section, the indeterminacy thesis does not involve the claim that there is any epistemologically significant difference between the canons of linguistic, or psychological methodology and those of the natural sciences. So, there must be another, metaphysical reason for the claim that only what is determined by the physical facts genuinely constitutes facts at all. For Quine, as I interpret him, this claim is entailed by the thesis of physicalism. Physicalism, as Quine describes it, is the view that there is "no difference without a physical difference".[7] This view is elaborated in his article entitled 'Facts of the Matter' (Quine, 1978). Every event — a change in a state of affairs — is, on the physicalist thesis, a physical event. What is it to be a physical event? It is to consist in a change of at least one among a set of fundamental properties, or magnitudes, in some region of space-time. What properties are included in this set is to be discovered through the progress of physics. The heart of the thesis is not the designation of a branch of science called 'physics' as the basis for the set of fundamental properties. Rather, it is the belief that all objects — macro, micro, mental and nonmental — undergo changes of state determined (to the extent that there is determination) by the same fundamental properties. This doctrine is not intended to be primarily a philosophical one. Rather, this is seen as the working hypothesis of current physical theory. As Quine puts it, the physicist can be viewed as seeking an answer to the following question: ". . .what minimum catalogue of states would be sufficient to justify us in saying that there is no change without a change in those states?" (1978, p. 164).

He then goes on to summarize his argument for the indeterminacy thesis as follows:

> . . . I speak as a physicalist in saying there is no fact of the matter. I mean that both [translation] manuals are compatible with the fulfillment of just the same elementary physical states by space-time regions. (1978, p. 167)

Now I think Quine's opponents can be assumed to be physicalists as well. In particular, they probably agree that it is the business of physics to describe the world's causal machinery — to determine which changes in properties and magnitudes are those in virtue of which the events of this world take place. Thus the causal primacy of the physical is not at issue. The locus of disagreement must be in the inference from the causal primacy of the physical to its ontological hegemony. Why should one conclude from the fact that everything that happens in this world does so in virtue of changes in the fundamental physical properties, that only facts uniquely determined by those changes are hard, honest-to-God facts of the matter?

Perhaps from the very way the last question was phrased some find the answer obvious. Whatever metaphysical slack is left over after the causal determinants have been fixed — which includes not only every physical event which does take place, but also every physical event which could take place — has no way to be taken up. A complete ontological 'catalogue' would include the facts which are, as it were, pregnant with the counter-facts (by way of the causal determinants constitutive of the facts themselves), and that's all. I think Quine believes this is obvious, given that he goes no further than I have so far in defending his view. But I think at least a little more can be said, and needs to be said, to make the view as plausible as it is.

First of all, most mentalists who are also physicalists these days, believe that intentional states are kinds of functional states. Functional states are abstract states — defined in terms of inputs to and outputs from a particular functional system, and in terms of each other — realized in the physical states of a functional system. The difference between a functional state and the physical state which realizes it can be characterized as the difference between a second-order and a first-order state. This way of capturing the essence of functional states (or properties) is presented in Putnam's 'On Properties'. He says there:

> To say that a finite automaton has a certain machine table is to say that *there are*

> *properties* . . . which the object has . . . , and which succeed each other in accordance with a certain rule. Thus . . . [it] . . . is a *property of having properties which* . . . , it is of "second order" in the old Russell-Whitehead sense . . . . (1979, p. 313)

For a second-order attribution to be true of an object, then, there must be some, but it doesn't matter which, first-order attributions true of it, such that they satisfy the conditions defining the second-order predicate. In the case of functional definitions for mental states, the definition states that there must be some set of physical states (if we're physicalists) which causally relate to each other and stimuli and behavior (or just output) in a particular way. Except for mentioning the inputs and outputs, the condition the definition lays down for satisfying the predicate is purely second order. So, then, we see that all a functional (i.e. because second-order) predication requires to be true is for it to be realized in the object of which it is predicated.

But now look at the situation facing the cognitive/functionalist psychologist. We have been assuming that it is possible for two conflicting psychological models to be realized in the same physical system (biological organism, robot, or whatever). We have also seen that, insofar as psychological models are second-order (in Putnam's sense), they are truly ascribed to a person if the person's physical states (and causal relations to her environment) realize them. If you put these two claims together, they *entail* that there is an ontological indeterminacy about the facts described by the psychological models. One might see it this way. If functional states are causal roles, as many describe them, what else but the facts concerning the causal determinants in nature could render a functional ascription an honest-to-God matter of fact? This is how I interpret Quine's appeal, quoted above, to the "whole truth about nature."

There is one more argument I want to consider. Suppose someone countered by denying that psychological states were *merely* second-order states. True, their causal relations are second-order, in that it is the causal features of the realizations of psychological states which account for their causal efficacy. Nevertheless, they are *more* than causal roles picked out by second-order predicates. This 'something more' involves the suitability of the psychological description as a whole for the task of facilitating explanation and prediction of human behavior. Some second-order, psychological models are better suited to this purpose than others — they provide especially useful tools — and this is the heart of the canons of good methodology referred to earlier. On this view, to count as the content of someone's belief is to be described as such by the most workable functional theory available.

This response works only if one is willing to drop realism, which I assumed at the beginning of this section. Insofar as Quine himself describes theories generally as 'useful tools,' and the canons of scientific methodology as the pragmatically best organizers of our sense experience, he has no ground for the ontological distinction he makes between the facts of physics and the facts of psychology. This much was admitted at the end of Section II. But let's assume, what may ultimately turn out to be incoherent itself,[8] metaphysical realism about the facts claimed to obtain by physical theory. What does it mean to be a realist about these facts; that is, to what is it opposed? At the beginning of this paper I put it this way. I said that the soft, non-objective facts were those that obtained, to a crucial extent, in virtue of the interpretation given them, the way they are described, by a rational subject. While the objective, hard facts are those that obtain independently of, or invariantly across, the ways they are interpreted by rational subjects. Now, if we want to base the genuine factuality of psychological ascriptions on something about the way *we*, as rational enquirers, can deal with the theory making the ascriptions, it seems this could only be self-defeating. For isn't that to make the alleged facts relative to, and not independent of, the interpretive stance we take toward them? So long as we deny that the epistemological criteria we employ for choosing among *physical* theories are in any way constitutive of their truth, we can't both say otherwise with psychological theories and also claim that psychological facts enjoy the same ontological status as physical facts.

## IV

Let's take stock. We have seen that those of Quine's arguments for the indeterminacy thesis which derive from his empiricist and behaviorist leanings, do not hold up to scrutiny. Insofar as he lays down *a priori* constraints on the sorts of things a respectable psychology can talk about, he is violating his own (1969) strictures. That is, if there is no first philosophy, if epistemology becomes empirical, then we should follow psychological theory wherever it leads. If it leads us to positing a language of thought, then we should have no more scruples about this than we do about positing quarks. This attitude would be in the true spirit of 'Epistemology Naturalized'.

However, we also saw that a realist attitude about the posits of an intentional psychology can run into conflict with a realist attitude about the posits of physics. If, and this we saw could not be decided, it is possible for there to be two conflicting psychological theories which are both compatible with all the relevant physical facts, there is good reason to say that there is no fact of the matter as to which one is right. This is the central core of Quine's argument that remains after one strips off what I called his 'residual dogma of empiricism'. This is not intended to be an *a priori* truth, and it relies on no questionable epistemological premise. It is simply a matter of what follows from a commitment to realism and physicalism. To the extent that these commitments seem reasonable, and to the extent the posits of intentional psychology seem to be necessary for an adequate explanation of our behavior, we are faced with a dilemma. Do we take these posits seriously or not?

The force of this dilemma can be brought out as follows. Why do we feel compelled to adopt a realist attitude toward the theoretical entities of physics, anyway? It is that the best explanation for the phenomena we observe involves reference to the workings of these theoretical entities. If they are to actually explain the phenomena we observe, they must actually exist. Now, human behavior exhibits properties which, it is alleged, require us to posit internal, cognitive states to adequately explain them. How can we maintain the force of these explanations if we turn around and deny reality to the cognitive states upon which they depend?

If one accepts Quine's argument, as I have reconstructed it, it turns out that, among apparently conflicting psychological accounts of human behavior, all consistent with the same physical facts, there is no genuine distinction in fact. This applies not only to descriptions of internal psychology, but also to descriptions of behavior insofar as these are not couched in the language of the physical sciences. That is, our interpretations of the data to be explained — how stimuli and behavior are categorized in the first place — is indeterminate within the same constraint of consistency with the physical facts as is our interpretation of the internal goings-on. (This is because descriptions of behavior usually derive from hypotheses concerning its mental causes.) Thus, the ontological force of psychological explanations remains as strong as the ontological force of the facts standing in need of explanation. Now, one may insist that it is still the *physical* properties of behavior which, ultimately, stand in need of psychological explanation. But, by hypothesis, all of the apparently conflicting psychological theories in question can account for the physical events. So nothing, objectively speaking, remains unexplained.

If indeterminacy is in fact true, it will mean that the physical facts determine equivalence classes for psychological theories much wider than what consideration of the theories themselves would suggest. What we normally would consider clear cases of a difference in fact, and not merely a difference in manner of description, would turn out not to be cases of metaphysical difference at all. Should this situation make any difference to the way psychological investigation is carried on? I think not. It would be a great feat if we could come up with even one, relatively complete psychological model, much less many, that worked. It is clear that we will never be able to explain and predict human behavior without recourse to something like a cognitive theory (unless, of course, someone actually shows us how). It is just that we would have to realize that what appeared to us to be a significant difference between two candidate theories might not really be one after all. In particular, if the two candidate theories were descriptions of our epistemological competence, it could turn out that what appear to be clearly conflicting standards of the rational assessment of beliefs, are equally attributable to us as the standards we employ. Even if this had no effect on the practice of psychology — which, again, I say, it shouldn't — the philosophical implications would be enormous. For if, from the point of view of what there really is, apparently significant distinctions among attributions of mental states, and among standards of rationality, are collapsed, this would deeply undercut our conceptions of what it is to be a rational being, a being with a mental life.

There is another effect of indeterminacy as well, and this one has to do with physics, not psychology. The argument of Section III took for granted a realist attitude toward physics, and derived indeterminacy from there. But, if there is really indeterminacy about the contents of our mental states, is there a fact of the matter about what we are talking about in physics? Of course, as Quine is the first to admit, there can't be. So, while there are, metaphysically speaking, physical facts which determine all that is the case, we would, on Quine's thesis, never be in a position to say what they are. Or, one could put it this way. If theories are treated as abstract objects, formal systems together with their interpretations, it turns out that the physical facts determine some to be true and some false. However, among a significant subset of these theories, the physical facts don't determine which one it is that we believe.

One final word. One might find comfort in the possibility that there aren't, in the end, two rival psychological theories that are compatible with all the relevant physical facts (even ones we don't know about).[9] But two things should be kept in mind. First of all, I wouldn't bet on it. To repeat again, as the accumulator case makes vivid, there are many, many possible interpretations, the vast majority of which we haven't the imagination to consider. But even if the complete stock of relevant physical facts did determine a unique, correct assignment of contents to our psychological states, wouldn't the ontological firmness thereby gained be only derivative from that of the physical facts? Perhaps that wouldn't matter. Perhaps that is all we can ask for. Still, it does seem to vindicate at least the spirit of Quine's thesis; namely, that physicalism entails that our mental lives cannot metaphysically stand on their own.

*Bates College*

## NOTES

[1] The importance of these notions for psychology has been emphasized by Chomsky. See especially Chomsky (1975a, Ch. 1; 1984). Note that the sense of 'module' intended here is not quite the same as the sense intended by Fodor (1983).

[2] Cf. Horwich (1982, Note 5) where he too attributes this view to Quine.

[3] Cf. Antony (1982) for a discussion of the relation between holism about confirmation and holism about semantics. My discussion here relies upon the points made there to a significant extent.

[4] I take it that, at least minimally, realists must believe that there are states of affairs that constitute the truth conditions of theoretical sentences, and that these sentences are not merely inseparable components of the entire theory, the truth conditions of which are its observational consequences. However, granting this minimal realism, it is still an open question whether, when confronted with two formally incompatible but observationally equivalent theories, one does or does not admit the possibility that they conflict in fact, as well as in form. A maximal realist would allow this possibility, where the minimal realist might say that two such theories are really two different ways of saying the same thing. That is, as Quine (1975a) puts it, they could be considered two 'formulations' of the same theory. Paul Horwich (1982) argues that if there is a predicate mapping $f$ from one of the two theories, call it $T_1$, onto the other, call it $T_2$

(i.e. a mapping of the open sentences of $T_1$ onto those of $T_2$), such that if $T_1$ includes the sentence '*Fx*', then $T_2$ includes the sentence '$f(F)x$', we should consider $T_1$ and $T_2$ mere notational variants (I oversimplify somewhat here). Quine (1975a) seems to espouse a similar position, though he is more equivocal about even minimal realism. On this view, whether or not the ultimate theoretical conflict envisaged by the maximal realist is possible, depends on the possibility of constructing the requisite predicate mapping. If one thinks such a mapping always exists, one will not admit the possibility of such an ultimate theoretical conflict. As far as I can tell, nothing I say in this paper depends on whether 'realism' is to be taken in its minimal or maximal sense. Therefore, I will continue to use it ambiguously.

[5] However, notice that earlier in that paragraph he seems to qualify this 'realistic point of view' somewhat. The tension between his realist and positivist sympathies reveals itself here, as elsewhere.

[6] Of course these two sorts of indeterminacy cannot be kept strictly separate. This is because a block term's role in logical/computational operations will be part of the determination of its semantic interpretation. For instance, we can decide, because of the way two apparently different terms behave in these operations, that they are really two notational variations of the same term. When we leave the block world and deal with neurophysiological structures, this consideration looms large.

[7] Although Quine has often used this phrase in lectures, I cannot find it anywhere in print. However, he says (Quine 1978a), ". . . nothing happens in the world, not the flutter of an eyelid, not the flicker of a thought, without some redistribution of microphysical states." Also, see the quotation below (Quine 1978b).

[8] For arguments along these lines, see Putnam (1978).

[9] This seems to be Michael Friedman's (1975) attitude.

## REFERENCES

Antony, Louise: 1982. 'Realism and the Theory of Meaning.' Ph.D. Dissertation, Harvard University.

Chomsky, Noam: 1959. 'A Review of Skinner's *Verbal Behavior*,' *Language* **35**:1, 26–58.

Chomsky, Noam: 1975a. *Reflections on Language*. New York: Pantheon Books.

Chomsky, Noam: 1975b. 'Quine's Empirical Assumptions.' In Davidson and Hintikka (1975), pp. 53–68.

Chomsky, Noam: 1984. *Modular Approaches to the Study of the Mind*. San Diego: San Diego State University Press.

Davidson, D. and J. Hintikka (eds.): 1975. *Words and Objections; Essays on the Work of W. V. Quine*. Rev. ed. Dordrecht: D. Reidel.

Fodor, J. A.: 1975. *The Language of Thought*. New York: Thomas Y. Crowell Co.

Fodor, J. A.: 1978. 'Propositional Attitudes,' *Monist* **61**:4, 501–523.

Fodor, J. A.: 1980. 'Methodological Solipsism Considered as a Research Strategy in Cognitive Psychology,' *Behavioral and Brain Sciences* **3**:1, 63–72.

Fodor, J. A.: 1983. *The Modularity of Mind*. Cambridge, Mass.: Bradford.

Fodor, J. A.: (unpublished) 'Off the Slippery Slope, or Why Paramecia Don't Have Mental Representations.'

Friedman, Michael: 1975. 'Physicalism and the Indeterminacy of Translation,' *Noûs* **9**, 353—373.

Harmon, Gilbert: 1975. 'An Introduction to "Translation and Meaning", Chapter 2 of *Word and Object*.' In Davidson and Hintikka (1975), pp. 14—26.

Horwich, Paul: 1982. 'How to Choose Between Empirically Indistinguishable Theories,' *Journal of Philosophy* **79**, 61—77.

Putnam, Hilary: 1978. 'Realism and Reason.' In *Meaning and the Moral Sciences*. London: Routledge and Kegan Paul.

Putnam, Hilary: 1979. 'On Properties.' In *Philosophical Papers*, Vol. 1. 2nd ed. Cambridge: Cambridge University Press, pp. 305—322.

Quine, W. V. O.: 1960. *Word and Object*. Cambridge, Mass.: MIT Press.

Quine, W. V. O.: 1969. 'Epistemology Naturalized.' In *Ontological Relativity and Other Essays*. New York: Columbia University Press, pp. 69—90.

Quine, W. V. O.: 1970. 'On the Reasons for Indeterminacy of Translation,' *Journal of Philosophy* **67**, 178—183.

Quine, W. V. O.: 1975a. 'On Empirically Equivalent Systems of the World,' *Erkenntnis* **9**, 313—328.

Quine, W. V. O.: 1975b. [Reply] To Chomsky. In Davidson and Hintikka (1975), pp. 302—311.

Quine, W. V. O.: 1978a. [Review of Nelson Goodman's *Ways of Worldmaking*] *New York Review of Books* **25**:18, p. 25.

Quine, W. V. O.: 1978b. 'Facts of the Matter,' *The Southwestern Journal of Philosophy* **9**:2, 155—169.

Rey, Georges: 1980. 'The Formal and the Opaque,' *Behavioral and Brain Sciences* **3**:1, 90—92.

Rorty, Richard. 1972. 'Indeterminacy of Translation and of Truth,' *Synthese* **23**, 443—462.

ABNER SHIMONY

# COMMENT ON LEVINE

Most of my comments will concern the first section of Levine's paper, in which he agrees with Quine's assimilation of epistemology to psychology but differs from him by proposing that the appropriate psychological theory for this purpose is cognitive rather than behavioral. Levine says in effect (p. 262 and Note 1) that if the stimulus-response analysis of behaviorist psychology is inadequate to account for verbal behavior, then *a fortiori* it is inadequate to account for the intricacies of proposing hypotheses, gathering data, searching, assessing, and inferring that constitute the epistemic activities of ordinary life and of scientific research. I agree with him on the relative merits of behaviorist and cognitive psychology and believe that an adequate naturalistic epistemology must make extensive use of the latter. I do not agree, however, that a naturalistic point of view requires the *assimilation* of epistemology to *any* kind of psychology.

Epistemology is a normative discipline, which tries to formulate criteria for assessing human claims to knowledge. It is not the generic trait of being normative, however, that makes epistemology resistant to assimilation by psychology, but rather the specific character of the norms which it hopes to achieve and the specific limitations of human capacities. Electrical engineering is a normative discipline, and yet there is a clear sense (despite Polanyi's objections) in which it can be assimilated to physics: once the desiderata for devices are stated, good engineering consists in finding controllable boundary and initial conditions which yield the desired outputs in strict accordance with physical law. An analogous relation of epistemology to psychology does not hold, for reasons given below.

Of the two epistemological problems which Levine considers, only the second — "How are (scientific) claims confirmed?" — evokes his ideas on how there can be some normative content in an epistemology which is assimilated to psychology: "In particular, two notions are crucial in this regard . . . the competence/performance distinction and the notion of modularity. . . . Thus, a theory of the competence of the human belief-confirmation capacity would serve as a normative



*Abner Shimony and Debra Nails (eds.), Naturalistic Epistemology*, 291—294.

standard for the evaluation of the reasonableness of belief." Levine's notions are well chosen, in two respects. First, they refer to teleological aspects of human activity, and obviously some reference to teleology is essential if one is to account for the normative character of epistemology. Second, the notions are fruitful — the first in linguistics, the second in the analysis of a whole range of complex systems, from computers to vertebrates.

Nevertheless, the two notions and the theses in which they are embedded are seriously deficient for the purposes of epistemology. If one accepts the conception of distinguishable cognitive modules, then presumably one would say that the mind as a whole performs well when these modules are integrated and subordinated to some overarching teleology of the person. What might this overarching teleology be? Personal survival is a candidate, but there are plentiful additional and sometimes competing candidates, depending upon circumstances: satisfaction of curiosity, religious exaltation, security for one's family, military victory, etc. (For present purposes there is no need to discuss the sociobiological thesis that all these *teloi* are subordinated to the ultimate grand aim of propagating one's gene pool.) Judiciousness in assessing the truth of propositions is of different value to different ones of these overarching aims (e.g., "In war the first casualty is the truth."). And even if some degree of sensitivity to the truth is indispensable in pursuing any of these aims, that sensitivity can be highly selective.

Since the integration of the cognitive modules does not provide the normative standards for evaluating epistemic claims, perhaps we should look in the opposite direction — at specific and specialized modules, ideally unencumbered by the passions or the propensity for wishful thinking. Such specialized modules resemble Piaget's cognitive schemata, and also the classical faculty of reason. I agree that there are processes of inference, to a large extent genetically determined, which operate with remarkable accuracy in the circumstances of ordinary life. However, even if one concedes the utmost to genetic determination and plays down the contribution that environmental regularities on the one hand and culture on the other hand make to the development of reasoning, one still does not have a module sufficient to achieve the aim of epistemology. The short-fall is most evident in sophisticated scientific inference. The controlled deployment of analogical reasoning, of the hypothetico-deductive method, of probability theory, of error theory, of criteria of simplicity, of theoretical constructs, and of mathematical analysis may perhaps be an exercise of our belief-confirmation capacity, but it goes far beyond whatever is genetically determined in that capacity.

Far more promising than Levine's assimilation of epistemology to psychology is the recognition of both a biological and a cultural matrix of inquiry, as proposed by one of the pioneers of naturalistic epistemology, John Dewey (1938). For cultural, historical, and biographical reasons the search for the truth about nature became the overarching aim of a few remarkable individuals and, eventually, of a few institutions within the larger society. The remarkable individuals exercised their innate belief-confirmation capacity with great effectiveness, always of course in the context of some antecedent achievements, and they greatly refined this innate capacity. The articulation, the enrichment, and the finetuning of the procedures of scientific inference were cultural accomplishments with no definite historical beginning, but occasionally with discernible historical advances. These advances usually resulted from reflection upon exemplary scientific discoveries or occurred in tandem with them. It may be claimed, of course, that these discoveries only illustrated the genetic competence of one of our mental modules and provided the occasions and the stimuli for articulating the underlying rules of operation of that module. Such a claim is implicit, for example, in the program of formulating scientific inference within the framework of personal probability theory. I am very skeptical of such claims, however, and have argued in my contribution to this volume and elsewhere that there is an uneliminable *a posteriori* element in scientific methodology, which is learned in the actual world and is applicable in the actual world but not in all possible worlds.

Even if my doubts can be answered, and normative standards for evaluating epistemic claims can really be found in the rules governing a specific module of the mind, there still would remain a sense in which epistemology would resist complete assimilation to psychology. It would be a task of epistemology to exhibit the primacy of the belief-confirmation module over all the other modules which have competing aims. Since all the modules are equally subsystems of the mind, operating by rules and principles which are equally natural, the primacy could only be established *a posteriori* — by investigating which of the competing propensities of the mind is most conducive to the aims of epistemology. But in the course of this investigation, the epistemologist must make many epistemic claims, both about the environment and about the mind itself. Consequently, the enterprise of epistemology is inevitably self-referential and dialectical, in a way that normal sub-disciplines of psychology are not.

The bulk of Levine's paper is an examination of the compatibility of intentional psychological statements with the metaphysical thesis of physicalism. He focuses upon one of the desiderata of a naturalistic epistemology: the 'closing of the circle' of epistemology and metaphysics, so that each meshes with and complements the other. Levine agonizes over the dangers to closing the circle on his own terms. A scenario that worries him is

> that the physical facts determine equivalence classes for psychological theories much wider than what consideration of the theories themselves would suggest .... In particular, if the two candidate theories were descriptions of our epistemological competence, it could turn out that what appear to be clearly conflicting standards of the rational assessment of beliefs, are equally attributable to us as the standards we employ. (p. 287)

I am skeptical of the fruitfulness of scenario-writing of this kind, when the entire investigation is conducted at a high level of abstraction — when the class of physical theories, the class of psychological theories, the class of total bodies of evidence, etc., are deployed so sweepingly that one has only the vaguest idea of the extensions of the classes. Rather than agonize that "the philosophical implications would be enormous" if the scenario should turn out to be true, I recommend the philosophical strategy of Crossing That Bridge When We Come To It. And I shrewdly suspect that we shall never come to it, because of the discrediting of physicalism, if for no other reason. In view of the richness of cognitive psychology as compared with behavioristic psychology, the breakdown of a program of reconciling cognitive psychology with physicalism could be construed as one more *reductio ad absurdum* argument against physicalism.

*Boston University*

## REFERENCE

Dewey, John: 1938. *Logic: The Theory of Inquiry*. New York: Holt.

JOSEPH AGASSI

# COMMENT ON LEVINE

I approach Levine's paper with a very mixed response-pattern, and this fact makes me as nervous as any Pavlov dog. In my disposition to respond with pleasure to Quine's writings but with suspicion to any behavioristic writings, I was naturally conditioned to gloss swiftly over behaviorist passages in his writings.. I always responded, then, with surprise and incredulity, to friends and colleagues citing Quine's behaviorism.[1] The displeasure incurred by finding myself not well acquainted with the works of an author I have invested some effort in an attempt to master should now disappear with my reading of Joseph Levine's coherent interpretation of Quine's philosophy as inherently physicalist and (thus) as inherently behavioristic[2] (although without the ability to justify science as a whole except scientifically). It is still a fact that any detailed reduction of any theory of scientific activity to psychology is repellent to me on account of the fact that psychology is universal and science is a product of some cultures and so is not universal. Here, then, is the analysis of my response-pattern to Joseph Levine's paper. It is for me both anxiety reducing and anxiety raising.

Why should readers of Quine or of Levine care about my resonse-pattern as applied to my reading of Levine on Quine? Comments are supposed to be critical, not psychological self-observations.

What, however, is criticism? From the standpoint of behaviorism, it is a stimulus pattern that induces, when valid, an expected characteristic response pattern. Hence, to make my response into criticism, I should expect that as stimulus it will elicit in many readers of these lines a response-pattern recognizable as the disposition to assert that it possibly is a valid criticism. I do so expect. To be more precise, I do so expect on the assumption that my reader will go on reading the series of responses to Joseph Levine's paper presented here.[3]

The attempts to apply behaviorism to itself, in the intent to validate it, as well as in the intent to criticize it, are to date inconclusive. This means, behavioristically, that one conditioned to study behavior behavioristically is not deconditioned when studying, and that one conditioned to respond negatively to behaviorism is not deconditioned



*Abner Shimony and Debra Nails (eds.), Naturalistic Epistemology*, 295—298.

when facing the same situation either. Nevertheless, as long as behaviorism was not applied to the study of the history of science and as long as no attempt in this direction in this way was rewarded, but other, non-behavioristic, theories had some measure of reward, behaviorists should have recognized this reward and should either have supported such researches or approached them with a mixed attitude. A behaviorist aware of past failures of attempts to apply behaviorism to the history of science and of past successes to apply non-behaviorist theories to the history of science may try to apply behaviorism to the history of science. He may try harder than his predecessors. What, however, is the behaviorist equivalent of the non-behaviorist 'try harder'? We must know, since trying harder is a regular ingredient in research. Yet trying harder may be explained behavioristically as pitching a stimulus to action with a stimulus to non-action or to alternative action, e.g. by raising the expected reward. Is the reward of the successful application of behaviorism to scientific research in general, and to behavioristic research in particular, highly rewarding? I do not know. Yet the rewards of applying non-behaviorism to the same must be rewarding enough, since it is performed all the time. In other words, although the application of behaviorism to itself is not yet an argument one way or another, the application of behaviorism and of non-behaviorism to the history of science frustrates behaviorism and confirms non-behaviorism.

This argument is not conclusive and may very will be invalid. For one may attempt to translate all successful psychological studies of science and its history into the behaviorist system. And this is not difficult at all. Thus, incentives for scientific research may be stimuli. Query: are these universal? If not, why not? Assume that they are universal. Why, then, are most known cultures (magical and thus) not scientific? Answer, the stimuli accumulate slowly and the counter-stimuli [superstitions; prejudices; systems of thought competing with science such as religion and metaphysics] are also countered slowly. Assume that they are not universal. How then is their specificity explained with the use of the universal theory? Clearly, the difference must lie in the difference in initial conditions, the stimuli and the counter-stimuli rate of accumulation and the dependence of these rates on local physical conditions. Hence, whether we assume the incentives to scientific research specific — on a superficial level — or universal — on a more basic level — the answer is the same.[4]

This discussion establishes two very significant corollaries. The first, the use of psychology in the study of the phenomenon of science — whether as a cultural phenomenon, as a personal (biographical) phenomenon, or as the history of specific events in the growth of science — is comforting to the behaviorist because behaviorism should be applicable to all human phenomena or else it is not true. No effort in this direction is detrimental to behaviorism, even if it is the successful application of a competitor to behaviorism: the behaviorist can always try harder. The second corollary concerns the scientific status of behaviorism when considered (as a naturalistic epistemology) as a theory applicable to the phenomenon of science, and may be disturbing for the behaviorist. It is that at least for that purpose behaviorism is quite metaphysical and thus non-empirical, as the comment on the first corollary indicates.

Joseph Levine will not disagree with the second corollary as far as Quine is concerned. He shows that Quine's behaviorist view of science rests on his metaphysical views, such as physicalism, an empiricist theory of meaning, and more. As suggested here, this is a general truth which a behaviorist cannot deny. Whether he will agree, I cannot say.

Behaviorists, gratified by what they consider successes of behaviorism and frustrated by their failure to explain science as a phenomenon are as free to respond — they may respond the one way or the other, or throw a neurotic fit like Pavlov's dogs or get stuck like Buridan's ass. Once freed to make a decision, they may feel freed of the fetters of their stimuli.[5]

*York University* and
*Tel-Aviv University*

## NOTES

[1] This response accords with the theory that the same stimulus — a sentence from the works of Quine — may act differently on the same organism — myself — in different contexts: citing a passage is a context which enhances it as a significant stimulus.

[2] This displeasure accords with the theory that the removal of anxiety is a reward — *à la* Pavlov — and that finding patterns that can be firmly established or reinforced in further experiences is anxiety removing — *à la* Pavlov and *à la* Skinner.

[3] This expectation accords with the behavioristic view of science as a merely more complex than usual behavior pattern, but otherwise not outside the stimulus-response pattern of the behaviorist's usual lore. Also, it accords with Skinner's view of verbal behavior.

[4] It is that the same stimulus can elicit different responses in different contexts.
[5] See Agassi (1977) for the proof that behaviorism can explain no verbal behavior. See Fried and Agassi (1983) for a discussion of the holistic character of behaviorism (i.e. the high context-dependence of the response).

## REFERENCES

Agassi, J.: 1977. *Towards a Rational Philosophical Anthropology*. The Hague: Nijhoff.
Fried, Y., and J. Agassi: 1983. *Psychiatry as Medicine*. The Hague: Nijhoff.

ABNER SHIMONY

# INTEGRAL EPISTEMOLOGY*

This essay is both an appreciation of the epistemological contributions of Donald Campbell and a statement of an epistemological program which is different from his in several respects.

In a lecture to the Boston Colloquium for the Philosophy of Science in 1977, he said:

> What I am doing is "descriptive, contingent, synthetic epistemology." . . . I make a sharp distinction between the task and permissible tools of descriptive epistemology on the one hand and traditional, pure, analytic, logical epistemology on the other. Descriptive epistemology is a part of science rather than philosophy, as that distinction used to be drawn by philosophers. It is science of science, scientific theory of knowledge, were those terms not too pretentious for the present state of the art.
>
> While I want descriptive epistemology to deal with normative issues, with validity, truth, justification of knowledge — that is, to be epistemology — descriptive epistemology can only do so at the cost of presumptions about the nature of the world and thus beg the traditional epistemologist's question [Campbell, 1977a, p. 1].

Campbell's resolute restriction of his investigations to descriptive epistemology is both his great strength and his weakness. It is his strength because it frees him, at one stroke, from the slow-paced type of inquiry that dominates the literature of analytic epistemology; for example, "When I see a tomato there is much that I can doubt" (Price, 1932, p. 33). His investigations lead out of the study into the open air. There is a wonderful sweep in his survey of the stages of cognitive development (Campbell, 1974, pp. 422—434). In the perspective that Campbell offers, nature is in no way subservient to humans; however, because of the sequence of adaptations to nature which occurred in the human phylogeny, we are able to achieve something approaching objective knowledge.

* The research for this essay was supported in part by the National Science Foundation. The essay was first published in *Scientific Inquiry and the Social Sciences: A Volume in Honor of Donald T. Campbell*, ed. by M. B. Brewer and B. E. Collins (San Francisco, Washington, London: Jossey-Bass Publishers, 1981, 98—123). It is reprinted here, with small deletions and changes, with the kind permission of Jossey-Bass Publishers.



*Abner Shimony and Debra Nails (eds.), Naturalistic Epistemology*, 299—318.

The weakness in Campbell's program is that the traditional problems of analytic epistemology continue to be haunting, especially when answers to some of them are postulated as preconditions for his own investigations. He postulates an ontology of independent real objects: that there are entities in the universe which do not depend for their existence upon their being perceived or known by human beings. He accepts without argumentation the correspondence theory of truth: that a sentence is true if and only if the state of affairs which it expresses is the case, so that the truth of a sentence does not depend upon its being believed, or upon the utility of believing it, or upon the existence of evidence supporting it. He postulates a causal theory of perception: that one can understand the content of perceptual experience only by taking into account the causal relations which link the perceiver to objects existing independently of the perceiver's faculties. If Campbell's descriptive epistemology entirely abstained from normative questions, then there would be nothing wrong in principle with such postulation. It would be analogous to putting one's trust in the mathematicians and assuming the correctness of useful mathematical theorems without checking the proofs oneself. Since, however, Campbell wishes to deal with normative questions, especially with the justification of knowledge, he cannot avoid the problem of justifying his postulates.

This essay will propose an *integral epistemology*, in which certain methods of descriptive epistemology (which Campbell espouses) and certain methods of analytic epistemology (from which he abstains) are combined for the purpose of rationally assessing claims to human knowledge. It is anticipated that the results of scientific investigation about human beings will shed light on the reliability of human cognition, and reciprocally that adequate justification can be given for the presuppositions of scientific investigations. The proposed integral epistemology is unequivocally naturalistic, following Campbell not only in his general thesis that a necessary condition for understanding human cognition is to see man's place in nature but also in his insistence that detailed attention to the sciences is indispensable for solving epistemological problems. The proposed approach differs from his in its envisagement of a dialectical structure of epistemology and in its resort to methodological, decision-theoretical, and semantic analysis.

A disclaimer should be made at the outset with regard to novelty, not just because of the usual obligation to acknowledge intellectual indebtedness but because an integral epistemology is a synthesis by its conception. A naturalistic view of human knowledge is at least as old as Aristotle's *De Anima*, though it has been greatly expanded by applications of the theory of evolution; and the thesis that epistemology has a dialectical structure goes back, of course, to Plato. Sustained attempts to incorporate naturalistic epistemology into a dialectical framework are, however, uncommon. Among classical philosophers, Peirce seems to come closest to the integral epistemology which I envisage, and the contemporary philosophers who approach it most closely – notably, Quine (1969) and Rescher (1977) – acknowledge their affinity to him. Rescher explicitly characterizes his epistemology as dialectical, but Quine (1974, p. 137) does not seem to like the word, in spite of his espousal of a philosophical methodology which seems to me unequivocally dialectical. A dialectical naturalism is also outlined in Shimony (1970).

## SOME DIALECTICAL THESES

A dialectical method offers a reasonable resolution to a difficulty which is intrinsic to the enterprise of epistemology. If epistemology is a comprehensive study of human cognition, then the question of the validity of its own procedure falls within its domain. How, then, does the enterprise rationally begin? One radical type of answer to this difficulty is to identify incontrovertible initial propositions, guaranteed, for example, by clarity and distinctness, by self-certification, or by intuition; a radical answer at the opposite extreme is skepticism. In the work of a number of philosophers who call themselves "dialectical," one finds several characteristic features which permit them to navigate with control between these two extremes. Tentative suppositions are accepted at the beginning of inquiry, but they are subject to criticism and may be revised and refined on examination. Fundamental principles are, then, the end rather than the beginning of inquiry. The terms in initial suppositions are often vague, clarification of meaning being inseparable from the assessment of propositions. Most important, the dialectic is open, with no foregone conclusions and no suppositions that are so entrenched that they cannot be critically evaluated. In particular, the openness applies to the dialectical method itself. Inquiry may show that the dialectical method is in need of supplementation or that it is only a temporary expedient on the way to a deeper method.

Because it is characteristic of dialectical methods to examine the initial suppositions of inquiry and to reflect on the dialectical process itself, there is a widespread opinion that these methods entail a commitment to a coherence theory of truth — the theory that the truth of a sentence consists in its properly fitting into a system of assertions. The opinion is correct concerning some historically important versions of dialectic — notably, Hegel's — but it is not correct as a generalization. The integral epistemology envisaged in this paper agrees with Campbell's recommendation "to accept the *correspondence meaning of truth and goal of science*, and to acknowledge *coherence* as the major but still fallible *symptom of truth*" (1977a, p. 20).

In the subsequent discussion of connections between description and justification, four procedural or organizational theses will be deployed. For the most part, these theses are implicit in the general features of dialectical methods which have already been noted.

(1) *Commonsense judgments about ordinary matters of fact have a prima facie credibility, which should not be discounted without clear positive reasons.* The distinctions between truth and falsity, between reliability and unreliability, between appearance and reality are habitually made in ordinary discourse, which consists largely of expressions of the kinds of judgments mentioned in thesis 1. Perhaps anticipations of these distinctions can even be found in stages of individual development before ordinary discourse is mastered, but one can hardly doubt that a secure achievement of a commonsense view of the world is a precondition to any kind of sophisticated examinaton of these distinctions. Consequently, a sweeping skepticism about the reliability of commonsense judgments undermines the rough-hewn contrast between, for instance, true and false judgments or real and apparent properties of things, with the danger that the dialectical process of refining these epistemological distinctions will be aborted at an early stage. It must be emphasized that the presumption of truth for the specified class of judgments does not imply a full commitment to them.

(2) *The road to inquiry should not be blocked.* This thesis is Peirce's "supreme maxim of philosophizing" ([1897] 1931, p. 135). The thesis is implicit in the openness of the dialectic, which was noted previously, but it is a reminder that openness is not preserved without some effort. In a later section (headed "Evolutionary Considerations"), this thesis will be applied in the course of evaluating criteria of meaningfulness.

(3) *Epistemology and knowledge of the world* (*a vague phrase, intended to refer both to the sciences and to metaphysics*) *should mesh*

*and complement each other.* This thesis really has the status of a desideratum, for there is no a priori guarantee that we are capable of "closing the circle," in the sense of giving an objective account of the knowing subject as an entity in the world in a manner that adequately explains its cognitive capacities. Nevertheless, thesis 3 is reasonable, and its failure to hold in a particular philosophical system is prima facie evidence that something is amiss, unless convincing reasons are given why it should fail (as Kant, of course, tries to do in the *Critique of Pure Reason*).

(4) *A vindicatory argument — that is, an argument to the effect that a certain method, M, will yield good approximations to the truth in a certain domain if any method will, so that nothing is lost and possibly something is gained by the use of M — is an acceptable form of epistemological justification.* This thesis has a decision-theoretical character. It is similar to recommending a strategy that performs as well as or better than all other strategies in a game. The thesis has been used by Hans Reichenbach, Herbert Feigl, Wesley Salmon, and others in discussions of the justification of inductive inference. It provides, incidentally, a good example of a somewhat vague concept that becomes clarified as a result of dialectical examination. In common sense there is a vague concept of justification, the explication of which has been a major philosophical problem. The retreat, under the critical scrutiny of Hume and others, from attempts to provide a justification in a strong sense for inductive inference has constituted a remarkable extended dialogue, and the resort to vindication has been a way of salvaging something from the epistemological optimism of an earlier epoch. An important warning is that, despite its reasonableness as stated, thesis 4 is difficult to apply in actual epistemological argumentation, because one seldom can be confident that the condition "nothing is lost and possibly something is gained" is satisfied.

Although these four theses are obviously in need of further discussion, it may be more illuminating to aim at clarification in the course of applying them to concrete epistemological problems than to analyze them further in an abstract way. The following three sections will examine the epistemological relevance of three different bodies of factual information about human cognition, together with some non-descriptive considerations. The section headed "Empirical Psychology and the Reliability of Perceptual Judgments" will discuss the reliability of perceptual judgments in the light of psychological studies of perception and cognition. The section on "Evolutionary Considerations" will consider the implications of evolutionary biology for the scope of human knowledge. And the section headed "Inductive Inferences and Vindicatory Arguments" will outline a program for the justification of inductive inference, involving among other things some reliance on the history of science (which is entitled to be recognized as part of descriptive epistemology). It must be emphasized in advance that answers to epistemological problems cannot be anticipated simply as corollaries of scientific results. An appropriate deployment of factual information is impossible without analysis, if only because of the considerable differences in the questions typically posed by scientists and by epistemologists. This warning is not a revocation of claims for the power of a naturalistic point of view but is, rather, a reassertion of the conception of an integral epistemology, in which factual information is indispensable but not self-sufficient.

## EMPIRICAL PSYCHOLOGY AND THE RELIABILITY OF PERCEPTUAL JUDGMENTS

An index of the significance of empirical psychology for epistemological purposes is the breadth of the overview that it provides of some of the traditional debates of analytic epistemology. The strong points on opposite sides of debates are often seen to fall into place; and, even more impressively, the naturalistic point of view is often completely at ease with conclusions which analytic epistemologists seem to have reached reluctantly, with a sense of abnegation and retrenchment. I shall illustrate these claims by examining the epistemological problem of the reliability of perceptual judgments. (Another illustration, concerning the extent to which observation is theory laden, is given in Shimony, 1978.)

Within empiricism there has been a major debate (Swartz, 1965) between sense data theorists (such as Price and C. I. Lewis) and their critics (such as Firth and Quinton). The issue can be posed more generally than it usually is: Are there intermediate steps between the physical stimulation of the sense organs and the final perceptual judgment which justify or support the perceiver's final judgment? The candidates for the intermediate steps are not restricted to sense data. Few, if any, authors dissent from the proposition that there are unconscious causal links, at least partly neural in character, in the total process of perception, but there is disagreement about the significance of these links for claims of knowledge.

Among the arguments presented by sense data theorists, there is one that generically favors epistemic mediation, without specifying its character. The argument is that a perceptual judgment, of the sort that can be expressed in a statement about physical objects, has a content that could not be summarized in any finite body of observational data; the judgment implies, for example, what would be seen if the identified object were viewed under normal conditions from any of an infinite number of vantage points. But a judgment of such strength must be an extrapolation from the actual observational base. That sense data are the entities that mediate between physical stimuli and final perceptual judgments can be demonstrated, according to sense data theorists, in an indirect way by consideration of illusions. Since an imitation often is classified perceptually as a real object of a certain kind, something common must be presented to the observer in the two cases. A reasonable candidate for the common element is *appearance*, which is interpreted as a network of sense data.

This reasoning has frequently been criticized. Quinton (1965), for example, argues that the identification of the locution "This appears to be $\phi$" with "There is a $\phi$ sense datum" is mistaken. He claims that the primary usage of "appears" is not to signal a definite claim to a meager bit of knowledge about the state of the subject but, rather, to signal a tentative belief in a proposition about ordinary objects. He does, however, admit that there is a secondary sense of "appears," which is applicable when persons turn their attention away from things and toward their own awareness — as, for example, in painting impressionistically or taking audiological tests. This possibility of switching to what he calls a "phenomenological frame of mind," however uncommon its occurrence may be, seems at first to support the claim that sense data are revealed directly by introspection. The revelation occurs when one performs upon oneself the operation called "perceptual reduction," in which one attempts to eliminate all elements of experience that are not unequivocally "given" — in particular, all conceptual contributions (Firth, 1965, p. 235).

The critics of sense data theories insist, however, that the phenomenological frame of mind is incompatible with the normal object-perceiving state and that characteristics discerned in the former by introspection cannot legitimately be imputed to the latter (Firth, 1965, pp. 236—237; Quinton, 1965, p. 507). Firth (1965, p. 214) asserts, furthermore, that "sensations do not occur as constituents of perceptions, but at most only as complete and individual states of mind." And Quinton (1965, p. 510) says, "When we transfer our attention from objects to experience, an enormously richer awareness is obtained. We then suppose that we were in fact having experiences of as complex and detailed a kind while attending to the objects, although we were unaware of the complexity and detail. This move is not inference supported by recollection, but a convention."

It follows from this sharp dichotomizing of mental states that perceptual judgments cannot be interpreted as the results of unconscious inference from premises concerning sense data. This conclusion does not require the rejection of unconscious mental processes. Rather, it rests on a sharp distinction between causes and reasons. Whatever the causal importance of the unconscious processes, they cannot be supposed to have the status of premises in the epistemic system of a subject who is operating in the normal mode of perception. Quinton (1965, p. 525) softens this conclusion somewhat: "Perception is an intelligent activity (not an infallible reflex). . . . But not all intellectual processes are types of reasoning." Unfortunately, however, he provides no explanatory details. I suggest that one of the important ways in which empirical psychology can illuminate traditional epistemological debates is by providing some detailed information about the "intelligent activity" at which Quinton hints.

I shall now list five propositions, expressing experimental results and theoretical ideas of recent psychology, which are relevant to the problem of epistemic mediation in perception. It should be noted in advance that the five propositions are far from being straightforward matters of fact. Each of them contains one or more terms — like "information" in proposition 1 and "expectancies" in proposition 4 — which should be clarified by more conceptual analysis than there will be space for in this paper. That these propositions are drawn from different schools is not merely opportunistic, for I am convinced by a synoptic view (see Neisser, 1976, p. 24) which combines elements from information-processing theories, from hypothetico-deductive models (especially Bruner and Gregory), from Campbell's (1956) analysis of perception as substitute trial and error, from theories which emphasize the richness of stimulus information (J. and E. Gibson), and from developmental psychology (especially Piaget).

(1) There are stages in the process of perception which are fairly well defined temporally and describable in such terms as the selection, transformation, encoding, storing, and use of information. Especially relevant for the debate between sense data theorists and their critics is the experimental evidence for a very brief period of iconic storage, in which an image is presented as a whole, as if offered for further examination, and a longer period, in which features are analyzed and the resulting information is stored in a short-term memory (Neisser, 1967, pp. 89—90, p. 301).

(2) The experimental data concerning visual and auditory pattern recognition favor a hierarchically arranged feature-analysis mechanism, rather than, for example, a template mechanism (Neisser, 1967, ch. 3).

(3) Feature analysis is not merely a matter of making decisions but requires a constructive activity of synthesizing a perceptual object: "The mechanisms of visual imagination are continuous with those of visual perception" (Neisser, 1967, pp. 94—95).

(4) The perceptual process is one of active search, in two different ways. First, there is an internal search for appropriate features and appropriate concepts (Bruner, 1957). Second, there is a search of the environment for additional stimulus information to confirm or disconfirm expectancies aroused by earlier perceptual activity (Gibson, 1950; Campbell, 1956; Neisser, 1976).

(5) The generic trait of search for significant stimulus information is innate, since it is exhibited in newborn infants (Bower, 1966). However, the ability to search with increased control, purposefulness, and sensitivity to salient features of the physical stimuli is a skill, which is developed naturally by a growing child interacting with the environment but which can be greatly enhanced by training (Gibson, 1969).

The epistemological importance of these results and theoretical ideas goes beyond adjudicating between the sense data theorists and their critics, though one can score points for and against both sides. It seems to count against the sense data theorists that the stage of perception which is closest to the physical stimulus and least modified by subjective processing is epistemically labile: "After a tachistoscopic exposure in Sperling's . . . experiment, the subject feels that he 'saw' all the letters, but he cannot remember most of them. The uncoded ones slip away even as he tries to grasp them, leaving no trace behind. Compared with the firm clarity of the few letters he really remembers, they have only a marginal claim to being called 'conscious' at all" (Neisser, 1967, p.

301). On the other hand, the occurrence of a stage of feature detection as part of normal perception of objects is a point in favor of the *general* position of the sense data theorists that epistemic mediation occurs in perception, with the qualification that the identification of a feature may not be a conscious occurrence. In spite of this qualm, however, the evidence for a temporal fine structure to perception at least indicates greater continuity between the "phenomenological frame of mind" and the normal state of object perception than the critics of sense data theories admitted.

What empirical psychology is beginning to provide is a view of perception as an intelligent activity in a way that none of the opposing schools within empiricism explicitly formulated (though the quotation from Quinton shows that some epistemologists vaguely foresaw the possibility of an extension of the conception of rational process beyond ordinary inference). In this view, the remarkable reliability of perception under normal circumstances — reliability in the sense of agreement on the whole between any confident perceptual judgment and further judgments based on reexamination of the scene; in the sense of consilience among normal observers with similar access to a scene; and in the sense of helpful guidance in practical activity — is the result of an immense amount of critical scrutiny. The scrutiny occurs on two entirely different time scales. One is the time scale of seconds, on which whatever may be called an act of perception is consummated. (In speaking of this time scale, I make no commitment to the view that perception occurs in discrete units; this view has been justly criticized by Gibson and Neisser, and earlier by Dewey; see Ratner, 1939, p. 958 and p. 962.) It is the scale on which both the internal search for relevant features of the stimulus information occurs, and on which there is search of the environment for information which confirms or disconfirms expectations. The other is the time scale of years, during which skills in perceptual search have been developed, to some extent by neural maturation but certainly in large part as a result of trial and error.

The fact that perception involves search, with resulting confirmation or disconfirmation, and often with consequent refinement of expectancies on the short time scale and of search strategy on the long one, can be construed as a vindication of those epistemologists who maintain that perception is an epistemically mediated process. The price of the vindication, however, is a considerable transformation of the conception of mediation. Furthermore, those epistemologists who maintain the opposite view regarding mediation can find some comfort in the fact that perception, like other skills which are improved by practice, has the property that its analytic components are usually smoothly integrated and suppressed from consciousness.

One cannot stop at this point in adjudicating, however, since there is at least one other constituent in the "intelligence" of perception. The smooth integration which normally occurs when expectations are fulfilled can be interrupted by disconfirmations or by various kinds of challenges, such as the one posed by a painter who is trying to replicate an optical effect or by an epistemologist who is trying to understand the structure of his knowledge. The ease of such interruption and the ease of resumption of the integrated sequence are further evidence that the phenomenological frame of mind and the normal state of object perception are not as disjoint as the critics of sense data theories maintained. (It may be relevant to note that a common pedagogical technique for developing skills other than perception is to exhibit the elements of a temporal sequence separately — for instance, musical phrases or parts of a tennis stroke — and then to coach the student into smoothly integrating them into a larger unity.) The possibility of consciously verifying whether it is appropriate to attribute a certain feature to the stimulus information and of consciously searching the environment for confirming or disconfirming information concerning a tentative identification is one of the major ways in which perception is distinguished from reflex activities. In this way a kind of control is exercised by the subject as a whole — whatever that may mean — over the perceptual process, with some similarity to the control exercised in inference and problem solving, which are considered exemplary instances of intelligent activity.

This is a good place to point out that my invocation of empirical psychology for epistemological purposes differs in important respects from Quine's. According to Quine (1969, pp. 82—83), "Epistemology . . . studies a natural phenomenon, *viz*., a physical human subject. This human subject is accorded a certain experimentally controlled input — certain patterns of irradiation in assorted frequencies, for instance — and in the fullness of time the subject delivers as output a description of the three-dimensional external world and its history." Quine regards the observer as a black box whose perceptual processes are reliable if his reports are correct and his behavioral responses adequate, as judged by third persons who take for granted the scientific description of the physical world. This way of construing naturalistic epistemology is inseparable from a commitment to behaviorism in psychology and seems to me excessively narrow.

The cognitive psychology which I have invoked, in order to show that the reliability of ordinary perceptual judgments is well grounded in the critical scrutiny exercised by normal subjects, aims at "getting inside the black box," at least in the minimal sense of offering a "subpersonal theory" (see Dennett, 1978) analogous to microphysics, and perhaps in the further sense of linkage with introspection. In giving an account of the structure of my knowledge, I wish to be free to switch back and forth between a first- and third-person point of view. The ability to take a third-person point of view of myself is not an achievement of epistemological sophistication but, rather, the by-product of socialization and of the mastery of a public language; and it is, of course, a most remakable thing, which we are far from completely understanding. The ability to take a first-person point of view is so fundamental that it would seem not to need a defense, except for the challenge to its scientific respectability by various forms of behaviorism; and the best defense on the same plane as the challenge is to refer to a cognitive psychology, which is a well-articulated and fruitful alternative to behaviorism.

## EVOLUTIONARY CONSIDERATIONS

The discussion of the "closing of the circle" of epistemology and knowledge of the world can be carried out at different levels, with different degrees of fineness of description. At one level — the level at which one studies a human being functionally, as an organism interacting with its environment and endowed with strategies for pursuing vital goals — the closing of the circle is quite well established. We can see, from a variety of psychological studies, *that* human cognitive strategies are effective for practical purposes and, to some extent, *how* they are effective. It is possible to discuss these questions without inquiring into the genetic question of how human beings came to be endowed with such effective cognitive strategies and with apparatus for carrying them out. The genetic question obviously lurks in the background, however; indeed, one reason why it was not obtrusively disturbing is that almost all readers, and certainly all who follow

Campbell, take for granted an evolutionary answer. *Some* answer to this question — either the evolutionary one or a surrogate — is necessary if we are to know where human beings fit in nature and therefore to accomplish the "closing of the circle" at a level deeper than that of practical functioning. (I should add, however, in view of the difficulty posed by the mind-body problem, that an evolutionary answer to the genetic question does not ipso facto "close the circle" at the deepest level. Instead, we are confronted with the problem in a new guise: how is it that creatures endowed with the consciousness that we know ourselves to possess could have evolved from ancestors apparently not so endowed?)

Another reason why the question of the origin of human cognitive faculties did not arise obtrusively in the previous section is that we were concerned with the reliability of perception, which is a practical matter, or certainly can be so understood, given the customary usage of the word *reliability*. A perceptual judgment is reliable if on the whole it is confirmed by further experience, if on the whole it agrees with the judgments of other well-placed and competent observers, and if on the whole it can serve as a basis for actions which favor the interests of the perceiver. That perception is reliable in this sense is a commonplace, shared by a number of philosophers, including some who are not naturalistic. In particular, philosophies which offer radically different answers to the genetic question may concur regarding the reliabilty of perception in the sense stated. Berkeley, for example, repeatedly insists that his immaterialism does not undermine the commonsense view of the world, and indeed he claims to provide a better explanation of the accessibility of the commonsense world to knowledge by perception than do the advocates of a mysterious material substratum.

Nevertheless, different answers to the genetic question are linked with different epistemological positions, but the linkage concerns other problems than those considered so far; for example, the extent to which perception reveals properties of material objects which are independent of the cognitive relation, and the extent to which the structure of experience is imposed by the mind of the subject. I agree with Campbell that, in assessing the various solutions offered to these problems, one must take into account the overwhelming evidence for an evolutionary history of human cognitive faculties. This evidence makes very implausible the Kantian thesis that the mind legislates to nature. Furthermore, the evolutionary point of view permits a naturalistic explanation of the pervasiveness of the concepts of substance and causality in experience, contrary to Kant's thesis that we can recognize the operations of the understanding but cannot give a causal explanation for these operations. (According to Kant, an attempt to give such a causal explanation would be an instance of applying the category of causality to the things in themselves.) As Lorenz has argued, with Campbell's enthusiastic concurrence (1974, pp. 441–447), a lineage of animals which evolves in a physical environment inhabited by relatively stable macroscopic objects and governed by macroscopic regularities would benefit from a propensity to apply the categories of substance and causality to the sensory content of experience. The general lines of this argument are independent of the detailed nature of the propensity — whether it is ontogenetically innate, or requires neurological maturation, or requires some empirical input to consummate the imprinting. The evolutionary point of view suggests a less rigid role of the categories of the understanding than the one to which Kant is committed because of his theory of the *constitutive* function of the understanding. Insofar as flexibility in applying the categories seems to be required not only to account for the heterogeneous intermingling of regularity and randomness in ordinary experience but also to fit the picture of the world provided by modern physics, the evidence seems to favor a naturalistic rather than a transcendental explanation of a priori elements in experience.

Our concern with the structure of epistemology, however, should make us wary of an easy victory for naturalism. It may reasonably be objected that a question has been begged when the theory of evolution and other theories with which it is linked, such as historical geology and parts of the physical sciences, are accorded a realistic interpretation. If scientific theories are interpreted nonrealistically, merely as means for "economy of thought" or as instruments for anticipating experience of the ordinary world or even one's personal sensations, then their significance for adjudicating epistemological arguments becomes negligible. Consequently, the appeal to evolutionary and other advanced scientific theories is illegitimate in epistemology without an antecedent analytic argument that these theories should be interpreted realistically. The objection can then continue that strong arguments raised by Berkeley, Hume, Mill, James, and the positivists suggest that a realistic interpretation of scientific theories is either meaningless, or else possibly meaningful but in principle undemonstrable by human beings, or else indistinguishable in cognitive content from sufficiently sophisticated phenomenalistic interpretations.

This issue is intricate, and evidently I cannot do justice to it in a part of a section of a paper which aims only at giving an overview. Fortunately, I feel that I can respectably abstain from a detailed discussion of the issue, since Skagestad examines it in this volume and I have written about it elsewhere (Shimony, 1971). A rough outline of my position is the following.

The arguments against realistic interpretations depend on the adoption of quite narrow criteria of meaningfulness and of evidential support. There are no strict and compelling reasons for the adoption of these narrow criteria, and there is a general metaphilosophical or dialectical reason for adopting broader criteria at least at the outset of inquiry: specifically, Peirce's maxim that the road to inquiry should not be blocked (thesis 2, presented earlier). Given broader criteria, one has the possibility of bringing relevant evidence to bear on realism, instrumentalism, and other rival interpretations of scientific theories. The possibility is also kept open that the initially broad criteria of meaningfulness and evidential support will be tightened as a consequence of inquiry, perhaps for reasons of scientific fruitfulness or of coherence (the "closing of the circle" of epistemology and knowledge of the world). If such tightening should occur, it would be nonarbitrary, in contrast to an initial stringency of criteria, which has the effect of hedging a favored philosophical position against evidence. The final step in the argument is that when realism, instrumentalism, and phenomenalism are placed on the same footing as candidates for the interpretation of scientific theories, none being excluded and none favored a priori, then realism is overwhelmingly supported a posteriori.

Needless to say, the foregoing analysis is excessively condensed, but at least it illustrates a major claim of integral epistemology: that naturalistic, analytic, and dialectical considerations must be intertwined if one is to do justice to the peculiar difficulties of the central problems of epistemology.

## INDUCTIVE INFERENCE AND VINDICATORY ARGUMENTS

Concern for the structure of epistemology requires that at least a few statements be made about inductive inference, even though they will be inadequate for a complicated subject. Campbell explicitly states that one of the presuppositions of his descriptive epistemology is the validity of inductive inference, and he accepts Hume's skepticism with regard to the possibility of justifying this presumption. One might have expected that he would attempt, as some other naturalistic epistemologists have done, to justify this presupposition by evolutionary considerations, along the following line: (1) Something like inductive inference has been practiced throughout the evolutionary process. (2) In particular, long before the articulation of the scientific method, the higher animals and primitive humans developed (by trial and error) the strategy of proposing hypotheses and testing them by observations. (3) The success in the biological sense of a population which has adopted this strategy shows that it is a good strategy to employ in the world as it actually is constituted, and this fact justifies inductive inference. Actually, Campbell does assert (1) and (2), but he refrains from (3). He does so, I believe, because he realizes that the transition to (3) is itself an instance of inductive inference, the validity of which is under examination. In other words, Hume's criticism of the attempt to justify inductive inference by its past successes applies equally well to arguments which appeal to evolutionary evidence.

If Hume is not to have the last word, several strategies are open. One is to make full use of the apparatus of probability theory and decision theory, which themselves are defensible on analytic or a priori grounds. Another is to concede that any kind of inductive justification of induction involves some circularity but to reason that some types of circularity are nonvicious, in the sense that they do not foreclose the issue under examination and do not hedge a favored position against adverse evidence. This strategy is clearly in the spirit of the dialectical approach to epistemology advocated as a meta-theory in an earlier section (headed 'Some Dialectical Theses') and followed elsewhere in this paper. The third strategy is to employ a vindicatory argument, as sketched in Thesis 4 of that section. My own opinion is that all these strategies are needed and must be used in tandem. I am not satisfied that any methodologist has worked out the details of how the meshing of these strategies is to be accomplished, but I have advocated a program along the following lines, which seems to me to be in the right direction (Shimony, 1970, 1976).

Briefly, I believe that the theory of inductive inference has an a priori component, consisting of personal probability theory and decision theory, and making free use of deductive logic. The personal probability evaluation should be qualified by a maxim of critical open-mindedness, which is called "the tempering condition"; that is, give sufficient prior probability to any seriously proposed hypothesis that it may achieve high posterior probability if favored by the data.

A vindicatory argument can be given as follows for this maxim. If we are not optimistic about human ability to make good conjectures, then seriously proposed hypotheses are coordinate with frivolously proposed ones and with merely logically possible ones that no one has ever singled out for mention. But concerning any nontrivial matter of interest, particularly concerning generalities, infinitely many rival hypotheses are logically possible, and it is hopeless to explore all of these. Our only hope of approaching the truth concerning nontrivial matters is to be optimistic about human abductive power, in the sense of permitting a seriously proposed hypothesis to be accepted after experimental scrutiny. This optimism precludes no particular method for assigning an order to hypotheses, since the proposal of an order would itself be a special case of abduction. Consequently, nothing is lost and possibly something is gained — in accordance with the general line of vindicatory arguments — if our subjective probability evaluations are made in accordance with the tempering condition. The tempering condition as stated, however, is manifestly vague, and attempts to eliminate this vagueness by a priori explications of the phrases "seriously proposed" and "sufficient prior probability" are likely to be arbitrary.

A more promising procedure is to use a posteriori considerations in explicating these phrases. By reflecting on exemplary scientific discoveries in the past, one can recognize types of hypotheses which deserve prima facie attention — meaning, of course, no more than a moderate degree of prior probability, remote from full credence. Circularity obviously enters into the proposed epistemological structure, since there is reliance on exemplary scientific achievements, and these in turn could have been identified only by inductive inference. The circularity is nonvicious, since a seriously proposed hypothesis of the past can be subjected to experimental scrutiny and as a result may acquire so low a posterior probability that it is discredited. Furthermore, there is an openness in the procedure in spite of the recourse of past achievement, for a radically new hypothesis can be accorded the status of "seriously proposed" for many reasons; it may, for example, satisfy the standards of exposition of exemplary hypotheses of the past, in spite of the novelty of its content.

This program for justifying inductive inference evidently needs much further clarification and exploration. However, space permits only two last remarks, which bear on the structure of epistemology. The first is to reiterate the indispensability of analytic considerations in an integral epistemology. Without probability theory and decision theory, and without the appeal to a vindicatory argument, there seems to be no alternative to skepticism regarding the justification of induction. The second remark is that what Campbell calls "descriptive epistemology" has a role in the justification of induction, as it does in other epistemological matters. But it appears that the term should be extended, so as to apply not only to results from the natural sciences but also to some historical information. As far as we can reconstruct the reasoning process of prescientific people, we find a mélange of critical and noncritical elements. Along with *modus tollens* and rudimentary probability theory, there is uncritical use of simple inductions and very uncritical reliance on analogies and anthropocentric explanations. If one argues from the biological success of Homo sapiens to the reasonableness of the genetic thought processes of man, then one has an undiscriminating justification of the whole package of primitive thought processes, critical and uncritical together. The a posteriori considerations which permit us to separate out different components of the package and to assess them are considerations concerning not the evolutionary development of the race but, rather, the history of science. The improvement of primitive inductive processes is thus a cultural achievement. It is, however, a very special cultural achievement, since it is the by-product of search for the truth about nature. There is no better illustration of a dialectical structure in epistemology than the refinement of inductive inferences by reflection on exemplary results of science, which themselves were achieved inductively.

*Boston University*

## REFERENCES

Bower, T. G. R.: 1966, 'The Visual World of Infants.' *Scientific American* **215**(12), 80–92.

Bruner, J. S.: 1957, 'Perceptual Readiness.' *Psychological Review* **64**, 123–152.

Campbell, D. T.: 1956, 'Perception as Substitute Trial and Error.' *Psychological Review* **63**, 330–342.

Campbell, D. T.: 1959, 'Methodological Suggestions from a Comparative Psychology of Knowledge Processes.' *Inquiry* **2**, 152–182.

Campbell, D. T.: 1974, 'Evolutionary Epistemology.' In P. A. Schilpp (Ed.), *The Philosophy of Karl Popper*. La Salle, Ill.: Open Court.

Campbell, D. T.: 1977a, Introductory comments to a lecture on Evolutionary Epistemology to the Boston Colloquium for the Philosophy of Science, February 22.

Campbell, D. T.: 1977b, 'Descriptive Epistemology: Psychological, Sociological, and Evolutionary.' William James Lectures, Harvard University.

Dennett, D. C.: 1978, 'Toward a Cognitive Theory of Consciousness.' In C. W. Savage (Ed.), *Minnesota Studies in the Philosophy of Science*. Vol. 9: *Perception and Cognition: Issues in the Foundations of Psychology*. Minneapolis: University of Minnesota Press, pp. 201–228.

Dretske, F.: 1971, 'Perception from an Evolutionary Point of View.' *Journal of Philosophy* **68**, 584–591.

Firth, F.: 1965, 'Sense-Data and the Percept Theory.' In R. J. Swartz (Ed.), *Perceiving, Sensing, and Knowing*. New York: Doubleday, 1965, pp. 486–496.

Gibson, E. J.: 1969, *Principles of Perceptual Learning and Development*. New York: Appleton-Century-Crofts.

Gibson, J. J.: 1950, *The Perception of the Visual World*. Boston: Houghton Mifflin.

Gregory, R. L.: 1970, *The Intelligent Eye*. New York: McGraw-Hill.

Heffner, J.: 1981, 'Causal Relations in Visual Perception.' *International Philosophical Quarterly*, **21**, 303–332.

Neisser, U.: 1967, *Cognitive Psychology*. New York: Appleton-Century-Crofts.

Neisser, U.: 1976, *Cognition and Reality*. San Francisco: Freeman.

Peirce, C. S.: 1931, 'Notes on Scientific Philosophy.' In C. Hartshorne and P. Weiss (Eds.), *Collected Papers of Charles Sanders Peirce*. Vol. 1. Cambridge, Mass.: Harvard University Press. (Original manuscript 1897.)

Price, H. H.: 1932, *Perception*. London: Methuen.

Quine, W. V.: 1969, 'Epistemology Naturalized.' In *Ontological Relativity and Other Essays*. New York: Columbia University Press, pp. 69–90.

Quine, W. V.: 1974, *The Roots of Reference*. La Salle, Ill.: Open Court.

Quinton, A. M.: 1965, 'The Problem of Perception.' In R. J. Swartz (Ed.), *Perceiving, Sensing, and Knowing*. New York: Doubleday, pp. 497–526.

Ratner, J. (Ed.): 1939, *Intelligence in the Modern World: John Dewey's Philosophy*. New York: Random House.

Rescher, N.: 1977, *Methodological Pragmatism*. Oxford: Blackwell.

Russell, B.: 1921, *The Analysis of Mind*. London: Allen and Unwin.

Shimony, A.: 1970, 'Scientific Inference.' In R. G. Colodny (Ed.), *The Nature and Function of Scientific Theories*. Pittsburgh: University of Pittsburgh Press, pp. 79–172.

Shimony, A.: 1971, 'Perception from an Evolutionary Point of View.' *Journal of Philosophy* **68**, 571–583.

Shimony, A.: 1976, 'Comments on Two Epistemological Theses of Thomas Kuhn.' In R. S. Cohen, P. K. Feyerabend, and M. W. Wartofsky (Eds.), *Essays in Memory of Imre Lakatos*. Dordrecht, Netherlands: Reidel.

Shimony, A.: 1978, 'Is Observation Theory-Laden? A Problem in Naturalistic Epistemology.' In R. G. Colodny (Ed.), *Logic, Laws, and Life: Some Philosophical Complications*. Pittsburgh: University of Pittsburgh Press, pp. 185—208.

Swartz, R. J. (Ed.): 1965, *Perceiving, Sensing, and Knowing*. New York: Doubleday.

# PART III

# CRITICISMS OF NATURALISTIC EPISTEMOLOGY

PAUL T. SAGAL

# NATURALISTIC EPISTEMOLOGY AND THE HARAKIRI OF PHILOSOPHY

(0) Bertrand Russell, in the early part of the twentieth century, labored vigorously on a proper foundation for mathematics. One of his efforts resulted in a set theory along with a theory of types, a theory of levels of discourse. The existence of sets was made relative to the level of discourse already employed in the construction of the theory. This is the beginning of what can be termed constructive set theory, the attempt to tie talk of mathematical entities to linguistic operations of a certain kind, and hence to provide a philosophical justification for mathematics. Having embarked on this constructivist path, Russell detoured because he thought it was necessary to talk about collections of sets across levels. Consequently, he introduced his famous *axiom of reducibility* and committed what Hermann Weyl called the *harakiri* of reason. This attempt to justify mathematics resulted in suicide.

To give up philosophy's traditional attempt to provide a systematic justification for human knowledge, especially for the fundamental principles of science, is to commit suicide on a grand level. And this is what naturalistic epistemology, or the naturalization of epistemology, is all about.

(1) For pre-Socratic times, the line between philosophy and science is difficult to draw. Was Thales the first philosopher, the first scientist or both? With Socrates and Plato the line gets drawn, and quite sharply. In Book VI of Plato's *Republic* in the figure of the divided line we get a powerful statement of the priority of philosophy over science, specifically mathematics or geometry. In mathematics, we work with axioms or principles which are assumed or hypothesized and we deduce the logical consequences from these. Philosophy gets behind or beneath the axiomatic method. It's the distinctive task of philosophy or the Platonic dialectic to justify the axioms. But in what can such a justification consist?

(2) Justification of axioms cannot consist in demonstration, in deduction from further axioms, since these axioms themselves would require justification. Philosophers from Plato through Descartes, Husserl and C. I. Lewis have taken the ultimate concern of philosophy



*Abner Shimony and Debra Nails (eds.), Naturalistic Epistemology*, 321–332.

to be epistemological. The primary question is: How can we justify the fundamental principles of human knowledge? If science, especially natural science, is taken to be paradigmatic of human knowledge, then it is the task of philosophy to justify science. Science cannot justify itself because it does not consist in questioning everything. Its method remains axiomatic. Fundamental concepts and principles are postulated, tested perhaps, but not epistemologically justified. The hypothetico-deductive method of science in its several varieties, what we sometimes term scientific method, itself requires justification.

(3) Plato saw that there was a remarkable relationship between concepts and principles. Definitions provide the bridge; hence the prominent role given to the search for definitions in Plato's philosophy. Later philosophers, especially the logical positivists, spoke as if we justified the employment of a concept through definition or partial definition (meaning postulate). The definition in turn was justified as being analytic, a tautology, in another idiom. Now even if we bracket Quinean qualms about analyticity, we can see that calling a statement analytic does not justify it. All definitions are analytic but some definitions are better than others. Einstein's definition of fundamental physical concepts is superior to Newton's. Einstein's concepts and definitions are superior, but his definitions are not anymore analytic than Newton's. Plato despaired of genuine knowledge in natural science apparently because natural science, physics, would have had to be empirical, and empirical knowledge was a contradiction in terms. Implicit in his thinking here seems to be recognition of the problem of induction. Empirical science would have to be inductive, but induction cannot itself be justified so neither can natural science, so natural science cannot be genuine knowledge. We can see why Plato treated mathematics and ethics on a par. Both were non-empirical. Both were axiomatic systems which required justification of fundamental terms and principles. In both, definitions play a major role. The key philosophical enterprise becomes the search for definitions. It is hardly surprising that Plato is so concerned with questions like what-is-virtue and what-is-justice. His is not a search for tautologies (they are only too easy to find), nor a search for well-confirmed hypotheses, for the concept of confirmation has its own philosophical problems.

(4) The dignity of philosophy, its own special importance, is that its task is to provide the foundation for all other intellectual disciplines, for science taken broadly in the sense of the German term *Wissenschaft*.

This means that philosophy must be self-critical. Philosophy must contain a meta-philosophy or a philosophy of philosophy to justify the fundamental concepts and principles of philosophy itself. Hegel's characterization of philosophy as thought comprehending thought appears to be an appropriate one.

(5) If Plato made an heroic effort to distinguish philosophy from science, and to give philosophy a most noble and special task, Aristotle represents a step in the other direction. Philosophy and science are brought closer together. The Aristotelian epistemological ideal is demonstrative science. Actually, Aristotle distinguishes two senses of science, one in which science consists in the logical consequences of the axioms but not in the axioms themselves, and the other in which the science includes the axioms also. How are the axioms justified for Aristotle? Dialectic in Plato's sense no longer plays the dominant role, though logical principles like non-contradition are provided with dialectical justification. For Aristotle, dialectic is inferior to demonstration, so axioms require some other kind of justification. Intuitive induction plays this role. We grasp essences which are expressed through definitions after suitable experience. The details of this method are found in the intricacies of Aristotle's rational psychology. While Aristotle's intuitive induction is different from the modern variety of induction which proceeds to unexamined cases from examined cases instead of grasping essences on the basis of limited observations, Aristotelian induction can be seen to have at the very least the same epistemological problems as the modern variety. While Aristotle provides a psychological account of the process of intuitive induction, he nowhere provides a justification for it. Was Aristotle a naturalist? Did Aristotle voluntarily and happily identify psychological or scientific explanation with philosophical justification? I doubt it; still some of the seeds of naturalism were planted. Certainly Aristotle has this in common with contemporary naturalistic epistemologists. He does not take scepticism seriously. If the sceptic need not be answered, but merely dismissed, much of the burden of traditional epistemology, with its roots in Plato, is lifted.

(6) The Middle Ages was not a period dominated philosophically by epistemology and the challenge of scepticism. Religion made it difficult to take scepticism seriously, though Augustine at the beginning of the period and Nicholas of Autrecourt towards the end of the period, among others, certainly made valuable contributions to tradi-

tional epistemology.[1] It is really with Descartes that philosophy makes a complete recovery, that is, epistemology as foundation of science, returns to center stage. Descartes rejects science without a proper foundation. He sought to reconstruct science from the ground up. In this quest, Descartes was super-ambitious: He actually tried to ground everything on a single principle, his famous *cogito*. I think Descartes is on the right track in emphasizing thinking, inquiry, and method. But the step from the *cogito* to the existence of God and beyond is more than a giant step; it is an unjustified leap. Descartes rejected the philosophy he learned at La Flêche but he never seemed to call into question the very language of scholastic metaphysics along with the general principles which formed its framework. How else do we explain Descartes' uncritical use of principles like 'there is at least as much reality in the cause as in the effect' which he emphasized in his first argument for the existence of God?

(7) Empiricism proved itself unable to provide a proper foundation for knowledge. It was unable to bridge the gap between private subjective experience and the public experience of science. And, of course, as David Hume demonstrated, it could not bridge the inductive gap from the observed to the unobserved. Empiricism could thus not provide a justification for fundamental principles. It is easy to see how Hume's critique of induction awakened Kant from his dogmatic sleep. According to Humean empiricism, all principles must concern either relations of ideas (they must be analytic) or matters of fact (they must be synthetic). If the latter, the justification of principles would require induction; hence there could be no justification. If the former, principles would be tautologies, trivial knowledge. But the fundamental principles of science, e.g., Euclidian geometry and Newtonian science, were not trivial, they were substantive. Kant saw that Humean empiricism was an epistemological dead end. Classical rationalism was no better. If epistemology was going to survive it would have to be provided with new foundations. It is this task that the Kantian critical philosophy set for itself. Kant asked how science is possible. What are the necessary and sufficient conditions for science? He assumed that science was possible in the case of Newtonian science because it was actual. But even if it were not actual, he could still work back from the idea or concept of science, a methodically disciplined inquiry leading to objectively justifiable principles. As is well known, Kant found the conditions of science in the workings of the mind. We can know about

nature what our minds put into nature. Our minds are structured in a Euclidian-Newtonian way and nature necessarily reveals itself to us as Euclidian and Newtonian. The details of the Kantian enterprise do not concern us. How exactly Kant would have responded to the discovery of non-Euclidian geometry and non-Newtonian physics is pure speculation. One thing I believe is clear is that he would have not given up the critical enterprise altogether. The view of epistemology we will sketch and defend is a variation of Kantian critical theory. It is Kantianism with a pragmatic and linguistic twist.

(8) No history of the theory of knowledge or traditional theory of knowledge, even a history as potted as this one would be complete without mention of the work of Edmund Husserl and his phenomenological school. Husserl's work can best be seen as an attempt to provide a pre-Kantian brand of radically empiricistic foundation for knowledge. Experience was broadened to include *a priori* or formal structure. For instance, the necessary connections that Hume couldn't find in experience Husserl found. Husserl sought a return to the primordial given in experience, *zu den Sachen selbst*, but what he found in the primordial given was a far richer content than what Hume and the traditional empiricists found. Both Kant and Husserl flirted too closely with psychology. It often appeared that empirical psychological questions were being given *a priori* philosophical answers. There remained the problems of transcending subjectivity, of getting from consciousness to the world. It's not difficult to see that epistemology done in what we could call the *psychological style* appears to have insurmountable problems. And that this could cause the faint of heart to give up the epistemological task altogether. It is basically this kind of epistemology which culminates in Husserl, Carnap's *Aufbau* and to some extent in the work of C. I. Lewis that is most explicitly rejected by the epistemological naturalists. We will look at a post-linguistic turn variation of Kantian epistemology that still holds promise for the traditional epistemological enterprise, and then examine Quine's classic advertisement for the naturalization of epistemology.

(9) Hugo Dingler studied under Husserl and David Hilbert among others. He translated the Kantian psychological idiom into that of pragmatism or operationism. Knowledge was to involve the construction; the making of statements in methodical order. Science is to be justified in terms of its contribution to technology, the construction of objects useful for life. The philosophical quest is essentially a quest for

certainty and security; it is a reflection of the same quest which gives us engineering and technology. Given the commitment to certainty, we can arrive at a conception of science to fit it. We can rationally and freely decide to employ certain concepts and hold on to them come what may. We decide to immunize these from experimental test; we are not compelled to do so *à la Kant* because of the structure of the human mind. Dingler advocated the method of decision and exhaustion. Principles are decided on *a priori* largely on the basis of technological and methodological need; when the principles fail to stand up to refutation, we do not modify or give up the principles, we exhaust the disturbing conditions, we postulate conditions which would explain the failure. When observations failed to jibe with Newtonian theory's predictions about the orbit of Uranus, we did not throw out Newtonian theory, we looked for disturbing conditions. And this is how the planet Neptune was discovered. Dingler considers this example paradigmatic of good scientific method.[2]

Contemporary German philosopher Paul Lorenzen and his *Constructive Philosophy* reflects the influence of Dingler along with that of mathematician Hermann Weyl. It is a living paradigm of a non-naturalistic theory of knowledge. I will sketch Lorezen's approach and then compare it with the naturalistic approach of Quine.

(10) Paul Lorenzen achieved a measure of academic recognition for his work in logic and the foundations of mathematics. He formulated an improved version of Gentzen's proof for the consistency of arithmetic. He successfully carried on the work of Hermann Weyl; he provided a constructive version of the continuum and hence the arithmetic of real numbers, a version which makes no use of impredicative principles, principles which assume the existence of objects belonging to a certain set in generating the existence of the set.

Lorenzen's *Constructive Philosophy* can be viewed as a generalization of, or extrapolation from, his constructive theory of mathematical knowledge. The program is to give all domains of knowledge, all intellectual disciplines, a constructive treatment, no circles or gaps are to be countenanced and the foundation itself must be more reasonable than any alternative. Lorenzen offers a new foundationalism, a constructive foundationalism, if you will, one with both empiricist and rationalist features. Constructive mathematics, for instance, is literally a step-by-step, non-circular construction of the language of mathematics. *Constructive Philosophy* comprises the construction of the fundamental portion of all intellectual disciplines all sciences in the sense of the German term 'Wissenschaft'.

*Constructive Philosophy* has this in common with Logical Positivism: it is an *ideal* rather than an *ordinary* language philosophy. For Logical Positivism, the ideal language was a language constructed to be adequate for all natural science, which was considered to exhaust the realm of cognitively significant discourse. This ideal language left out large portions of disciplined discourse, especially normative discourse. Ethics and related disciplines were banished to the land of the cognitively meaningless.

The Logical Positivists were not especially clear about how the ideal language was to be constructed or what metalinguistic devices were philosophically permissible. They tended to use some natural language, German or English as their meta-language. Lorenzen, on the other hand, is meticulous in his concern about the principles of the linguistic construction. Furthermore, the meta-language cannot itself be accepted uncritically. Meta-linguistic means should not go beyond acceptable linguistic apparatus already introduced in the object-language. For instance, Lorenzen wants us to be as critical in our discourse about natural science as we are in natural science itself. The scientist should be able to defend science 'scientifically', i.e., in a disciplined way.

(11) Let us turn to the constructive foundations of mathematics. We begin with logic and arithmetic. Let us take logic for granted, though Lorenzen has developed a constructive logic. Arithmetic is constructed as follows: We do not begin with some axiom system for arithmetic, e.g., Peano's Axioms. We begin with arithmetic practice, with calculation, with school arithmetic if you will. First rules are given for the production of numerals. The first numeral 0 or 1 can be produced thus: $\rightarrow$ 0 or $\rightarrow$ 1. All other numerals can be constructed from the first numeral by adding strokes, so we have n $\rightarrow$ n1. We have then a recipe for producing numerals. Recipes can also be provided for addition and multiplication. This is principally all we require for ordinary arithmetic. We arrive at an understanding of, we provide a foundation for, arithmetic by re-inventing, re-constructing arithmetic from arithmetical practice.

All the talk has been of numerals and not of numbers. A number, it is true, is an abstract object and as such it cannot be constructed in the concrete manner we have indicated above. But, after all what is a number but a numeral whose style considerations — whose 'looks' —

are unessential. Do the rules of arithmetic require any particular way of writing numerals? Arithmetical truth is a matter of what can be defended in terms of our arithmetical recipes. Arithmetic, of course, will not suffice for mathematics as a whole. The Greek response to this situation involved what could be termed a *geometrization* of arithmetic. The modern, or Post-Cantorian, response involves the grafting of set-theory onto arithmetic. In this way we will be able to get the higher arithmetic, or analysis. The problem with the modern approach is simply the problem with set-theory. Which set-theory? Intuitive or naive set-theory (Cantor's) led to contradiction. It seems we simply have a choice between alternative axiomatic versions of set theory. But how are we going to choose between these theories? How are we going to justity or legitimize any one of them? There are not even consistency proofs available. More basically, does an axiomatic system enable us to truly understand the subject matter? The axiomatic method was eschewed in connection with arithmetic; it is to be eschewed here too. We require a recipe for constructing sets just as we had a recipe for constructing numbers.

We want thus to construct sets much in the manner as we constructed numbers. Actually we constructed numerals, and obtained numbers or talk of number as a *façon de parler* through an abstraction procedure. Let me sketch a parallel construction for sets. We construct formulae, open sentences, sentences with variables, say *Fx*, where substituends for *x* comprise our previously constructed arithmetical numerals. At the next level our formulae would contain variables which would allow substituends including numerals and first level formulae. What we have then is a step-by-step recipe for the construction of formulae. Sets, the abstract objects, are obtained from formulae by means of the abstraction procedure. Two formulae *Fx* and *Gx* are said to represent the same set when for all *x, Fx* if and only if *Gx*. To say that something *z* is a member of a set of *F* is simply to say *Fz*. Without going into detail here, we can characterize the constructivist program as the construction of sets sufficient for the conceptual [ontological] needs of analysis, for the mathematical requirements of physics. If we suppose such a construction can be carried out, it is important to see how the constructive approach differs from the traditional axiomatic approach.

In the constructive approach the existence of sets is tied to our ability to construct certain formulae. In the traditional approach the existence of sets is established by *postulate*. Some version of the following existence or comprehension principle is employed: $(E\alpha)(x)$ $(x \in \alpha$ iff $\Phi x)$. A set exists, something is an element of a set, iff the element has some property $\Phi$, where no restriction is placed on the nature of the $\Phi$; $\Phi$ could involve all kinds of sets including the set defined by $\Phi$ itself. This principle leads to contradiction [where $\Phi$ is the property of not being a member of itself. This result is due to Russell]. Attempts at modification of this principle can be shown to avoid Russell's contradiction; but they cannot be shown to avoid contradiction altogether. The axiomatic alternative is the method of postulation and prayer [prayer for consistency].

(12) To appreciate the constructive enterprise in mathematics is to be well on the way to appreciating the constructive enterprise as a whole. Such an appreciation should help us focus on such questions as (1) Why do we need intellectual disciplines, e.g., mathematics, at all? (2) Why do intellectual disciplines require foundations? (3) According to what principle should foundations be chosen? (4) What principles of construction ought to be adopted? The questions themselves are fundamental and the answers should be adequate. Here we can only sketch the constructivist response. (1) Our practice needs to be improved. The goal is better living through intellectual disciplines, through science, if 'science' is construed in a suitably broad manner. We can count and measure without the disciplines of arithmetic and geometry; we can move heavy objects without a science of physics, but with the disciplines we can accomplish these tasks more effectively. (2) Unless the foundation is secure we cannot commit ourselves intellectually to the use of a discipline; we cannot even critically understand the discipline unless there are some first steps available for defending its intelligibility and adequacy. Foundationalism remains a compelling ideal. The quest for certainty and intellectual security should only be given up when all the serious attempts to provide foundations have been examined. Constructive foundationalism is not to be dismissed merely because it is foundationalist. Of course, it has to put up or shut up. (3) Theory arises out of practice, and the first theoretical steps must be rooted in practice. Theory may ultimately modify practice, but it cannot cancel out the initial practice from which the theory itself arose. Arithmetic begins with counting, grammar with speech, and physics with primitive engineering. Even Ethics can be said to begin with the practice of conflict resolution. (4) The construction must follow a methodical order. Each step can only be justified in terms of previous steps.

Circularity vitiates the principle of methodical order. Furthermore, *each* step in the construction must be justified. As we mentioned previously, there are to be no circles and no gaps. The answers to (3) and (4) are justified in terms of an ideal, the ideal of certainty, the ideal of intellectual security. From Plato through Descartes, Kant, Husserl and Dingler, the epistemological ideal has been the same. Of course, success in meeting the ideal, has left much to be desired. In our own century, we have even seen some give up the ideal; we have seen the *naturalization* of epistemology, the attempt to treat the theory of knowledge or science as itself a part of science.[3] The circularity involved in this has been called 'virtuous'. Paul Lorenzen's *Constructive Philosophy* is a contemporary attempt at realizing that *old* ideal which made Philosophy itself fundamental.

(13) W. V. O. Quine's view of philosophy reflects his judgment that traditional epistemology is dead. He does not mourn for it . . .

> I see philosophy not as an *a priori* propaedeutic or groundwork for science, but as continuous with science . . . in the same boat — a boat which, to revert to Neurath's figure as I so often do, we can rebuild at sea while staying afloat in it. There is no external vantage point, no first philosophy. All scientific findings, all scientific conjectures that are at present plausible, are therefore in my view as welcome for use in philosophy as elsewhere. (Quine, 1969b, pp. 126—127)

On this account, what happens to questions about justification and evidence? What kind of questions do these become? Do they become psychological questions? Do all questions of norms and justification get replaced by questions of scientific explanation? In 'Epistemology Naturalized' (Quine, 1969a) we appear to get affirmative answers to these questions. We are back to the psychologism that Husserl fought so fiercely against. There really is nothing very new in the naturalistic program. Husserl were he alive today would not have to read Quine since there were many Quines in his generation. Of course, this point hardly makes for a refutation of naturalism. If naturalism has not been frequently refuted it has, at least, been frequently rejected. In the twentieth century the traditional epistemological task came to be termed rational reconstruction and this was considered to be a peculiarly philosophical as opposed to scientific task. The aim of rational reconstruction was to provide an idealized substitute for systems of knowledge which exhibited the justification of claims to know. The naturalization of epistemology involves rejecting rational reconstruction.

> But why all this creative reconstruction, all this make-believe! The stimulation of his sensory receptors is all the evidence anybody has had to go on, ultimately, in arriving at his picture of the world. Why not just see how this construction really proceeds? Why not settle for psychology? (Quine, 1969a, p. 75)

But we are interested in the justification of the construction of sciences like psychology. How can we employ psychology in order to justify psychology? Of course, for Quine resting squarely in Neurath's boat, methodical circularity is no problem.

> Such a surrender of the epistemological burden to psychology is a move that was disallowed in earlier times as circular reasoning. If the epistemologist's goal is validation of the grounds of empirical science, he defeats his purpose by using psychology or other empirical science in the validation. However, such scruples against circularity have little point once we have stopped dreaming of deducing science from observations. If we are out simply to understand the link between observation and science, we are well advised to use any available information, included that provided by the very science whose link with observation we are seeking to understand. (Quine, 1969a, pp. 75—76)

Quine might be justified in giving up the empiricist version of epistemology, he seems to have in mind ('deducing science from observation'), but Hume showed the bankruptcy of this approach years ago. Just because there can be no justification of science in the traditional empiricist vein does not mean there is no justification of science *simpliciter*. Presumably, Quine would have to examine other options. But this he does not do in 'Epistemology Naturalized'. After all Kant understood Hume as well as Quine, but he did not give up epistemology. For this more argument than what Quine offers is necessary. We must not lose sight of exactly what Quine is asking us to give up. Isn't he asking us to give up philosophy as a distinctive intellectual discipline? If rational reconstruction goes, what is there left? Earlier in this century John Dewey (Quine admits to be a philosopher squarely in the Deweyan tradition) had similar ideas. But Dewey could not properly be said to reduce epistemology to psychology. His *Logic — The Theory of Inquiry* and his *Theory of Valuation* are works, perhaps difficult to classify; but they do not belong to psychology. Dewey appeared to recognize that his own practical philosophy required philosophical justification. And, of course, Quine's own epistemological naturalism itself requires justification! Unfortunately, or fortunately, for Quine, there is no room in his philosophy for such justification.[4]

*New Mexico State University*

## NOTES

[1] For some details see Sagal (1982).

[2] There is just about nothing by Dingler in English. Dingler (1964), with a preface by Lorenzen, is a good introduction to Dingler's work.

[3] For another discussion of some of these issues, see Sagal (1978).

[4] I feel compelled to add that whatever Quine is doing, philosophy, epistemology, psychology, or what have you, his work has been original, stimulating and of exceptional craftsmanship.

## REFERENCES

Dingler, Hugo: 1964. *Aufbau der Exakten Fundamental Wissenschaft*. Munich: Eidos-Verlag.

Kamlah, Wilhelm, and Paul Lorenzen: 1967. *Logische Propädeutik oder Vorschule des vernünftigen Redens*. B. I.-Hochschultaschenbuch, Bd. 227. Mannheim: Bibliographisches Institut. Reprinted, 1973.

Lorenzen, Paul: 1962. *Metamathematik*. B. I.-Hochschultaschenbuch, Bd. 25. Mannheim: Bibliographisches Institut.

———: 1965a. *Logik und Grammatik*. Duden-Beiträge, H. 26. Mannheim: Bibliographisches Institut.

———: 1965b. *Differential und Integral. Eine konstruktive Einführung in die klassische Analysis*. Frankfurt a.M.: Akademische Verlagsgesellschaft.

———: 1967. 'Methodical Thinking,' *Ratio*.

———: 1968. *Methodisches Denken. Gesammelte Aufsätze*. Suhrkamp Taschenbuch Wissenschaft, Bd. 73. Frankfurt a.M.: Suhrkamp. Reprinted, 1974.

———: 1969. *Normative Logic and Ethics*. B. I.-Hochschultaschenbuch, Bd. 236. Mannheim: Bibliographisches Institut.

———: 1974. *Konstruktive Wissenschaftstheorie*. Suhrkamp Taschenbuch Wissenschaft, Bd. 93. Frankfurt a.M.: Suhrkamp.

———: 1978. *Paul Lorenzen: Theorie der technischen und Politischen Vernunft*. Stuttgart: Reclam.

Lorenzen, Paul, and Oswald Schwemmer: 1973. *Konstruktive Logik, Ethik und Wissenschaftstheorie*. B. I.-Hochschultaschenbuch, Bd. 700. Mannheim: Bibliographisches Institut. Reprinted, 1975.

Quine, W. V. O.: 1969. *Ontological Relativity and Other Essays*. New York: Columbia University Press.

———: 1969a. 'Epistemology Naturalized.' In Quine (1969, pp. 69—90).

———: 1969b. 'Natural Kinds.' In Quine (1969, pp. 114—138).

Sagal, Paul T.: 1978. 'Epistemology De-Naturalized,' *Kant-Studien* **69**, 97—109.

———: 1982. 'Skepticism In Medieval Philosophy,' *Philosophical Forum* **14**:1, 80—93.

ABNER SHIMONY

# COMMENT ON SAGAL

Sagal criticizes naturalistic epistemology generically for falling short of the mission of justifying the sciences with "no circles and no gaps," and he outlines a program, essentially along the lines of Paul Lorenzen's constructive philosophy, which is intended to carry out that mission. I shall present some reasons for believing that the constructive program is unachievable. The answer which I shall then offer to Sagal's criticisms of naturalistic epistemology is based in large part upon the inevitability of curtailing our philosophical ambitions.

The part of constructive philosophy which Sagal presents in greatest detail, as a demonstration piece for its power, is philosophy of mathematics. He has correctly assessed the area of his greatest strength. It is indeed astonishing how much of mathematics, including most of the analysis which is used in the natural sciences, can be recovered constructively. Even a philosopher who is unwilling to reject on principle all of non-constructive mathematics can be impressed by the achievement of the constructivists and can admit that they have made a strong case that a large part of mathematics is an artifact of the mind.

But what reason is there for thinking that constructivism can be extrapolated beyond the domains which can plausibly be regarded as man-made (which include language, logic, perhaps mathematics, perhaps ethics, and not much else)? To be sure, Kant suggested a way in which the domain of the natural sciences might be understood as man-made — by regarding nature as experience subjected to law, and considering the laws in their general form to be imposed by the mind. If, however, one does not regard the mind as a transcendental legislator, but rather as an entity in nature — and Sagal appears to be a naturalist to this extent — then one cannot use the argument strategy for extending constructive analysis to the natural sciences.

There exists, however, a non-Kantian defense of constructivism in the natural sciences, presented by Hugo Dingler, Lorenzen, and other protophysicists: they insist that scientific theory is founded upon practice, and of course crafts and technologies are human accomplishments. Since Sagal does not recapitulate their arguments in any detail,



*Abner Shimony and Debra Nails (eds.), Naturalistic Epistemology*, 333—336.

and since my knowledge of the literature is limited, I cannot pretend to offer a decisive criticism, but I can state my reasons for skepticism.

(1) It is hard to see how any extrapolation of principles from practical situations in which they have been successfully applied to new situations can dispense with a procedure of induction, and therefore the extrapolation is no more justified without circles and gaps than are the procedures of inductive logic. In reply, some constructivists resort to a vindicatory argument, which Reichenbach and others outside their school have invoked to defend induction: we cannot be sure that the principles of protophysics will work, but we cannot dispense with them and therefore it is reasonable to commit oneself to them (Peter Janich, cited by Andreas Kamlah, 1981, pp. 220—1). But a gap in the argument remains, because one does not know that an alternative protophysics, as yet unformulated, will not work as well as the Euclidean-Galilean one to which the constructivists are committed. Kamlah says:

> *Angenommen nun, alle bisherigen Normensysteme der Längenmessung, welche nicht die euklidische Geometrie voraussetzen — für andere Bereich der Protophysik gilt Analoges — seien tatsächlich als zirkelhaft erwiesen, was wäre die Situation? Kann man eine Aussage also apriori gültig ansehen, deren Apriorität nur induktiv erwiesen werden kann, so dass sozusagen ihre Apriorität 'aposteriori' gültig ist.* (Kamlah, 1981, p. 223)

(2) By now we have seen many concrete examples of theories which are accurate in their predictions in a restricted domain but inaccurate beyond it. They are limiting or special cases of theories which are reliable in broader domains. Niels Bohr introduced the notion of a correspondence principle governing the relation between quantum physics and its classical limiting cases. An extension of his notion can be applied whenever a theory is honorably superseded by a more accurate and/or more general one. Is it inconceivable that the principles of a practically efficacious protophysics are the 'correspondence limits' of some more general principles? This suggestion is particularly attractive if one takes a naturalistic view of human beings and recognizes that human practical activity is limited in spatial extent, in temporal duration, in the range of the objects handled, and in the fineness of human sensory observation. Is it inconceivable that instruments constructed in a shop, in accordance with Euclidean and Galilean principles, will be capable of discriminations which lend evidential support to modifications of those principles? If one studies in detail the machine shop procedures for fabricating an interferometer, and sees why it can be used to make length measurements more

precisely than the micrometers which were employed in constructing it, then the program just now suggested becomes quite plausible.

(3) The special theory of relativity provides an exemplary instance of inferring a scientific theory which corrects the principles assumed in the machine shop. The extraordinary agreement between the crucial predictions of this theory and observation, its internal coherence, and its ability to account for its own genesis constitute very strong evidence against the constructive philosophy of the natural sciences. A good part of the constructivist literature is devoted to disputing the virtues which I have just attributed to the special theory of relativity — to exhibiting its alleged conceptual incoherence and to showing that all its successful experimental predictions could be achieved within a Euclidean-Galilean framework. This dispute is too complicated and technical to summarize here. I shall only point out that there is an excellent anthology by Joachim Pfarr (1981) which reprints some of the relevant critiques of relativity theory by the protophysicists along with some assessments of their arguments. A later paper by John Stachel (1983) is also relevant. My impression is that the critics of the protophysical variety of constructivism make overwhelmingly a better case than the defenders.

Sagal's assertion that naturalistic epistemology has fallen short of the mission of justifying the sciences in based almost entirely upon an analysis of a paper of Quine. I agree with much of what Sagal says about Quine, as can be seen from my "Integral Epistemology" in this volume and from my comments on the paper of Levine, but I do not see how his criticisms are decisive against the *generic* enterprise of naturalistic epistemology. In particular, if one does not simply identify epistemology with psychology, but requires that a part of its method is a dialectic whereby it criticizes its own starting point, then one is conforming to Sagal's maxim "that philosophy must be self critical." (I suggest, incidentally, that Quine is more dialectical in practice than either Sagal or he himself admits.)

A very important part of the task of self-criticism is to examine the concept of justification itself. The concept of justification which one entertains at the beginning of a philosophical investigation may turn out to be too narrow, too idealized, and too much borrowed from a favored discipline. When Sagal says that the task of philosophy "is to provide the foundation for all other intellectual disciplines" and when he speaks of construction with "no circles and no gaps," he conveys the impression that his concept of justification is borrowed from mathematics, much as Descartes' was in his *Discourse on Method.* But may not the standards of justification in mathematics be inappropriate or inachievable in the natural sciences, the social sciences, and in practical activities? There are several different ways to approach this question. One is an examination of the history of thought, to see how dedicated searchers for knowledge were forced to discriminate between mathematical and causal necessity, and to retreat from imposing the ideal of mathematical demonstration in other disciplines. Another way is to reflect upon the place of human beings in nature and upon the constitution of the human faculties, a reflection which will almost inevitably force us to recognize human limitations and fallibility. A third way is to engage in dialogue, and since Sagal is a follower of Lorenzen, who espouses a dialogical method, I suppose that this is the way he would prefer.

Since the version of naturalism which I propose in 'Integral Epistemology' has a dialectical component, there should not be an impasse between Sagal and me. I have found counterparts in the writings of the constructivists to all the four theses for organizing dialectic that were listed in my paper, including notably the recourse to vindicatory arguments. (This last point should contribute greatly to a consensus between us, since it constitutes a definite retrenchment from the ideal of mathematical justification.) The constructivists' emphasis upon initiating theoretical investigation in the realm of practice is cognate to my thesis that "commonsense judgments about ordinary matters of fact have a *prima facie* credibility, which should not be discounted without clear positive reasons". Where I sharply diverge from the constructivists, however, is in my conviction that clear positive reasons for discounting initial commonsense and practical judgments *have sometimes been found.* In my opinion, the dialectical method gains in strength rather than loses because of its flexibility.

*Boston University*

## REFERENCES

Kamlah, Andreas: 1981. 'Zur Diskussion um die Protophysik.' In Pfarr (1981).

Pfarr, Joachim: 1981. *Protophysik und Relativitätstheorie.* Mannheim, Vienna, Zürich: B. I. Wissenschaftsverlag.

Stachel, John: 1983. 'Special Relativity from Measuring Rods.' In R. S. Cohen and L. Laudan (eds.), *Physics, Philosophy, and Psychoanalysis: Essays in Honor of Adolf Grünbaum.* Dordrecht, Boston, and London: D. Reidel.

JOSEPH AGASSI

# COMMENT ON SAGAL

(0) Paul Sagal offers a compound thesis when he says naturalistic epistemology is the desertion of the traditional philosophical attempt to justify claims for knowledge, which is the suicide of reason. I do agree with him that the traditional activity of philosophy includes as a major item on its agenda the justification of claims for knowledge. I would like to agree with him that W. V. Quine's naturalistic epistemology consists in the desertion of this task. (As I have argued in my note on Abner Shimony, Shimony's view of naturalistic epistemology is meant to have the cake and eat it too, but correcting it amounts to either rejecting it altogether or agreeing with Sagal that it amounts to the desertion of the justification of all knowledge claims.) What remains, then, for me to debate is Sagal's claim that deserting justification of knowledge claims amounts to the suicide of reason. Or, to convert his claim, he says that to reason is to justify claims for knowledge; it is the theory of rationalism as justificationism, which he rightly deems traditional. To disagree with him is to declare non-justificationist rationalism possible, then. My claim is stronger: to stick with justification of knowledge claims, I say, is the suicide of reason. If rationalism is to survive, some non-justificationist, i.e. critical, version of it should be propounded. (The numbering follows that of Sagal.)

(1) Sagal rightly observes that pre-justificationist Greek philosophy, Presocratic philosophy, is both philosophy and science. For, let me add, it was non-epistemic and metaphysical. Parmenides, I should say, was the first justificationist, claiming his view was *logôn piston*. This, I should add, gave rise to both non-justificationism — Democritus: truth lies in a bottomless well — and irrationalism — Heraclitus: the way up is the way down. Plato, says Sagal, was a fully-fledged justificationist. I agree. We disagree about Socrates, but this is a big side-issue. Plato's justificationism was the philosophy of axiomatics.

(2) Axiomatic philosophy was shared by Plato, Descartes, Husserl. It is doomed to failure, because of the sceptic argument from infinite regress or circularity. The same holds for the hypothetico-deductive method, observes Sagal. Agreed.



*Abner Shimony and Debra Nails (eds.), Naturalistic Epistemology*, 337–340.

(3) Nor will definitions help, since they are verbal and so not epistemic. If they are epistemic they are either baseless axioms or else hypotheses. Agreed.

(4) Philosophy must, then, justify both science and itself, says Sagal. This, however, follows from Plato's demand for justification. Why should we accept his demand? Because, says Sagal, the dignity of philosophy is at stake. My, oh my.

(5) Aristotle tries to justify axioms [definitions] intuitively; it does not work. Even if his theory of intuition is true, he cannot show intuition to be justification. Agreed.

(6) Skipping fideist mediaeval philosophy — on which Sagal has a splendid paper elsewhere — we arrive at Descartes' revival of Greek justificationism. He failed. Right.

(7a) Parallel to Cartesian intuitionism came empiricism which was hit by the problem of induction. Right.

(7b) Kant declared Euclidean-Newtonian science to be psychologically necessary for all humans. The advent of non-Euclidean and non-Newtonian science refutes him, yet Sagal thinks he would have survived the onslaught with certain modifications, as his current influence does.

(8) Husserl went to pre-Kantian theory of knowledge, claiming that experience has *a priori* valid forms. This claim is in part psychology, in part question begging. Agreed.

(9) Hugo Dingler took a pragmatist way out, relying on the need — psychological need — for certainty, and demanding the justification of each step of science in a pragmatic way.

(10) Paul Lorenzen followed Dingler, with a vigorous execution of the construction, which applied to both (*a priori*) mathematics and (empirical) science. He thereby plays the intuitionistic analogue to the logical positivistic logicism: whereas the logical positivists deemed the language of science an ideal language akin to that of Whitehead's and Russell's *Principia*, Lorenzen followed Hermann Weyl's constructivism in mathematics and extended it to natural science. And, I add, Lorenzen fails like the logical positivists.

(11) Here Sagal outlines Lorenzen's view of arithmetics. He fails, since he prefers not to construct numbers, which are abstract, but numerals. What a numeral is he cannot say. A numeral is any sign similar to a given sign designating a number, of course. But this is troublesome. Among other things it includes the matter of classes, which, anyway, is necessary, as Sagal notes, so as to construct mathe-

matics out of arithmetic. Sagal follows Lorenzen and rather than construct sets, he speaks of formulas, open sentences, and closed sentences, which represent sets. What is a formula? It is any sign which has a given shape. What is a shape?

Constructivists may naively ignore my question: as long as one can write formulas and learn rules for writing more formulas, as long as one recognizes mathematical truths contingent on such rules, one can go on with the very exciting exercise. But whereas the exciting constructivist may ignore my question, the constructivist justificationist cannot enjoy this luxury: his aim is not to do exciting work but to quieten any sense of doubt by an appeal to reason. This makes me very nervous as I am not at all sure I know what is shape and how the shape common to all manifestations of a given formula permits them or makes them or enables them to represent the one and the same formula. Nor do I think it true, but I will not hike my doubt: it is Sagal's job to push it down.

(12) Why do we need math? Because, says Sagal, we need to improve. But the justified is certain and so is not improvable, say I. So better give up foundations. Why do intellectual disciplines require foundations? Because, otherwise, says Sagal, we cannot commit ourselves to any discipline. But damn commitment, say I. But then, retorts Sagal, we cannot even critically understand a discipline unless we can defend its intelligibility. Empirically false, retort I: a child finds something intelligible, e.g. arithmetic, yet is unable to commit himself to, much less defend, its intelligibility; and once a child intelligently comprehends arithmetic, she can learn to try to be critical of it, or at least of errors of arithmetics encountered here or there. Foundationism, concedes Segal, remains a compelling ideal. But if ideal then not reality, and if compelling, only to Sagal not to me, say I. The quest for certainty should only be given up, Sagal chides me, when all serious attempts to provide foundations have been examined. I reject the sermon: my Rabbis told me not to leave the faith until all my efforts to defend it have failed; my party theoretician told me to stay with the camp of the glorious future until all glory is gone; my physics teachers told me not to desert physics until I had used up all my potential for scientific work proper. Oh, no.

Theory, pontificates Sagal, rises out of practice, and must rest on it. No, say I, equally pontifically. Theory can rise out of the collapse of practice, as a form of dogmatic pragmatodikea, and theory can rise in rebellion against practice or in attempts to change practice. And it can be but idle speculation: lovely idle speculation. This is too much for Sagal, and he adds a pontifical edict: we must follow methodical order, each step justified by previous steps. Could Euclid work thus, or Newton, or Cantor? At least Einstein confessed he was an opportunist.

(13) And from the ode to Lorenzen's justificationism, Sagal moves to lament Quine's non-justificationism. Sagal quotes Quine's refusal to justify and then comments to say Quine's justification is psychological. This defies comment. Quine is then cited to say constructivism conflicts with psychology: why not settle for psychology, asks Quine (rather than for construction)? Sagal sees in this rhetoric question a plea for psychological justification. He considers another quote from Quine as proof that Quine is for circular justification: he recommends a bootstrap theory. But bootstrap theories are non-justificationist and so not justifications and so not circular justifications and not non-circular justifications. Is this hard for Sagal to imagine?

Thus, Sagal sees non-justificationism as psychological justificationism, and he says it cannot be defended. The boot is on the other foot: the Dingler-Lorenzen school is psychological, and rests on an empirically refuted psychology.

*York University* and
*Tel-Aviv University*

JOSEPH AGASSI

# NATURALISTIC EPISTEMOLOGY: THE CASE OF ABNER SHIMONY

## SHIMONY'S FRAMEWORK FOR NATURALISTIC EPISTEMOLOGY

Shimony's 'Integral Epistemology', in this volume takes as a point of departure Campbell's non-justificationist descriptive epistemology: when we say we know something, we make a conjecture, not offer a proof. Shimony has no objection to, and indeed supports, descriptive epistemology, but dissents from Campbell's view and declares possible analytic epistemology, proof or some other justification. His "purpose . . . is to advocate an integral epistemology, in which descriptive and analytic considerations are brought together for the purpose of rationally assessing claims of human knowledge". The operative words are 'rationally assessing'. He wants to "shed light upon the reliability of human cognition', so that "adequate justification can be given for the presuppositions of scientific investigation".

I find this highly disturbing, let me confess. Scientific investigation goes on in the real world, with real people applying for real public funding of the substantial expenses required for their conduct. There are canons of making proposals, and committees are supposed to follow these and grant the funds available to some and not to others. They do not of course, but this does not mean that they ignore the canons. Are these canons of "adequate justification . . . for the presuppositions of scientific investigation"? If yes, the canons cannot be improved; if not, they must improve. Between these impossibility and necessity a whole world of choice is lost. (See Agassi, 1981.)

Nor is this peculiar to Shimony. It is, I have argued in detail (Agassi, 1969) a necessary corollary of justificationism: once I justify my views as best possible, and my opponent does the same, then either I or he clearly wins; and he who refuses to acknowledge defeat is a knave or a fool. Even if my justification is only in my proof that in today's circumstances my own view is the most probable amongst available views, then today all my opponents are knaves and fools. This conflicts with the commonsense view that some but not all disagreements are rational, and more so when dealing with "the presuppositions of



*Abner Shimony and Debra Nails (eds.), Naturalistic Epistemology,* 341–351.

scientific investigations". Planck's investigations in the field of radiation theory, we remember, were based initially on the rejection of Boltzmann's presuppositions in that field, and then on the acceptance of the same presuppositions; then, as it happened, Planck tried hard to get rid of them again. Where was he rational and where not? How are we to judge his wavering as rational and justify each of his stances? Is this not too excessive?

Shimony offers as a framework or guidelines the following four maxims (quoted here *verbatim*):

(a) Commonsense judgments about ordinary matters of fact have a *prima facie* credibility, which should not be discounted without clear positive reasons.

(b) The road to inquiry should not be blocked.

(c) Epistemology and knowledge of the world (a vague phrase, intended to refer both to the sciences and to metaphysics) should mesh and complement each other.

(d) A vindicatory argument — that is, an argument to the effect that a certain method, M, will yield good approximations to the truth in a certain domain if any method will, so that nothing is lost and possibly something is gained by the use of M — is an acceptable form of epistemological justification.

I will argue that these four maxims are, as worded here, so wishy-washy, that no rational commentator can ascertain that they should be accepted or that they should be rejected:

(a) Shimony proposes that common sense should gain a *prima facie* acceptance. What is clear is that Shimony is aware that sometimes we refuse to accept the verdict of observation reports: no matter how often a researcher fails to find an effect, he may still try again. Faraday sought effects for decades, at times his search was crowned with success, and at times not, yet he stuck with all his hopes (e.g. to discover the Kerr effect and the Zeeman effect and some electrogravitational effect).

Did Faraday violate Shimony's first maxim? This depends on whether we agree to judge his reason 'positive': his contemporaries did not; we do. Hence, the maxim is question-begging. Elsewhere (Agassi, 1975) I said, if excuses for ignoring facts are at all permitted, then a promise of an excuse should suffice too — and I offer historical instances where such promises were judged acceptable. As long as an investigation is interesting, I say, the problems it faces, including its conflict with accepted observation reports, may be put on ice. If we identify Shimony's demand for positive reasons with my justification of a research as interesting, then our claims coincide. But then, obviously, my claim leaves no room for, nor any need for, epistemology as a justification. And I do not see why he says we need one.

(b) Shimony advocates that roads to research not be blocked. Now, there are active and passive ways to block research. Putting one in jail because one is doing research on the occult, for example, is by now opposed by everyone in our circles. Refusing a request for a grant in the aid of the project that later on happens to win the sweepstakes is still very common, yet hardly objectionable: for logical reasons alone members of a grants committee cannot forecast the future results of an investigation, as Henri Poincaré already noted. There is a borderline, where the editor's refusal to publish progress reports lies. Shimony hardly notices such matters, and I have dealt with them at length elsewhere.

(c) Theory of knowledge and knowledge should mesh, says Shimony. Unless — he qualifies this — unless good reasons are given. This leads me to repeat my comments on Shimony's first maxim as well as to press my question to him, what counts as knowledge? For example, the theory of knowledge accepted among scientists in the last century led them to declare that Newtons's mechanics can never be superseded; consequently, when Newton's mechanics was superseded, they all felt that their theory of knowledge must be changed. Nevertheless, two other alternatives were open to them. They could deny that Newton's theory was superseded. Lord Rutherford closed a presidential address of the British Association for the Advancement of Science (in 1913) by declaring the view that Einstein has overthrown Newton's theory a popular prejudice. As an argument, he gave the observation that Einstein's theory, on the contrary, broadened its base. This makes no sense to me: I literally cannot tell what the metaphor means here.

Thus, the questions are, what is knowledge and what is a good reason; without answers to them the third maxim of Shimony is of little use. Since this third maxim is very pertinent to my previous discussion I will return to it.

(d) Any research which promises no loss and possible gain is justified in Shimony's eyes. Is there anything to lose by pursuing astrological research? This depends on one's goals. If one is employed in a university, has tenure, enjoys astrology, and is not particularly ambitious, then the answer goes one way; if one is ambitious and wants some positive results, then perhaps the answer goes the other way: one loses precious time by studying astrology. Once time is considered a scarce commodity, Shimony's fourth maxim becomes inapplicable in principle, and so utterly useless.

When we apply Shimony's four maxims to any specific case we can see what they amount to. Consider Boltzmann and Planck again. First, as to Shimony's support of commonsense judgment on matters of fact: Is the second law of thermodynamics violated by the statistical reading of thermodynamics? Boltzmann was worried and tried to block the statistically rare but still possible violation of the second law of thermodynamics. At times he did so by applying statistics only to unobservable atoms. At other times he did so by preventing too many small effects from ever accumulating so as to produce a perceivable one. Planck disliked this, took thermodynamics as fundamental and statistical mechanics as conflicting with it and so as objectionable. He later changed his view and followed Boltzmann. Amongst students of these matters, Einstein was the first — temporarily — to violate Shimony's maxim: in 1905 he applied statistical mechanics to all bodies, large and small, while offering no "clear, positive reason" whatsoever. On its basis he predicted statistical fluctuations that violate the "accepted commonsense judgments about ordinary matters of fact" and identified them with Brownian motion. Einstein's predictions were confirmed, thereby supplying *ex post*, of course, the "clear, positive reason" that Shimony's maxim requires. And Plank admitted error soon after. (Boltzmann died soon after and what his response would be is a matter for speculation.) The maxim, therefore, is clearly good *ex post* but clearly not so good *ex ante*.

As to Shimony's second maxim (which objects to the blocking of the road for research), nobody could block either Boltzmann's or Planck's inquiries, but everyone who could tried to discourage them except for a small group of friends. Would Shimony have encouraged their wild researches? He has no maxim to that effect.

Since neither Planck nor Boltzmann worried at that stage about the integration of knowledge with theory of knowledge, Shimony's third maxim (which advocates such an integration) may be inapplicable to their case. Paul Feyerabend repeatedly complained that not enough attention has been paid to Boltzmann's liberal view of scientific research which he (Feyerabend) thinks has helped Boltzmann himself in conducting his researches more freely and boldly than his colleagues. He may be right — I am too ignorant to judge the complex historical case. Certainly Planck, in the eve of his life, showed increasing concern for philosophy and attempted an integrated view, scientific, metaphysical, methodological, and epistemic. He failed, or at least this is the common verdict. The word 'should' in Shimony's wording of his third maxim makes it unclear to me whether its application to Planck's case leads to commendation for attempt or to condemnation for failure. But the goal of the maxim is so high and ambitious that failure to live up to it is the common lot in any case.

The fourth maxim of Shimony (which vindicates hopeful research) is so vaguely worded that it must be successfully applicable to both Boltzmann and Planck if time is not considered wasted in research, or to neither if it is. If Shimony's fourth maxim is to be of any use it has to tell the difference between a good and a bad method. Does it?

## SHIMONY ON SCIENCE AND ON THE THEORY OF SCIENCE

Immanuel Kant had in his philosophy room for two kinds of psychology, empirical and transcendental, and they did not sit well with each other — in principle, not necessarily in point of fact. This is — in point of history, as far as I can judge — what caused the collapse of Kantian philosophy. The situation has greatly altered since: we do not quite know what is scientific knowledge, but we do know it is not what Kant said it is. We also do not know what is scientific research, but we try to present a theory of research which will enable us to explain current science as the outcome of past research — in some sense of 'outcome'. Hence, we still have to square our philosophic view of what science is and what research is with our historical knowledge of facts about science and about research, especially the current science and the research that has preceded it. In addition to this we have scientific theories about the nature of man, about the growth of knowledge of individuals, especially children, and about perception. In addition to the desire to harmonize all three — evolutionary biology, developmental psychology, and perception theory — we wish to harmonize these with our philosophy and history of science (yet it is unclear how harmonizing is to be effected).

No one denies this. What I deny are two claims which, I think, Shimony makes. First, that the three fields mentioned in the previous sentence are already harmonized, and, second, that harmonizing them successfully with the philosophy and history of science somehow vindicates or justifies philosophy. In addition to this denial, I deny that philosophy of science is harmonized with historical facts about science — particularly Shimony's philosophy of science.

Shimony's presentation includes some discussion of the harmony between the three scientific fields. In particular he emphasizes that we learn to acquire reliable information by perceiving things in order to adapt and survive. This claim is too vague, since survival is the sole criterion of the reliability of information and we know full well that common sense information which was adequate enough for the survival of humanity during millennia of human prehistory and history was shown utterly unreliable nonetheless as soon as modern science evolved. To claim that science is essential to human survival is fantastic, and to claim that if not essential it is at least conducive to survival is debatable and hotly debatable indeed. But, in fact, at least the claim is uncontested that the unscientific cultures are threatened by the scientific cultures. Hence, Shimony's claim cannot perform the task of vindicating science, which task he allots to it. Let me quote his claim verbatim.

Shimony claims, emphatically, that "perception is reliable because it is an activity of an organism constructed so as to extract from proximal stimuli significant information about distant objects. It is in this sense and at this level that the structural demand of meshing epistemology with knowledge of the world is satisfied." Let us take an example — and since Shimony speaks of "an organism" we may well take an insect, whose perception mechanism is much better explained by science, than, say, that of a simian. We can see a butterfly or a moth navigating flight at times successfully, at times not. The theory which describes the navigation mechanism is an explanatory success if and only if it explains both navigational success and navigational failure, of course, as well as the survival of the species; the theory must, then, include the assumption that the rate of success suffices to insure procreation. Similarly, a theory which supersedes a previous theory should have superior explanatory success, and therefore it should explain both its predecessor's explanatory success and its explanatory failure. But whereas the butterfly species' overall success is measured as an ability to procreate one year after another, how is the overall success of science to be explained? Perhaps as its ability to help mankind procreate and thus survive. But then, if we do manage to destroy our planet — with the aid of science, of course — then science must be judged a failure, even if the destruction of the planet — with the aid of science, still — will be effected not by scientists or by scientific ideas. There is something amiss here.

Nor does Shimony harmonize evolutionist biology, developmental psychology and perception theory. To do so he needs a canonic version of each, if not also an axiomatic wording — and this is not available. Things are in flux, and all we can say is that efforts to harmonize, e.g. evolutionary biology with perception theory, are very fruitful, but still controversial: the way James Gibson suggested to approach perception theory, i.e. ecologically, is still highly contested by most perception theorists. Now that he is dead, they may be more willing to pay his ideas the attention they deserve, as a reviewer of his posthumous book has gallantly proposed in *Science* (Haber, 1980).

What puzzles me most, however, is that Shimony sees no problem in his holding evolutionary dialectic epistemology on one hand and static inductivism on the other — not even when he demands that any "personal probability evaluation should be *tempered* by a maxim of critical open mindedness". What is this emphasized "should be *tempered*"? Which of the two — the temperer and the tempered — under what terms, gains the upper hand? Shimony explains: one should "give sufficient prior probability to any seriously proposed hypothesis that it may achieve higher posterior probability if favored by the data". This means that if Boltzmann or Planck or Einstein has a theory which is testable, and someone tests it and it is confirmed, then you should be sufficiently critically minded to waive your hostility to it and accept it. This is terrible on two counts. First, it is useless for the researcher; second, researchers must be allowed enough openmindedness to reject well confirmed theories, or else their research is too constrained. Faraday rejected the most confirmed views of his day and therefore tried to refute them — at times with success, at times without.

## SHIMONY'S TRANSCENDENTAL ARGUMENT

Whenever a justificationist runs out of ammunition he takes recourse to a transcendental argument: without justification science is impossible. Here is Shimony's transcendental argument:

> If we are not optimistic about human abductive power, then seriously proposed hypotheses are coordinate with frivolously proposed ones and with merely logically possible ones that no one has ever singled out for mention. But . . . infinitely many rival hypotheses are logically possible, and it is hopeless to explore all of these. Our only

> hope of approaching the truth . . . is to be optimistic . . . in the sense of permitting a seriously proposed hypothesis to be accepted after experimental scrutiny. This optimism precludes no particular method for assigning order [of a prior significance] to hypotheses . . . . Consequently, nothing is lost and possibly something is gained . . . . The tempering condition is manifestly vague . . . . A more promising procedure is . . . explicating [vague] phrases . . . . Circularity obviously enters . . . . The circularity is non-vicious, since a seriously proposed hypothesis can be subjected to experimental scrutiny and . . . discredited. Furthermore . . . a radically new hypothesis can be accorded the status of 'seriously proposed' . . . . This program for justifying inductive inference. . . .

I apologize for the long quotation and for the detailed analysis to come. It is passages like the one just quoted which usually make me prefer not to debate Shimony's ideas. For, the passage quoted is presented as programmatic, so that readers are called to be indulgent towards its sloppiness and put on it the best interpretation of it which they can find. And doing this I read it as a program to present a philosophy which has already been developed by Karl Popper and criticized — to my satisfaction at least.

Let me analyze the above passage in detail:

If we are not optimists, we may take equally seriously or unseriously a seriously proposed hypothesis and a silly one, says Shimony. I deny this. I am myself hardly an optimist, and when in a pessimistic mood I think I outdo Schopenhauer without even trying. Yet I am not aware of my taking equally seriously serious and unserious hypotheses. Indeed, if I am allowed to boast I would say I deem myself more discriminating than Shimony, his professed optimism notwithstanding. Since my claim to hold a discriminatory non-optimism clashes with Shimony's claim for optimism as a necessary condition for discrimination, at least one of these must be erroneous. Which option is better? Let us take the first: I am in error. Now, if I were non-discriminating, I would not care which option to choose. Since I present a discussion on it, I care about it, and since I care about it, I try to discriminate. Hence, I am an optimist despite myself. How can we examine this? Not by my professed non-optimism and not by my contesting Shimony's claim — Since both are under scrutiny. Perhaps the point can be analytically established: any discrimination between hypotheses, it may be contended, of necessity just is research oriented, future oriented, forward-looking; hence necessarily optimistic. A pessimist, by this contention, cannot explain his activities: he can explain his eating as a response to brute instinct, but not his stocking of food! This analysis is refuted by many cases of people acting under an overwhelmingly justified assessment and sense of utter despair, motivated by a sense of responsibility, by the desire to give others the benefit of doubt — by the duty of hope, to use the apt expression of Herbert Samuel. Now, if the duty of hope makes us optimists, then we may cut optimism out as the middle man and speak of the duty to discriminate between good and bad hypotheses. Hence, Shimony's claim is false.

But perhaps I am finicky. Let us agree, Shimony and I, that the assumption that possibly we may progress already obliges us to try, whereas assuming no progress possible leaves us in utter inactivity, so that on his fourth maxim we should try to progress. This makes all justification quite unnecessary. Moreover, justification of progress makes this maxim unnecessary. Unnescessary in the sense in which watering the flowers in the rain is unnecessary.

But let us proceed. Infinitely many hypotheses are basically possible. This worries Shimony. Why? Because we cannot examine them all. We cannot even articulate them all. Yet without articulating them he wants them to have prior probabilities so that some of them may have also posterior probabilities to justify our hopes. But this is highly dangerous: the most confirmed hypothesis may be not yet articulated! Today we can develop a model of the world in which a hypothesis is most confirmed and even has a name, yet is not yet worded! In brief, Shimony's method — it is really that of Harold Jeffreys and Dorothy Wrinch (and in these feminist days her name is still lost, though her hypothesis increasingly popular) — is more worrying than tranquilizing. To allay this worry, we should postulate that the already worded hypotheses have such initial priority that empirical evidence can never point away from all of them and in the direction of an unknown one.

But the history of science is full of examples of the kind just proposed by this postulate to be outlawed. These examples are called by Thomas S. Kuhn periods of crisis and they generate, he avers, sleepless nights in which the leaders of the scientific community conceive new hypotheses — new paradigms he calls them, since they give birth to a new optimism in the scientific community at large.

Hence, either Shimony's idea averts all crises and thus calls for optimism that is stagnating and hence quietist, or else his idea allows for a permanent crisis that is pessimistic in that it permits the indefinite stay in the dark night of the soul. What he would like to have, of course, is crises crowned with reliefs, with ever renewed joyous dawns following each bleak dark night. But the theory of Jeffreys and Wrinch which he adumbrates does not allow for such a complex phenomenon. It is no accident that Jeffreys explicitly denies that a hypothesis may be rejected before an alternative to it is offered (as does Kuhn); hence no dark nights and no sleepless nights for the justificationist.

This may be a technicality. If we accept the view of overall themes, of metaphysical guidelines, etc., we may claim that each theme offers a set of prior preferences, that when a theme dries up a crisis follows pregnant with a new theme. Fine. But then the pregnancy may be abortive. We do not know. And the transcendental argument will be now an inductive generalization: in the past the future did take good care of itself, so maybe in the future it will not do worse. Perhaps. It is incumbent on us to give Shimony the benefit of the doubt.

This is a very funny situation. If I give Shimony the benefit of the doubt, then possibly he justifies science. Is this an admission of defeat? Modal logic says, no: a possible defeat is not a defeat. Does he, *per impossibile* lose? No. Not even if a possible victory, as modal logic tells us, is also a possible defeat, since neither for him nor for me is a possible defeat a defeat. But since a possible victory is not a real victory, he has not won: he has not proven his case. But he claimed that we need a proof or else all is lost. Hence he has lost indeed. Pity: I'd love to know that as in the past so in the future, the future will take care of itself: pessimism is no fun.

The logic of all this is the logic of justification: it claims too much. Once a non-justificationist position is deemed not refutable, the whole justificationist philosophy has been superseded. Even the inability to refute the stupidest theory (e.g. Lewis Carroll's theory, 'the world is the Red King's dream') is the inability to prove. If justification is proof then I must be in error. If I may be in error but also not, then justification is an error.

But is justification proof? Can we not justify by giving the benefit of the doubt? If yes, then Shimony is in error when he claims to be contesting Donald Campbell's non-justificationism and demanding a justification. And here lies the rub. There is the necessary, the possible, and the impossible. Within the range of the possible all sorts of monsters lie. They may drive one to despair. Some of us want to exorcise these monsters by moving away from the merely possible: it has to be at least plausible or probable to gain respectability. Now plausibility or probability is either merely possible and so fails to exorcise all monsters, or necessary and thus exorcises all monsters but also destroys a hundred flowers. Otherwise plausibility or probability is merely plausible or probable and so begs the question all over again.

Conclusion: no monsters, no flowers.

Moral: Let us not allow the monsters to drive us to despair.

Corollary: With no despair there is no need for Shimony's transcendental argument.

Proposal: Let a hundred flowers bloom.

*York University* and
*Tel-Aviv University*

## REFERENCES

Agassi, J.: 1969. 'Unity and Diversity in Science.' In R. S. Cohen and M. W. Wartofsky (eds.), *Boston Studies in the Philosophy of Science*, Vol. 4. Dordrecht and Boston: Reidel, pp. 463—522. (Reprinted, with a new Appendix on Kant, in Agassi, 1975, pp. 404—468.)

Agassi, J.: 1975. 'Towards a Theory of Ad Hoc Hypotheses.' In *Science in Flux*. Dordrecht: Reidel, pp. 189—207.

Agassi, J.: 1981. 'Research Projects' and 'The Methodology of Research Projects: A Sketch.' In *Science and Society*. Dordrecht: Reidel, pp. 262—272 and 273—282.

Haber, R. N.: 1980. Review of James G. Gibson, *The Ecological Approach to Visual Perception* (Boston: Houghton-Mifflin, 1979), *Science* **209**, 799.

ABNER SHIMONY

# COMMENT ON AGASSI

In a brief space I can only reply selectively to Agassi's wide-ranging comments, but I shall try to do so as constructively as possible.

Although controversies are of no value for their own sake, they are of inestimable value as indices and as means. They are indices of the freedom to inquire and of the *élan* to pursue inquiry. They are means of assessing claims to the truth, and also for stimulating invention and insight. Agassi and I probably agree on these theses, despite differences of emphasis. I also partly agree with him concerning metaphysical controversies. They have indeed generated research programs which may give birth to scientific theories — a thesis which is one of his main contributions to philosophy. However, I am more sanguine than Agassi about the possibility of deploying scientific evidence and analysis to elucidate metaphysical controversies. For just this reason we are, in my opinion, living in a golden age of metaphysics. For example, the traditional metaphysical problem of the relation between the organic and the inorganic can be treated with more control than ever before — and above all in a more interesting manner — because of the discoveries of molecular biology, even though the problem is not completely solved.

Agassi classifies me in Sections 2 and 4 of his paper as a 'justificationist', who is therefore guilty of whatever faults are generic to "the whole justificationist philosophy." There is indeed a large and diverse class of philosophers who espouse the general theses that it is legitimate to ask why standard scientific inference (or a suitable refinement thereof) is reasonable and that the mission of philosophy is not completely fulfilled unless this question is given an illuminating answer. If this class is the extension of "justificationist," then I certainly belong to this class, but not if the term is used narrowly to apply to those who "appeal to universally valid proof" (Agassi, 1975, p. 422). I do not think that we can *prove* (in any sense that resembles the usual meanings of this word) that a theory which is well confirmed in accordance with scientific procedures is definitely true, or has greater verisimilitude than any rival theories, or bears a better correspondence relation to the truth



*Abner Shimony and Debra Nails (eds.), Naturalistic Epistemology*, 352—355.

than any of its rivals, nor can we *prove* that the indefinitely prolonged application of the scientific method will with certainty lead asymptotically to the acceptance of a theory with any of these virtues. My credentials as a fallibilist are in good order. Nevertheless, I maintain that the scientific method exhibits a kind of sensitivity to truth, and that refinements of scientific method — achieved by reflections upon scientific achievements — can enhance this sensitivity.

Although no one has given to my satisfaction a full and balanced exposition of this sensitivity, it does seem that the contributing elements can be discerned: notably, the recognition of the possibility of systematic errors in experimentation and the development of procedures for controlling them; the recognition of the possibility of random errors, and the use of the machinery of probability theory for controlling these; the recognition of the possibility of alternative explanations of phenomena, and the consequent need for precision of predictions and of observation in order to adjudicate among them; and the recognition of the difference between an *ad hoc* hypothesis and one which is 'natural'. I deliberately mention this last element in order to emphasize that the scientific method is vague and informal in certain important respects and therefore in need of further refinement. In my contribution to this volume and in the essays referred to there I have argued that the spectacular predictive successes of the natural sciences constitute phenomenological evidence for the claim that the scientific method as actually practiced manifests a considerable degree of sensitivity to the truth; but I also claim that we are able to peer inside the machinery of the scientific method to see why it works. Although I confess to sketchiness in the program that I have just outlined, I do not see what generic irremediable fault of justificationism has been committed and what in principle prevents the more and more complete realization of the program. Agassi has orally expressed agreement on this point.

Perhaps Agassi is arguing that the obstacle in principle to realizing my version of a justificationist program is the infinite richness of the possible: the infinity of alternative hypotheses compatible with any finite body of data (which Popper insists upon in his rejection of confirmation), the infinity of hybrid predicates like Goodman's 'grue', etc. As Agassi says eloquently at the end of his essay, "within the range of the possible all sorts of monsters lie." The instrument which I propose for dealing with the monsters of possibility is the vindicatory argument, schematically presented in maxim *d* of my contribution to this volume (to which Agassi objects). If nature is so intent upon hiding, or demons are so intent upon deceiving, or chance is so blindly perverse that a set of hypotheses is steadily confirmed by the most careful tests even though they are utterly remote from the truth, then there is nothing that we miserable human beings can do to discover the truth. But it is rational to set aside these scenarios in actual scientific investigation, since acting in accordance with them would leave us without guidance even in a world hospitable to our search. (It is also rational, however, to remember that they are indeed possible scenarios, in order to remind ourselves of the fallibility of our faculties.) My practical conclusion is not very different from Agassi's "Let us not allow the monsters to drive us to despair." The difference between us is that I give a rational account of the decision not to yield to despair. My account is recognizably rational from the standpoint of decision theory, even though it lies outside the mundane domain of standard applications of that discipline.

I admit that my maxim *d* is obscure when stated at a high level of abstraction, and that Agassi is right in saying we cannot judge whether "nothing is lost and possibly something is gained" unless we know a person's goals. I had in mind only the context of epistemology, in which the goal is to find reliable ways to assess claims to the truth or to closeness to the truth. There are other possible goals, perhaps in conflict with this one, and *a priori* one cannot rule out a value theoretical judgment that other human aims have primacy over the search for the truth. Epistemology can by-pass this problem of value theory and will have its hands full in working out the details of the hypothetical imperative, that if you wish to approach knowledge of the truth then you must follow such and such procedures. For this reason, I do not consider the fact that destructive technology is based upon the physical sciences to constitute a valid objection to their epistemic worth.

Agassi attributes to me a strategy for dealing with the monstrous infinity of possible hypotheses which I must explicitly disown: that of Jeffreys and Wrinch of ordering an infinity of possible laws and assigning to them prior probabilities which sum to one. Perhaps it is this methological proposal which he has in mind when he attributes to me a 'static inductivism'. In Shimony (1971, pp. 98—9, 131) I specifically say that any *a priori* ordering is arbitrary and that prior probabilities must be relativized to the set of seriously proposed hypotheses concerning the matter under investigation. Since this set depends upon social milieu, climate of ideas, biographical contingencies, and sheer inspiration, my methodological proposal is anything but static. Indeed, the intention was to break down somewhat the rigid distinction between the context of discovery and the context of justification, and to acknowledge the abductive powers of human beings even in the latter context. In making this proposal I was following Jeffreys (1961, p. 129) himself, who criticized his own earlier work with Wrinch. One can indeed see far by standing upon the shoulders of a giant, but only if one is sure-footed enough to maintain balance.

Agassi's criticism that my four maxims are 'wishy-washy' is penetrating and I intend to take it into account in my later work. A partial answer to him is that the maxims are not intended to be self-sufficient. They are not an entire epistemology, but are rather principles of conducting the dialectic which I believe to be indispensible to epistemology. They are enriched by engagement and by reflection upon the cognitive achievements of human beings thus far. Contrary to Agassi's representation of my views, I agree with his statement that "Things are in flux, and all we can say is that efforts to harmonize, e.g., evolutionary biology with perception theory, are very fruitful but still controversial." We can, nevertheless, sensibly inquire about the meshing of our account of human beings as knowers with our account of the nature of things, even when our knowledge of both is provisional. (Heffner's paper in this volume is an admirable example of such an inquiry.) Epistemology, as I conceive it, is the study of how to be rational when all is in flux, including the method of investigation itself.

*Boston University*

## REFERENCES

Agassi, J.: 1975. *Science in Flux*. Dordrecht and Boston: D. Reidel.

Jeffreys, H.: 1961. *Theory of Probability*. 3rd ed. Oxford: Clarendon Press.

Shimony, A.: 1971. 'Scientific Inference.' In R. Colodny (ed.), *The Nature and Function of Scientific Theories*. Pittsburgh: University of Pittsburgh Press.

MARX W. WARTOFSKY

# EPISTEMOLOGY HISTORICIZED

## 1. INTRODUCTION

Historicaly, epistemology has been ahistorical. That is to say, epistemologists have sought to fix the universal and necessary conditions of any knowledge whatever, or to establish the essential nature of the human mind. Thus, whether empiricist or rationalist, realist or phenomenalist, traditional epistemologies have shared a common essentialism. What made such epistemologies different were alternative accounts of what are the fixed, essential modes of the acquisition of knowledge, or what are the universal and unchanging structures of the human mind. Essentialism merged with foundationalism or justificationism just to the extent that putatively descriptive accounts of the modes of the acquisition of knowledge were tacitly normative (i.e., in that it was *knowledge*, and not mere belief, or habit, or phantasms of the imagination which resulted from the proper working, or the proper correction, of these cognitive procedures.) Or else, essentialism was openly justificationist, in the converse procedure of constructing just those accounts of, say, perception or cognition such that the normal exercise of these faculties would generate knowledge, or make truth attainable.

Things have changed recently. Over the past century, and more strikingly in the last thirty years, epistemology has been given a rough time. It has been biologized, evolutionized, naturalized, socialized, structuralized, geneticized, behaviorized, linguicized, computerized and most recently ostracized (or, as one may say, Rortyfied.) What started out as a difference over the locus of justification, in essentialist terms, has ended as an attack on justificationism itself, substituting Popperian 'hope' for Cartesian certainty, Feyerabendian anarchism for Platonistic ascent, Kuhnian descriptive sociologism for Kantian normativism, and whatnot.

Tempting though it may be to pursue a critical history of the demise, dismemberment, and transfiguration of epistemology, that is not my object here. I want to speak in defense of epistemology, but by way of an anti-essentialist argument. I also intend a critique of traditional



*Abner Shimony and Debra Nails (eds.), Naturalistic Epistemology*, 357—374.

epistemology, but by way of a proposal for replacement, not elimination. Though I will consider a number of reformist and revolutionary moves that have marked recent discussion, I will find them all wanting. Instead, I will propose an *historical* epistemology which starts from the premises that the acquisition of knowledge is a fundamental mode of human action; that it is therefore inseparable from other forms of human practice; and that it is also inseparable from the historicity of these other modes, that is, from their historical change and development.

Several things follow from these premises, as I hope to show. *First*, the appropriate domain for the study of human cognitive practice is not the abstract and relatively featureless domain of the 'human mind', whether *tabula rasa*, or packed full of innate ideas or faculties; but rather the concrete, many-featured and historical domain of human practices — social, technological, artistic, scientific. I am proposing a methodological shift of no small proportions: we may be said to learn what the human mind is, or what human knowledge is, or how it is acquired, by studying the modes of action and the embodiments of these actions in artifacts. This shift doesn't eliminate the epistemological subject, but it suggests that we come to know the subject via its embodiments of cognitive activity. The *second* consequence is that cognition proceeds by way of representation — but not, in the first place, by internal representations, 'in the mind' or 'in the brain'. Rather, it proceeds by taking things – first of all artifacts – as *symbols*, whether linguistic, pictorial, gestural, or theoretical. The third consequence is crucial: since human practices have a history, and since representation itself has a history, I will argue that modes of cognitive practice, perception, thought, ways of seeing and ways of knowing, also have a history; in short, that modes of cognition change historically in relation to changes in modes of social practice, and in particular, in relation to historical changes in modes of representational practice. I claim, for example, that modes of visual perception change with historical changes in dominant or canonical scientific world-views. I am not arguing for the evident case that the *content* or our knowledge or of our perception changes historically, that we believe or that we know different things; but rather for the view that *what constitutes knowing*, or the mode of cognitive activity, changes historically. The third consequence entails that there is no fixed, species-specific or essential form or structure which all human knowledge has; but rather than human beings create and transform the human mind in the course of their manifold historical practices; that, in effect, the mind is a changing and developing artifact which we ourselves make; and that it is, in this sense, historically self-transcending.

The general project of an historical epistemology has been laid out in rather bald and sketchy terms, without arguments, and in a shamelessly assertive way. Clearly, what I can present here will be a fragment of the entire project.

Within the compass of this paper, I will focus on three questions: *first*, in what way is such an historical epistemology situated between the naturalized, socialized or evolutionized epistemologies of, say, Dewey, Quine, Piaget, Popper, Campbell, on the one hand; and relativistic epistemologies of Carnap, Kuhn, or Goodman on the other? *Second*, what is 'representation' on my view, why should it be said to play an essential role in cognition, and how does it relate to the notion of internal representation? *Third*, what is the full import of the notion of historical change in modes of cognition, or in modes of perception? In answering this third question, I will limit myself to the specific example of visual perception in its relation to changes in canons or modes of pictorial representation.

One last word, by way of introduction: it has sometimes occurred to me that the project of an historical epistemology might be dramatically presented in the following form: What would have happened, if Nelson Goodman had read Karl Marx seriously? I leave you to ponder the question, and hope to shed some light on it as I proceed.

## 2. EPISTEMOLOGY SOCIALIZED AND NATURALIZED

One of the fascinating transformations in the history of philosophy was the emergence of a socialized version of empiricism out of the 'first-person' empiricism of Locke and Hume. A fragment of this history will suffice for our purposes. On a question of fact the crucial witness before the court is the eye-witness — moreover, an eye-witness who speaks only in terms of the simplest, least-interpreted, most immediately available evidence of the senses. Or so it seemed. The philosophical unpacking of this starting point, however, led quickly to complications.

Carnap's *Protokolsätze* (Protocol sentences), which were to provide the reduction base for an empirical philosophy of science, were not sense-data. These he disclaimed, in favor of what he called (following the lead of his Gestalt-psychologist friends) *Erlebnisse* ('lived experiences' is the usual English translation). These too were not enough, since it wasn't simply the *experience* which operated in the logical reconstruction, but the sentence describing it. But such sentences — the Protocol sentences — had to be in *some* language; and the choice of a language in which to describe such experience was, according to Carnap, a matter of expediency and of practical convention. Thus, Carnap's 'principle of tolerance' in the choice of linguistic frameworks, in which to describe the relevant empirical evidence, presented no intrinsic or necessary specification of the reduction-base for any given science. Thus Carnap relativized the crucial epistemological point of reference to what was, in effect, a social or a cultural choice: *chacun à sa réduction.* The choice of empirical reduction base not being a matter of whim, it becomes instead a matter of social convention, dictated by pragmatic or instrumentalist criteria of adequacy, in Carnap's view.

Similarly, Popper's empirical test — this time, not for the verification, but the falsification of theories — is the 'basis statement' (*Basissatz*), and here too, what constitutes something as a basis statement is a cultural or social convention — what the practitioners of a given mode of inquiry *take to be basic*. The falsificationist test of theories turns for Popper, after all, on social psychology.

The social relativization of the empirical reduction base, or of the test for verification or falsification of scientific statements, is an unacknowledged commonplace of the logical empirical program. It is more obvious in Wittgenstein, for whom the constitution of meaning is itself part of a form of life, the social reality of a language-game constituted by its rules. The relativity of frameworks which yield alternative perceptual or cognitive judgments, whether duck-rabbit, or Ptolemaic-Copernican, or Aristotelian-Galilean, finds its expression later in Hanson, Kuhn, Feyerabend, in various ways, all of which appeal to one or another social construction as the source of the alternative framework: whether an alternative theory, differentially loading observation; or an alternative paradigm, characterizing a scientific subculture; or a different politics, a different interest, a different rhetoric or different articles of faith.

Finally, we have a strange fellowship among the epistemologists of the *consensus-gentium*, for whom agreement among the members of the community of inquiry becomes the test of truth, or rather, the test of what will *count as* the test of truth. The strange bedfellows here are Quine and Habermas, and the locus of their affinity is just this appeal to consensus among the speakers of a language. It is not simply that intersubjective agreement on what's what reduces truth to consensus, but rather that there exists a 'meta' consensus – either tacit, or reconstructible as the precondition for the existence of a speech community in the first place – which marks the ultimate reference point for truth and objectivity. *It is what we all agree to agree on* that finally grounds reference, the observation statement, or the ultimate or last court of epistemic appeal, on this view.

Quine (1969) writes, (in 'Epistemology Naturalized'):

> An observation sentence is one on which all speakers of the language give the same verdict when given the same concurrent stimulation. (p. 87)

and later he continues:

> It is no shock to the preconceptions of old Vienna to say that epistemology now becomes semantics ... also that epistemology merges with psychology as well as linguistics. (1969, pp. 89–90)

Aside from the creeping circularity of this approach (e.g., how will "the same concurrent stimulation" be determined to be "the same" except by appeal to that same concurrence among speakers of a language that is involved in the verdict which determines some utterance as an observation sentence in the first place?) Quine's appeal to the linguistic community is strikingly like Habermas' appeal to the ideal speech situation or the ideal of undistorted communication. For both, truth is seen to be grounded in the intrinsic norms of language, or of communicative action, which, according to Habermas, have the condition of free (hence, undistorted) agreement among all the speakers of a language as their end.

To sum up briefly, the socialization of epistemology in modern philosophy, especially in philosophy of science, proceeded on empiricist grounds from the notion of intersubjective testability, to the relativization of the empirical reduction-base, to social choice, language convention, or expediency: thence, to modes of public or professional consensus, as courts of last resort in matters of epistemic authority. It is not that agreement on matters of fact grounds claims to truth, but rather that agreement on method — i.e. on what will count as matters of fact, or grounds for agreement — is the warrant of knowledge-claims.

Epistemic authority thus changes its locus from the 'internal' or 'mental' contexts of traditional epistemology — what Rorty has described as 'privileged representations' (e.g., the clear and distinct ideas, the indubitables and incorrigibles, the unmediated 'givens' of sense-datum theory; or what it is 'impossible to conceive of as otherwise' of an earlier rationalism) – to the externalized and socialized contexts of the behavior of a community of speakers. Justification shifts its locus from the eye to the mouth (so to speak). Thus, Rorty can say (commenting on Quine and Sellars): ". . . if assertions are justified by society rather than by the character of inner representations they express, then there is no point in attempting to isolate privileged representations," and further:

> Explaining rationality and epistemic authority by reference to what society lets us say, rather than the latter by the former, is the essence of what I shall call "epistemological behaviorism," an attitude common to Dewey and Wittgenstein. (1979, p. 174)

Epistemology socialized thus seems to end up as epistemology behaviorized, or semanticized, since the modes of socialization that seem to be canonical here seem limited to a 'society' of all those commonly conditioned by schedules of reinforcement to agree on 'observation sentences', or all those who participate in the same language game, or all those whose community of discourse is necessarily grounded in the 'Universal Pragmatics of Language' which presupposes 'coming to agreement' as the (quasi) transcendental ground of a speech community. Epistemology socialized joins hands with epistemology naturalized if the modes of socialization in a community are dominated in certain respects by the methods, results, and institutions of the natural sciences, as is the case in a significant segment of our own society. Quine, of course, is the outstanding advocate of an epistemology which is simultaneously naturalized and socialized.

This reduction of epistemology to behavioristic psychology or semantics or pragmatics (*cum* appeal to neurophysiology and the purported physicalism of 'irradiated nerve endings') seems to me a poor substitute for 'philosophical' or essentialist epistemologies it is to replace, not because these older views are adequate, but because the newer ones seem to me to be no less inadequate.

The inadequacy is not simply because the *consensus gentium* seems a poor substitute for sense-certainty or rational necessity, or because the socialized epistemologies suffer the dangers of total relativism (though both of these considerations *are* relevant); but rather because the socialization is so *thin*, its conception of the social so abstract and ahistorical, and the idea of human activity (even in speech alone) so impoverished, that the human knowledge which such models would support is a stripped-down caricature of human knowledge.

Nor does it seem to me that Rorty's heroic critique of ahistorical and asocial epistemologies does the trick. For if we end up with physicalist behaviorism *cum* hermeneutics, we haven't solved the problem, but have rather restated it in sharper terms. Further (as I have argued elsewhere), Rorty's move to eliminate the old 'privileged representations' by a wholesale attack on the notion of 'representation' seems to me to throw out the baby with the bath water.

So if 'epistemology naturalized', *à la* Quine, won't do; if 'epistemology socialized', in terms of alternative conventions of language, or theoretical frameworks, won't do; and if 'epistemology ostracized', *sans* representation, won't do either, what will?

I have proposed, in a number of papers (some published, some not), the alternative of an epistemology historicized, i.e. an account of cognitive activity which takes it to be historically changing in the sense that modes of cognition and of perception change with changes in the forms and norms of the whole range of human practices, social, technological, scientific, economic, artistic, etc. Here, I can focus on only one aspect of that general program, but I will try to show how, in respect to that aspect, namely *representation*, the sense of the historicity of cognition may be understood.

## 3. REPRESENTATION AND COGNITION

By historical changes in modes of cognition, I mean changes both in what constitutes 'knowledge' as a category, and in what constitutes ways of knowing or ways of perceiving. I want to argue that such changes are brought about by the creative transformation of a given mode of cognition by means of human activity itself, in the pursuit of those essentially cognitive practices which characterize human action. With philosophers as disparate as Dewey, Wittgenstein, Polanyi, and Goodman (and with special reference to Marx and Cassirer), I characterize 'cognitive' activity not simply as intellection, or 'having thoughts', but as including also the range of feelings, skills or competences in forms of practice; and including also those more or less *tacit modes of judgment* that are involved in moral, aesthetic, scientific and technical contexts, such as assertion, approval or disapproval, enjoyment, dissatisfaction. This is a rather broad domain to include under the rubric 'cognitive', but there is good precedent for it, even in so rationalist a thinker as Descartes. ("By thinking", says Descartes in the *Meditations*, he means "doubting, understanding, conceiving, affirming, denying, willing, refusing, imagining, perceiving" (*Med.* II).)

I want to propose further that all such cognitive activity is intrinsically normative, i.e. that 'knowing' (whether knowing-that or knowing-how) is an achievement-verb, and that learning, in general, presupposes some norm. Human cognitive activity does not take place in some context-free way in consciousness in general, but is typically inseparable from various forms of practice, e.g. hunting, production, social interaction, story-telling, war, politics. Practices then become the concrete modes of cognitive activity; and theory may (too briefly) be characterized as the self-conscious reflection on the norms of a practicer; or most generally and abstractly, of practices as such.

By a *practice*, I will mean a relatively coherent set of human actions, characterized by a commonly understood object, or good — i.e. a *telos*; and by a socially structured and commonly accepted repertoire of means, whether instruments or skills, which are understood as means to the end or good of that practice — i.e. a *techne*; as well as by commonly understood *norms*. These *norms* include the explicit *rules* or conditions under which the practice is carried out, the exemplary *models of the procedures or products of the practice* (as embodiments or exemplifications of the realized norms of the practice), and the tacit but recognized *modes of practical judgement* or *casuistry* which interprets the rules or the exemplary moders of the practice in terms of their application to specific instances of the practice. Note that all the constituents of a practice are intentional: they are constituted as what they are by being taken to be those things, or by being understood as such by the practitioners, and by the society which supports the practice, or benefits from it. It follows therefore that nothing can be a practice without being constituted as such by the understanding of conscious agents; and also, that no one can be a practitioner except in the context of both self-recognition and recognition by the relevant others *as* such a practitioner. Thus, practices are not extensionally definable, nor do they exist as self-subsistent roles or structures, though they may be reified as such, ideologically.

The relation of such a concept of a practice to Aristotle's notion of the Practical Intellect, and to Polanyi's or Kuhn's 'paradigms', or to Popper's 'third world' entities or to Habermas's 'norms of action', needs to be discussed, but I cannot do that here. It should be remarked, however, that the constitutive or constructive genesis of a practice, involving as it does reciprocal recognition in the acts of understanding by agents, has its sources in the dialectical phenomenology of Hegel and Feuerbach and their successors; and moreover, in Marx's insistence on the social and historical contexts which concretely define and differentiate such practices; and in Heidegger's notions of *Zeug* and *techne*.

In such contextually-bound cognitive activity, the norms of what will count as right or appropriate or adequate, as the correct way of doing things, the right answer to a question, the relevant behavior, may be characterized as the norms of that practice. Further, such practices are teleological, i.e. they involve some human purpose or aim, so that the norm of a practice is to be understood as involving what we may call the good of that practice. The notion of 'the good of a practice' (which I adapt from Alasdair MacIntyre) is clearly a social and, indeed, an historical notion, though its original sources in Greek philosophy, especially Aristotle, formulate the idea in essentialist terms.

Now I want to distinguish such teleological activity, involving the good of a practice, or its norm, from what we may characterize as goal-oriented or teleonomic behavior by non-human animals. Only in the human case is the purposiveness of the activity such that it is extrinsically embodied in norms, i.e. in rules, laws, methods, maxims, rules of art — prescribed ways of proceeding which are given in the socially understood forms of a practice. A practice embodies its rules in various ways: the model artifacts, which are the ones to be copied or reproduced in the activities of production (e.g., tools); in verbal or gestural instructions, commands, imperatives; in model-narratives, which describe appropriate modes of behavior, or point to a moral, e.g. parables, fables, myths; in rituals and rites; in dramatic pre-enactment or re-enactment of actions taken or to be taken, as rehearsals for the 'real thing,' e.g., hunting dances. One may say, generally, that a practice embodies its norms in the artifacts of that practice — the external, created and objective entities which come to be what they are, or are understood as being what they are, only within the system of practices in which these artifacts are made and used. Thus, something is a hunting-spear, or a knitting needle, or a shoe, or an appropriate response to a greeting (or, in fact, a 'greeting' in the first place) only within some system, where such a system involves a practice and the norms which define it.

What makes the practices cognitive, on my view, is the fact that the norms of the practices can be *represented*, whether initially *in* the embodiments of the practice themselves (in its products, or its 'right procedures') or in explicitly representational artifacts (words, pictures, diagrams, dramatic re-enactments) which are *about* the practice. It is the move to explicit representation of a practice, i.e., to the representation of one thing by another, or even in the reflexive mode by that 'same' thing, now taken to refer to itself, so to speak, or to be 'about' its use, or intentionality, or function — that on my view marks the genesis of cognitive activity. For then, an activity is conducted with respect to some norm which is consciously understood, and which guides the activity with respect both to its proper mode, and its proper end.

In this sense, no rote repetition of a task, or a course of action, however 'correct', can count as a cognitive practice, if the agent does not understand what he/she is doing in terms of the norms of that practice. But 'understanding' here is an intentional category, not reducible to, e.g., 'acting in accordance with a rule', which in fact would require no more than habit-formation, or indeed, the 'right' wiring, or stereotypic instinctive responses to cues. Thus, neither Myna birds nor apes 'speak', though they may learn to utter speech-like sounds (in the first case) or to form sequences of tokens (in the case of apes). To understand what one is doing, then, requires having learned it not simply by repetition, or by constraint (as one may be physically forced to perform a certain sequence of motions), but in terms of some representation of the action to be engaged in: Understanding it 'in the representation', or 'in the model' means, then, to have established some mode of reference between something, *taken to represent* a mode of action, and the action itself; thus, it entails taking something as a representation of something else, or realizing its 'aboutness'.

Nothing *is* a representation as such. It comes to be a representation by being *taken* as a representation, or *made* as a representation. And, as Goodman points out, *anything may* be taken as a representation of anything else, though in fact, not just anything is. There develop what I have called canons of representation, as social constructions, or conventions, which restrict the range of acceptable representations. Examples here, of special interest to me, are *Styles* in Art, and *Theories* in Science, both of which mark the boundaries of the acceptable representations within some symbol system. As Styles and Theories change historically, so does the range of what are taken to be, or constructed as, representations. A rough analogy to such canons of representation in Goodman's work is 'projectible Predicates' or 'categories', which he bases (too easily and abstractly) on familiarity, on habit, on 'entrenchment'. This leaves untouched the question of *why* certain predicates are entrenched, or projectible, rather than others; and stops short of an account of the concrete social, historical, technological, linguistic, etc., practices in which such entrenched predicates, categories, schemata develop. On the other hand, Goodman insists that such canons *are* generated by practices, and moreover, by *choice* (though not arbitrarily). They are thus not dictated by essential or fixed structures, and this leaves his systematic formal constructivism open to an historical interpretation; and equally important, to an historical interpretation which is not *historicist*, in Popper's pejorative use of that term.

Two things follow from this formulation of cognitive practice. *First*, as modes of practice change, and as the norms of these practices change, so too will the modes of representation. *Second*, as modes of representation develop, certain canonical forms of representational artifacts, namely, pictorial representations and linguistic representations, come to be dominant (for good reasons).

I conjecture that it is these forms of symbolic representation in particular tend to generate correlate 'images' in consciousness, which are taken to be 'internal representations'. Thus, in what traditional epistemology characterized as the faculty of Imagination, there arises out of human practice itself a capacity for reflective consciousness, an internalization of outward modes of activity, by which a differentiated and articulated 'mental life' develops. Much more needs to be said about this conjecture, and I have said some small part of it elsewhere. But the upshot of such an approach is that it does not eschew 'mentalism', but attempts to see it in some systematic relation to forms of social and historical practice. Further, it makes 'the human mind' a self-creating and self-transcending aspect of that practical agency which proliferates in all human activities. Finally, it sees the historicity of cognitive activity in relation to the historical changes in modes of representation; and the history of ways of knowing as a history of norms, or of the genesis and transformation of norms.

The sort of 'mentalism' that I am proposing may be characterized as a *constructive mentalism*. As against the essentialist view that epistemology (or psychology, or linguistics, for that matter) seeks to discover the fixed, universal, and necessary structures or features of the mind, a constructive mentalism suggests that epistemology inquires into the historical processes of the formation of the mind, in the context of the historical changes in modes of cognitive practices. If, by mentation, we mean the distinctively reflective or internalized processes of imagination and thought, then we may adopt as a slogan, "No mentation without representation", but in a rather qualified sense. Mental representations, or what are usually distinguished as 'internal representations', do not come to have cognitive priority until the *external* forms of representation, the pictures, narratives, rules of art or of practice, models, have been developed as active constituents of various practices, as norms of making and doing, embodied in the activity. Such forms of external representation then may come to be 'internalized', or mapped into those contexts of consciousness in which they become objects of reflection. The technical account of this possibility involves the mechanisms of memory and of image-formation in neurophysiological and psychological terms. But it is plain that we do not yet have an adequate theory here, though there are several interesting hypotheses. What is important, for my account, is that such an 'internalization' of outward forms of norm-guided practice is not limited at all to what is genetically given in the neural or perceptual structures, or to what may be said to be 'hard-wired'.

What is 'hard-wired' is the range of instinctual or innate, i.e., biologically *a priori*, actions or responses which characterize the stereotypic modes of animal behavior or of neural response. We may also say that genetic structure maps the typical modes of species-activity, or maps the life-world of the species, into its biological mechanisms. Thus, what the Frog's eye tells the Frog's brain is a function of a retinally-centered sensory-motor system evolved for the efficient response to predator-prey configurations. The Frog's eye maps the Frog's functional world or life-world or '*Umwelt*' into its neurophysiology, where this life-world may be said to be 'internally represented'. The long-term process of its coming to be 'internally represented' in this genetically-determinate way is the process of adaptive biological evolution; and the mechanisms of this evolution are the neo-Darwinian ones of natural selection from a variable gene-pool with random mutation (or perhaps some variant of this with weaker or stronger neo-Lamarckian tinges). Schopenhauer's expression of this biological mapping into organic structure of the life-conditions of a species is still vivid: Fangs and claws, he says (in *The World as Will and Representation*), are 'objectified hunger'.

Such functional 'internalization' into genetic and organic structure of modes of life-activity is, however, totally different from the sort of 'internalization' of forms of cognitive activity, in terms of their representation; and the difference is crucial to a constructive mentalism. The modes of representation which come to be internalized are not simply the modes of practical action themselves, but rather the ways in which such practices are understood, i.e. the ways in which they are represented, in the referential relation of 'aboutness'.

The virtue of the representational artifact — that is, of the representation made *as* representation, and not simply embodied in the reflexive mode — is that it is detachable from the contexts of the ongoing practice itself: it has a permanence which goes beyond the limits of the immediate practice, and therefore, also, a range of reference which goes beyond the limits of the given instances of its embodiment. The model artifact, the story, the ritual, the diagram, the picture, the rule, all transcend the particular instances of their genesis, or of their past application. This detached representational artifact is therefore a means of preserving, storing, and also of transmitting the norms of a practice. It begins to have the characteristics of a universal, but of a constructed universal; an 'essence', so to speak, which will change or vary with changes in the historically contingent norms and forms of human action or of social practices.

It is this sort of representational artifact which, I am proposing, is the source of that 'internal representation' which we may refer to as an 'internalized rule', or indeed, as a 'mental image'. That is to say, we don't have 'mental images' of 'things', *tout court*; but rather, we form mental images of the modes of our activity or practice; and images in this sense are laden with the intentionality or the teleology of these practices; they are normative images. But then, the perceptual image of an object is, in the first place, the image of a range of possible actions, 'possible' in the sense of normatively admissible or conceivable, given the conditions of representation of the norms of those practices which involve the perceived object. The act of perception then becomes an act of virtual or projectible practices, and the object of perception is perceived as the object *it is* — i.e., it is recognized — in terms of such a framework of virtual actions.

Crucially, human 'internal representations' offer two opportunities which biological or genetic internalizations of animal life-worlds lack. *First*, changes in modes of action, or in the norms of a practice, are directly transmissible by cultural means, i.e. by learning, or by the acquisition of these norms *in* social interaction. Such changes do not require the process of genetic selection by biological means, in order to be transmitted. *Second*, and of equal importance, the domain of 'external representation' (in representational artifacts — e.g., in language, in art, in rules and laws, in theories) acquires an independence from the constraints and conditions of direct, first-order practices. One may construct, reconstruct, invent and play with models of modes of action, and perform variations and innovations *in* the representation itself, which may in effect radically change its original constraints. This invention and creation of representations which are, so-to-speak, 'off-line' is open to an imagination of alternative modes of action beyond the limits of given forms or norms of a practice. This form of representational construction begins to approach that condition of '*zwecklose Zweckmässigkeit*' (purposeless purposefulness) which Kant describes in *The Critique of Judgement*. Such constructive activity is, in this respect, what we may call 'free construction' (within the limits of *chosen* rules of construction, of course) and feeds back into the modes of first-order practice by imagining or proposing alternatives.

What is the effect of these two opportunities which representational practice affords? First of all, changes in modes of practice, or acquired variations in norms of a practice may be transmitted from one generation to the next, or even within the span of generational life. This means that cultural evolution — taken here as the *historical* evolution of modes of cognitive activity or practice — is Lamarckian, not Darwinian. Acquired characteristics are transmissible, and not by means of the genes, but rather by means of the artifacts produced by a given historical form of life; and in particular, by its representational artifacts, its pictures, theories, histories, i.e., those whose function it is, specifically, to communicate norms of cultural and social practice.

The acquisition of such cognitive modes of action takes place by the process of coming to understand them, i.e., by learning to read the representations *as* representations: to *take* them as representations, or to *make* them as representations. This is a conventionally acquired skill; or to put it differently, it is the skill of acquiring or mastering a convention.

The historicity of this process is clear; in effect, it *is* history, and not the *result* of history. It *is* history in the sense that it is the self-transformation of modes of human action, and the preservation and transmission of these transformed modes by means of their embodiment in artifacts, and by means of the acquired ability to read back from such artifacts what it is they represent. One may then say that the artifact, insofar as it is understood representationally, is to cultural evolution what the gene is to biological evolution, except that this cultural evolution is Lamarckian while the biological evolution is Darwinian.

The internalized representation, as 'mental image', is initially parasitic on the external representation, or the activity of representing. But with its initiation, such representational activity develops an autonomy of its own, wherein the typically linguistic, pictorial or gestural forms of external representation come to be abstractly representable in what we may then call 'inner speech', 'inner picturing' or 'inner gesture' or in so-called 'mental schemata'.[1]

An historical epistemology proposes that these activities of external and internal representation *constitute* cognitive activity, and that what has traditionally been called the human mind is itself a construction which arises out of these activities. To put it in perhaps too radical a way, I would suggest that it is we who *invent* or *create* the mind, construct it, and transform it historically; that the human mind is, so to speak, the arena of our social and historical practice, insofar as this practice is cognitive, i.e., insofar as it understands itself in the mediated form of a representation. Modes of cognition, ways of knowing, then, are not simply different ways in which a given structure, called the mind, functions; rather, they are what the mind *is*, they constitute that normative category of reflective consciousness which we call knowledge; and further, constitute it as an historically changing category.

## 4. EPISTEMIC WARRANT

It remains to be seen how such an historical and cultural conception of mind, a constructive mentalism, is to be interpreted epistemologically, i.e., in terms of a theory of knowledge and not simply in terms of a sequence of historically variable norms or practice. Suppose, then, the mind to be a socio-cultural construction deriving from the various norms of human practices, as these come to be represented and understood. How does this conception provide anything but an historically relativized conception of 'knowing', or a descriptive account of the history of knowledge-claims? If, indeed, one takes the route of a socialized epistemology, then 'knowledge' reduces to the contingent and socially delimited convictions of '*unser Kulturkreis*'. And if one puts the burden of epistemic proof, so to speak, on the variable successes of different modes of practice (whether in straightforward pragmatic terms, or in socialized and historicized Marxist terms), then 'knowledge' seems to be caught in a 'praxiocentric predicament', larger in scope but systematically not unlike what R. B. Berry described as the 'egocentric predicament' two generations ago.

The appeal to some transcendent norm, of 'knowledge as such', is blocked by the refusal of essentialism. The accession to a radical relativism, whether social, historical, linguistic, is blocked by the fact that such a relativism eschews epistemology altogether, substituting, at best, a *description* of alternative norms, choice among which seems arbitrary, or parochial. What *could* possibly count as epistemic warrant for an historicized epistemology?

We could try to answer by discovering (or constructing) what can be plausibly taken to be invariants through historical transformation. Unfortunately, Goodman has scuttled many a promising argument of this sort, simply by pointing out that any constructed invariance is itself a world-version, subject to the limits of world-versions; namely, that there are others, right for other reasons, in alternative systems of construction or construal. So be it; there is no appeal to a versionless world. However, though there may be a plurality of world-versions, suiting our different purposes, interests, needs, or even our temperaments or talents, we still could answer that world-versions are not *ad hoc* (and certainly not for a painfully systematic constructivist like Goodman). They are bound by rules of construction, by norms, without which they wouldn't be versions at all, but merely piles or heaps or fragments of world-constituents. (Poincaré once said "a pile of bricks is not a house"). But if they are not *ad hoc*, and are not arbitrary, they are yet conventional. Given a variety of alternative conventions, there is, for Goodman, a range not merely of possible worlds, but of actual ones, with no necessary 'one world' of which they are a set of versions. What confers epistemic authority, at least on those world-versions which we do actualize – whether in physics, or in ethics, or in social theory? Presumably, some norm of 'correctness' relative to the rules of construction of that world-version. But this presumption yields the radical relativism of Goodman's theory, and that is what makes people queasy about it. I conjecture that this very condition of a relativized world-by-world authorization, each on its own terms, so to speak, *does* yield, if not a universal norm for the construction of worlds, then certainly a universal heuristic for the construction of norms for the construction of worlds, or what one may call a norm of norms.

What *is* invariant through transformations, world-by-world, is the requirement of systematicity. This would leave us, ultimately, within the circle of fire of a coherence theory of truth; and that would reduce ultimately to more or less sophisticated versions of the *consensus gentium*, which I have previously rejected as an abstract socialized epistemology. That we can all be wrong together strikes me not only as possible, but as historically well-evidenced, for any given historical community. But that very possibility, and the actual history of its actualizations — whether in the history of science, or of ethics, or of political theory — constitute the negative side of what may be formulated positively. We may, in some cases, all be right, though we cannot know for certain that we are. What it is we may all be right *about* is what is at issue, of course. But we are not all right *because* we agree, or come to agree at the limit. There is no limit, at which point, to use Peirce's phrase, we are all 'fated to agree' by following the 'right' method. There is, however, the *conception* of such a transcendental limit, which however transcendentally it is posited, is nevertheless posited immanently, by one or another concrete, historical human commuity.

*Baruch College of CUNY*

## NOTES

[1] In psychology, the notion of the internalization of outward modes of action (and of both motor activity and speech activity) is developed very fully by Piaget and Wallon, and also by Vygotsky and Luria. The philosophical correlates of this model of internalization (or of 'interiorisation', to use Piaget's term) are very rich, and are conceptually developed in various ways in great detail in the dialectical tradition, especially in Hegel, Feuerbach and Marx, and in some forms of dialectical phenomenology. They are also at the root of much of Dewey's more systematic thought, e.g. in *Experience and Nature*, and also Mead. But this is a separate matter which I cannot pursue here.

## REFERENCES

Quine, W. V. O.: 1969. 'Epistemology Naturalized.' In *Ontological Relativity and Other Essays*. New York: Columbia University Press, pp. 69—90.

Rorty, Richard: 1979. *Philosophy and the Mirror of Nature*. Princeton, New Jersey: Princeton University Press.

ABNER SHIMONY

# COMMENT ON WARTOFSKY

At one point Wartofsky summarizes his historical epistemology by saying, "To put it in perhaps too radical a way, I would suggest that it is we who *invent* or *create* the mind, construct it, and transform it historically." His suggestion is too radical because, by his own admission elsewhere, there are "mental and perceptual structures" which the inventive and creative processes must respect. Wartofsky seems to be struggling towards a wide-ranging epistemology that judiciously acknowledges both the genetic and the acquired, and both the neural and the mental aspects of cognition (I hope without conflating these two dichotomies, but I cannot be sure from his text). Unfortunately, the pendulum swings between exaggerated claims and cautious hedging are deleterious to a clear exposition of his epistemology.

Clarification is most urgently needed concerning the interface between the genetic and the acquired. Wartofsky's explicit statement on this matter is that "What is 'hard-wired' is the range of instinctual or innate, i.e., biologically *a priori*, actions or responses which characterize the stereotypic modes of animal behavior or of neural response." The selection out of this 'range' is, in his opinion, a historical matter, and he emphasizes that history is the "self-transformation of modes of human action, etc." But for this historicist thesis to make sense at all, there must be some genetically determined constitution of the mind which makes it possible for history to operate, to leave its trace, and to effect a selection from the range of human possibilities. Whether this constitution should be characterized in terms of faculties, schemata, strategies of learning, plasticity for habit formation, or something else is still unsettled, but whatever it is, its *modus operandi* must be extraordinarily subtle.

What I am asking of Wartofsky's epistemology may be expressed by a comparison with immunology. The antibodies which a mammal acquires in the uterus and through the mother's milk last only a few months, and therefore the stock of antibodies of an adult mammal is the outcome of its personal history — a history of invasions of its body by bacteria, viruses, and other foreign agents, and of the struggle to



*Abner Shimony and Debra Nails (eds.), Naturalistic Epistemology*, 375—377.

counter them. Instructions to create proper antibodies for a particular antigen (foreign agent) are not encoded in the organism's DNA; in Wartofsky's terms, the production of appropriate antibodies is not 'hard-wired', and the armory of antibodies is 'invented' or 'created'. But the science of immunology does not rest at this point. It attempts to explain, in terms of the genetically determined processes of the organism, *how* this armory is created; and although many problems remain, much progress towards an explanation has been made. It is fairly certain that antigens do not instruct the B-cells (the antibody producing cells) on what antibodies they should make; rather, if the antibodies which a specifically programmed B-cell produces happen to be appropriate for a given antigen, then their interaction with the antigen indirectly stimulates the B-cell to proliferate and generate a large number of clones of itself. Some understanding has also been achieved of the rapid trial-and-error process whereby a great variety of differently programmed B-cells are produced, in order to yield a large supply of various antibodies, among which there may be one appropriate for the antigen. This summary is crude and oversimplified, but it suffices to indicate how subtle is the 'hard-wired' strategy whereby the immune system of an individual organism can develop historically within its own lifetime. I cannot believe that the 'hard-wired' part of the human cognitive apparatus, which permits a person not only to accumulate a vast body of singular facts about the environment, but also to restructure the schemata of perception, representation, and conception, is any less subtle than its immunological analogue.

A reply may be made that the analogy which I have just developed is fundamentally fallacious, because the basic strategy whereby the mind invents and creates itself is itself subject to reconstruction, and hence is not 'hard-wired'; whereas the operation of the B-cells and of other elements of the immunological strategy (such as the contribution of the T-cells) is programmed in DNA, and hence unequivocally 'hard-wired'. This proposal cannot be defended *a priori*, however, and Wartofsky has not deployed enough psychological, linguistic, historical, etc. data to constitute an *a posteriori* defense. My own conjecture is that when the requisite data are deployed, which may require further development of the relevant sciences, they will point in the other direction.

In summary, I am very hospitable to a historical component in epistemology, but I believe that it will be incomplete, and indeed incomprehensible, without a naturalistic component. It is no accident that John Dewey, whose influence Wartofsky gratefully acknowledges, has a chapter in his *Logic* entitled 'The Existential Matrix on Inquiry: Biological' preceding his chapter 'The Existential Matrix of Inquiry: Cultural'.

## REFERENCES

Alberts, B., Bray, D., Lewis, J., Raff, M., Roberts, K., and Watson, J. D.: 1983. *Molecular Biology of the Cell.* New York and London: Garland Publishing, Inc., pp. 951–1012.

Dewey, John: 1938. *Logic: The Theory of Inquiry.* New York: Henry Holt and Co.

# INDEX OF NAMES

## BOSTON STUDIES IN THE PHILOSOPHY OF SCIENCE



1. Marx W. Wartofsky (ed.), *Proceedings of the Boston Colloquium for the Philosophy of Science 1961–1962.* 1963.
2. Robert S. Cohen and Marx W. Wartofsky (eds.), *In Honor of Philipp Frank.* 1965.
3. Robert S. Cohen and Marx W. Wartofsky (eds.), *Proceedings of the Boston Colloquium for the Philosophy of Science 1964–1966. In Memory of Norwood Russell Hanson.* 1967.
4. Robert S. Cohen and Marx W. Wartofsky (eds.), *Proceedings of the Boston Colloquium for the Philosophy of Science 1966–1968.* 1969.
5. Robert S. Cohen and Marx W. Wartofsky (eds.), *Proceedings of the Boston Colloquium for the Philosophy of Science 1966–1968.* 1969.
6. Robert S. Cohen and Raymond J. Seeger (eds.), *Ernst Mach: Physicist and Philosopher.* 1970.
7. Milič Čapek, *Bergson and Modern Physics.* 1971.
8. Roger C. Buck and Robert S. Cohen (eds.), *PSA 1970. In Memory of Rudolf Carnap.* 1971.
9. A. A. Zinov'ev, *Foundations of the Logical Theory of Scientific Knowledge (Complex Logic).* (Revised and enlarged English edition with an appendix by G. A. Smirnov, E. A. Sidorenka, A. M. Fedina, and L. A. Bobrova.) 1973.
10. Ladislav Tondl, *Scientific Procedures.* 1973.
11. R. J. Seeger and Robert S. Cohen (eds.), *Philosophical Foundations of Science.* 1974.
12. Adolf Grünbaum, *Philosophical Problems of Space and Time.* (Second, enlarged edition.) 1973.
13. Robert S. Cohen and Marx W. Wartofsky (eds.), *Logical and Epistemological Studies in Contemporary Physics.* 1973.
14. Robert S. Cohen and Marx W. Wartofsky (eds.), *Methodological and Historical Essays in the Natural and Social Sciences. Proceedings of the Boston Colloquium for the Philosophy of Science 1969–1972.* 1974.
15. Robert S. Cohen, J. J. Stachel, and Marx W. Wartofsky (eds.), *For Dirk Struik. Scientific, Historical and Political Essays in Honor of Dirk Struik.* 1974.
16. Norman Geschwind, *Selected Papers on Language and the Brain.* 1974.
17. B. G. Kuznetsov, *Reason and Being.* 1987.
18. Peter Mittelstaedt, *Philosophical Problems of Modern Physics.* 1976.
19. Henry Mehlberg, *Time, Causality, and the Quantum Theory* (2 vols.). 1980.
20. Kenneth F. Schaffner and Robert S. Cohen (eds.), *Proceedings of the 1972 Biennial Meeting, Philosophy of Science Association.* 1974.
21. R. S. Cohen and J. J. Stachel (eds.), *Selected Papers of Léon Rosenfeld.* 1978.
22. Milič Čapek (ed.), *The Concepts of Space and Time. Their Structure and Their Development.* 1976.
23. Marjorie Grene, *The Understanding of Nature. Essays in the Philosophy of Biology.* 1974.

24. Don Ihde, *Technics and Praxis. A Philosophy of Technology.* 1978.
25. Jaakko Hintikka and Unto Remes, *The Method of Analysis. Its Geometrical Origin and Its General Significance.* 1974.
26. John Emery Murdoch and Edith Dudley Sylla, *The Cultural Context of Medieval Learning.* 1975.
27. Marjorie Grene and Everett Mendelsohn (eds.), *Topics in the Philosophy of Biology.* 1976.
28. Joseph Agassi, *Science in Flux.* 1975.
29. Jerzy J. Wiatr (ed.), *Polish Essays in the Methodology of the Social Sciences.* 1979.
30. Peter Janich, *Protophysics of Time.* 1985.
31. Robert S. Cohen and Marx W. Wartofsky (eds.), *Language, Logic, and Method.* 1983.
32. R. S. Cohen, C. A. Hooker, A. C. Michalos, and J. W. van Evra (eds.), *PSA 1974: Proceedings of the 1974 Biennial Meeting of the Philosophy of Science Association.* 1976.
33. Gerald Holton and William Blanpied (eds.), *Science and Its Public: The Changing Relationship.* 1976.
34. Mirko D. Grmek (ed.), *On Scientific Discovery.* 1980.
35. Stefan Amsterdamski, *Between Experience and Metaphysics. Philosophical Problems of the Evolution of Science.* 1975.
36. Mihailo Marković and Gajo Petrović (eds.), *Praxis. Yugoslav Essays in the Philosophy and Methodology of the Social Sciences.* 1979.
37. Hermann von Helmholtz, *Epistemological Writings. The Paul Hertz/Moritz Schlick Centenary Edition of 1921 with Notes and Commentary by the Editors.* (Newly translated by Malcolm F. Lowe. Edited, with an Introduction and Bibliography, by Robert S. Cohen and Yehuda Elkana.) 1977.
38. R. M. Martin, *Pragmatics, Truth, and Language.* 1979.
39. R. S. Cohen, P. K. Feyerabend, and M. W. Wartofsky (eds.), *Essays in Memory of Imre Lakatos.* 1976.
42. Humberto R. Maturana and Francisco J. Varela, *Autopoiesis and Cognition. The Realization of the Living.* 1980.
43. A. Kasher (ed.), *Language in Focus: Foundations, Methods and Systems. Essays Dedicated to Yehoshua Bar-Hillel.* 1976.
44. Trân Duc Thao, *Investigations into the Origin of Language and Consciousness.* (Translated by Daniel J. Herman and Robert L. Armstrong; edited by Carolyn R. Fawcett and Robert S. Cohen.) 1984.
46. Peter L. Kapitza, *Experiment, Theory, Practice.* 1980.
47. Maria L. Dalla Chiara (ed.), *Italian Studies in the Philosophy of Science.* 1980.
48. Marx W. Wartofsky, *Models: Representation and the Scientific Understanding.* 1979.
49. Trân Duc Thao, *Phenomenology and Dialectical Materialism.* 1985.
50. Yehuda Fried and Joseph Agassi, *Paranoia: A Study in Diagnosis.* 1976.
51. Kurt H. Wolff, *Surrender and Catch: Experience and Inquiry Today.* 1976.
52. Karel Kosík, *Dialectics of the Concrete.* 1976.
53. Nelson Goodman, *The Structure of Appearance.* (Third edition.) 1977.
54. Herbert A. Simon, *Models of Discovery and Other Topics in the Methods of Science.* 1977.

55. Morris Lazerowitz, *The Language of Philosophy. Freud and Wittgenstein.* 1977.
56. Thomas Nickles (ed.), *Scientific Discovery, Logic, and Rationality.* 1980.
57. Joseph Margolis, *Persons and Minds. The Prospects of Nonreductive Materialism.* 1977.
58. Gerard Radnitzky and Gunnar Andersson (eds.), *Progress and Rationality in Science.* 1978.
59. Gerard Radnitzky and Gunnar Andersson (eds.), *The Structure and Development of Science.* 1979.
60. Thomas Nickles (ed.), *Scientific Discovery: Case Studies.* 1980.
61. Maurice A. Finocchiaro, *Galileo and the Art of Reasoning.* 1980.
62. William A. Wallace, *Prelude to Galileo.* 1981.
63. Friedrich Rapp, *Analytical Philosophy of Technology.* 1981.
64. Robert S. Cohen and Marx W. Wartofsky (eds.), *Hegel and the Sciences.* 1984.
65. Joseph Agassi, *Science and Society.* 1981.
66. Ladislav Tondl, *Problems of Semantics.* 1981.
67. Joseph Agassi and Robert S. Cohen (eds.), *Scientific Philosophy Today.* 1982.
68. Władysław Krajewski (ed.), *Polish Essays in the Philosophy of the Natural Sciences.* 1982.
69. James H. Fetzer, *Scientific Knowledge.* 1981.
70. Stephen Grossberg, *Studies of Mind and Brain.* 1982.
71. Robert S. Cohen and Marx W. Wartofsky (eds.), *Epistemology, Methodology, and the Social Sciences.* 1983.
72. Karel Berka, *Measurement.* 1983.
73. G. L. Pandit, *The Structure and Growth of Scientific Knowledge.* 1983.
74. A. A. Zinov'ev, *Logical Physics.* 1983.
75. Gilles-Gaston Granger, *Formal Thought and the Sciences of Man.* 1983.
76. R. S. Cohen and L. Laudan (eds.), *Physics, Philosophy and Psychoanalysis.* 1983.
77. G. Böhme et al., *Finalization in Science*, ed. by W. Schäfer. 1983.
78. D. Shapere, *Reason and the Search for Knowledge.* 1983.
79. G. Andersson, *Rationality in Science and Politics.* 1984.
80. P. T. Durbin and F. Rapp, *Philosophy and Technology.* 1984.
81. M. Marković, *Dialectical Theory of Meaning.* 1984.
82. R. S. Cohen and M. W. Wartofsky, *Physical Sciences and History of Physics.* 1984.
83. E. Meyerson, *The Relativistic Deduction.* 1985.
84. R. S. Cohen and M. W. Wartofsky, *Methodology, Metaphysics and the History of Sciences.* 1984.
85. György Tamás, *The Logic of Categories.* 1985.
86. Sergio L. de C. Fernandes, *Foundations of Objective Knowledge.* 1985.
87. Robert S. Cohen and Thomas Schnelle (eds.), *Cognition and Fact.* 1985.
88. Gideon Freudenthal, *Atom and Individual in the Age of Newton.* 1985.
89. A. Donagan, A. N. Perovich, Jr., and M. V. Wedin (eds.), *Human Nature and Natural Knowledge.* 1986.
90. Carl Mitcham and Alois Huning (eds.), *Philosophy and Technology II.* 1986.
91. Marjorie Grene and Debra Nails (eds.), *Spinoza and the Sciences.* 1986.
92. Stephen P. Turner, *The Search for a Methodology of Social Science.* 1986.

93. I. C. Jarvie, *Thinking about Society: Theory and Practice.* 1986.
94. Edna Ullmann-Margalit (ed.), *The Kaleidoscope of Science.* 1986.
95. Edna Ullmann-Margalit (ed.), *The Prism of Science.* 1986.
96. Gyorgy Markus, *Language and Production.* 1986.
97. F. Amrine, F. J. Zucker, and H. Wheeler (eds.), *Goethe and the Sciences: A Reappraisal.* 1987.
98. Joseph C. Pitt and Marcella Pera (eds.), *Rational Changes in Science.* (in preparation).
99. O. Costa de Beauregard, *Time, the Physical Magnitude.* (in preparation).
100. Abner Shimony and Debra Nails (eds.), *Naturalistic Epistemology: A Symposium of Two Decades.* 1987.